VOLUME THREE

Advances in NEUROTOXICOLOGY

Role of Inflammation in Environmental Neurotoxicity

VOLUME THREE

ADVANCES IN NEUROTOXICOLOGY

Role of Inflammation in Environmental Neurotoxicity

Edited by

MICHAEL ASCHNER

Albert Einstein College of Medicine, Bronx, NY, United States

LUCIO G. COSTA

University of Washington, Seattle, WA, United States; University of Parma, Parma PR, Italy

ELSEVIER

ACADEMIC PRESS

An imprint of Elsevier

Academic Press is an imprint of Elsevier
50 Hampshire Street, 5th Floor, Cambridge, MA 02139, United States
525 B Street, Suite 1650, San Diego, CA 92101, United States
The Boulevard, Langford Lane, Kidlington, Oxford OX5 1GB, United Kingdom
125 London Wall, London, EC2Y 5AS, United Kingdom

First edition 2019

ISBN: 978-0-12-815717-6
ISSN: 2468-7480

Publisher: Zoe Kruze
Acquisition Editor: Sam Mahfoudh
Editorial Project Manager: Peter Llewellyn
Production Project Manager: Vignesh Tamil
Cover Designer: Alan Studholme

Typeset by SPi Global, India

Contents

Contributors

Matthew J. Campen
Department of Pharmaceutical Sciences, College of Pharmacy, University of New Mexico, Albuquerque, NM, United States

Yu-Chi Chang
Department of Environmental and Occupational Health Sciences, University of Washington, Seattle, WA, United States

Gabrielle M. Childers
Neurotoxicology Group, National Toxicology Program Laboratory, National Institute of Environmental Health Sciences, National Institutes of Health, Durham, NC, United States

Jacki Coburn
Department of Environmental and Occupational Health Sciences, University of Washington, Seattle, WA, United States

Toby B. Cole
Department of Environmental and Occupational Health Sciences; Center on Human Development and Disability, University of Washington, Seattle, WA, United States

Lucio G. Costa
Department of Environmental and Occupational Health Sciences, University of Washington, Seattle, WA, United States; Department of Medicine & Surgery, University of Parma, Parma, Italy

Khoi Dao
Department of Environmental and Occupational Health Sciences, University of Washington, Seattle, WA, United States

Jesse L. Denson
Department of Pharmaceutical Sciences, College of Pharmacy, University of New Mexico, Albuquerque, NM, United States

Amedeo D'Angiulli
Department of Neuroscience; Institute of Interdisciplinary Studies, Carleton University, Ottawa, ON, Canada

Omar El Jordi
Department of Pharmacology and Toxicology, Indiana University School of Medicine, Indianapolis, IN, United States

Nikolay M. Filipov
Department of Physiology and Pharmacology, College of Veterinary Medicine, University of Georgia, Athens, GA, United States

Jacqueline Garrick
Department of Environmental and Occupational Health Sciences, University of Washington, Seattle, WA, United States

Consuelo Guerri
Department of Molecular and Cellular Pathology of Alcohol, Príncipe Felipe Research Center, Valencia, Spain

Michelle Guignet
Department of Molecular Biosciences, University of California-Davis, Davis, CA, United States

Gaylia Jean Harry
Neurotoxicology Group, National Toxicology Program Laboratory, National Institute of Environmental Health Sciences, National Institutes of Health, Durham, NC, United States

Huajun Jin
Parkinson Disorders Research Laboratory, Department of Biomedical Sciences, Iowa State University, Iowa Center for Advanced Neurotoxicology, Ames, IA, United States

Anumantha G. Kanthasamy
Parkinson Disorders Research Laboratory, Department of Biomedical Sciences, Iowa State University, Iowa Center for Advanced Neurotoxicology, Ames, IA, United States

Arthi Kanthasamy
Parkinson Disorders Research Laboratory, Department of Biomedical Sciences, Iowa State University, Iowa Center for Advanced Neurotoxicology, Ames, IA, United States

Jane Kasten-Jolly
Wadsworth Center, New York State Department of Health, Albany, NY, United States

David A. Lawrence
Wadsworth Center, New York State Department of Health, Albany; Department of Environmental Health Sciences, University at Albany School of Public Health, Rensselaer, NY, United States

Pamela J. Lein
Department of Molecular Biosciences, University of California-Davis, Davis, CA, United States

Emir Malovic
Parkinson Disorders Research Laboratory, Department of Biomedical Sciences, Iowa State University, Iowa Center for Advanced Neurotoxicology, Ames, IA, United States

Christopher A. McPherson
Neurotoxicology Group, National Toxicology Program Laboratory, National Institute of Environmental Health Sciences, National Institutes of Health, Durham, NC, United States

Carmen M. Mitchell
Department of Pharmacology and Toxicology, Indiana University School of Medicine, Indianapolis, IN, United States

María Pascual
Department of Molecular and Cellular Pathology of Alcohol, Príncipe Felipe Research Center; Department of Physiology, School of Medicine and Dentistry, University of Valencia, Valencia, Spain

Souvarish Sarkar
Parkinson Disorders Research Laboratory, Department of Biomedical Sciences, Iowa State University, Iowa Center for Advanced Neurotoxicology, Ames, IA, United States

Bryan K. Yamamoto
Department of Pharmacology and Toxicology, Indiana University School of Medicine, Indianapolis, IN, United States

Tamara L. Young
Department of Pharmaceutical Sciences, College of Pharmacy, University of New Mexico, Albuquerque, NM, United States

Katherine E. Zychowski
Department of Pharmaceutical Sciences, College of Pharmacy, University of New Mexico, Albuquerque, NM, United States

Preface

The Third Volume of *Advances in Neurotoxicology* is devoted to discussions on the role played by inflammatory processes in environmental neurotoxicity. Neuroinflammation refers to inflammatory responses in the central nervous system, which can be acute or chronic. In recent years, neuroinflammation has become a dominant theme in neuroscience as inflammatory processes in the central nervous system have been associated with neurodevelopmental and neurodegenerative disorders. This volume represents a timely addition to the literature as it discusses major issues related to neuroinflammatory mechanisms associating environmental exposures to neurodevelopmental and neurodegenerative disorders. The volume starts with a general overview on the nature of neuroinflammation, highlighting its main features, cell types involved, and principal mediators. It then focuses on several environmental neurotoxic agents known to cause neuroinflammation, such as metals (lead, manganese), pesticides (organophosphates), drugs of abuse, ethanol, and air pollution. Also discussed are the specific relevance of neuroinflammation in developmental neurotoxicity, and the issues of central vs peripheral inflammation and the role played by the brain–blood barrier.

This series fills a gap in the growing field of neurotoxicology, and the volumes should be of interest to academic and industrial researchers, graduate students, and practicing toxicologists. The series Editors will be happy to entertain proposals for specific topics to be the subject of future volumes, and for guest editing.

Michael Aschner
Lucio G. Costa

CHAPTER ONE

An introduction to innate immunity in the central nervous system

Gaylia Jean Harry*, Gabrielle M. Childers, Christopher A. McPherson
Neurotoxicology Group, National Toxicology Program Laboratory, National Institute of Environmental Health Sciences, National Institutes of Health, Durham, NC, United States
*Corresponding author: e-mail address: harry@niehs.nih.gov

Contents

1. Immune privilege in the CNS

Immune cells found throughout the body mediate pro- and anti-inflammatory responses and are major sources of soluble molecules, cytokines, hormones, and neuropeptides. This provides them with the tools to sense, process, and relay physiological signals beyond their canonical roles. In the periphery, dendritic cells (DC) function to detect the presence and nature of pathogens. DC take up protein antigens, process them into small peptides and then display these peptides on their surface with major

Advances in Neurotoxicology, Volume 3
ISSN 2468-7480
https://doi.org/10.1016/bs.ant.2018.10.002

histocompatibility complex II (MHCII) molecules. In general, DCs migrate through afferent lymphatic pathways to reach the draining lymph nodes. This allows for the presentation of antigen to naïve or memory T cells. DCs educate T cells on the pathogen threat to facilitate targeting of T cells to the specific pathogen site. Upon the localized expression of T cell cytokines, tissue resident or infiltrating immune cells respond to the signal to coordinate a focused attack on the pathogen. It is considered that pathogen ingress into the CNS involves movement from a peripheral site of entry that first elicits a response in the draining lymph nodes or spleen. While DC cells have not been detected within the brain parenchyma, cells expressing DC surface markers are present in the meningeal coverings and the choroid plexus. There is, as yet, no evidence that DCs with these classic capacities reside in the healthy CNS parenchyma (Galea et al., 2007). The blood-brain barrier (BBB) mechanically separates the CNS from the circulation by the presence of specialized endothelial cells tightly attached to each other via tight junctions and adherens junctions (Hawkins and Davis, 2005; Hermann and Elali, 2012). It excludes plasma proteins and many of the peripherally derived innate and adaptive immune cells as well as their associated inflammatory molecules (Banerjee and Bhat, 2007; DeSalvo et al., 2011) and by this blocking mechanism, serves to restrain CNS inflammation. The BBB actively contributes to the immune response of the CNS (Ifergan et al., 2008; Muldoon et al., 2013). While there is limited infiltration of peripheral immune cells into the CNS in physiological conditions, in the inflamed CNS parenchyma with disruption of the blood-brain barrier, cells expressing DC surface markers (e.g., CD11b, CD11c) can be observed. After physical injury, neutrophils, eosinophils, T lymphocytes, monocytes, and others can be found in the CNS parenchyma. Under these conditions, it is likely that the related cells are peripherally derived. With injury to the blood-brain barrier, blood-borne monocytes can enter the brain parenchyma and assume a brain macrophage phenotype (King et al., 2009; Mildner et al., 2007, 2009; Vallieres and Sawchenko, 2003). It is suggested that infiltrating cells are involved in severe inflammatory injuries as compared to a predominant focus of resident microglia in tasks related to maintaining tissue homeostasis (Ginhoux and Jung, 2014). In comparison, the luminal side of the BBB is in contact with leukocytes that are constantly surveying the barrier. This immune surveillance likely functions via the binding of endothelial cells of the BBB using adhesion molecules.

In the periphery, the lymphatic vasculature drains interstitial fluid (ISF) allowing for the clearance of toxic waste products, immune surveillance, and recirculation of immune cells in tissues (Louveau et al., 2015a,b). The CNS was long considered to lack a functional lymphatic drainage system similar to what is observed in the periphery and the absence of such a system was thought to contribute to the immune privilege state of the CNS. In 2012, Iliff et al. (2012) demonstrated that ISF is cleared from the brain parenchyma through a paravascular pathway. This pathway allowed for the drainage of ISF into the cerebral spinal fluid (CSF) and thus removal from the parenchyma. Given the dynamic nature of this system and the role that glial cells contributed to the clearance process, this ISF drainage system was termed the "glymphatic system." It is suggested to be responsible for the drainage of macromolecules and hydrophilic compounds into the CSF (Jessen et al., 2015). Early work demonstrated that CSF can drain to deep lymph nodes in the lymphatic system, however, the exact mechanism remained unclear (Foldi et al., 1968; Kida et al., 1993; Laman and Weller, 2013). While suggestive lymphatic vascular structures have been previously identified in the cranium (Kida et al., 1993; Laman and Weller, 2013), actual demonstration that these served as an immune surveillance drainage pathway to the deep cervical lymph nodes has been elusive. In 2015, two studies reported on a functional lymphatic vasculature system in the cranium that serves as a pathway for macromolecules, T-cells, MHCII cells in the cerebrospinal fluid (CSF) to be drained from the meningeal space (Louveau et al., 2015b) and the brain parenchyma (Asperlund et al., 2015) to deep cervical lymph nodes. This system now provides a route for the outward migration of CNS innate immune cells that can signal in the lymphoid organs as well as a second step in the drainage of the ISF from the brain.

2. Innate immunity

The function of innate immune cells is to recognize and respond to microbial pathogens. This occurs in the periphery and with significant modification in the CNS (Finsen and Owens, 2011; Ransohoff and Brown, 2012; Waisman et al., 2015). The lack of resident DCs and the predominance of inflammation-suppressive factors, e.g., transforming growth factor beta (TGF-β), interleukin (IL)-10, and gangliosides, in the CNS, likely contributes to a muted innate immune response. That being said, innate immune cells are responsive to primed T cells and released cytokines. With a T cell-mediated inflammation in the CNS, peripheral innate immune cells are allowed into the brain as a protective host defense against infection and for repair following stroke or physical brain trauma (Chen et al., 2001;

Hendrix and Nitsch, 2007) with multiple sclerosis (MS) or it's animal model, experimental autoimmune encephalomyelitis (EAE), serving as prime examples (Hauser and Oksenberg, 2006; Mix et al., 2008). In MS and EAE, it is possible that T cells are initially activated in peripheral lymphoid organs where they undergo cell differentiation and expansion. In the CNS, a microglia-T cell interaction is possible via the expression of IL-23 and IL-1β promoting T cell production of GM-CSF (Li et al., 2003) which can also serve as a proliferative factor for microglia.

These basic host defense mechanisms exist in microglia and astrocytes. The process begins with the recognition of pathogen signatures with the warnings mediated by pathogen-associated molecular patterns (PAMPs) such as bacterial, viral, and protozoal products (protein lipid, nucleic acid, carbohydrate). PAMPs are recognized by Toll-like receptors (TLRs) on the plasma membrane or in the endosomal compartments (Kawai and Akira, 2010). This then increases transcription of IL-1 family cytokine genes via activation of the transcription factor NF-κB. In a sterile inflammatory response, immune cells are activated in the absence of microbial compounds. Under these conditions, immune cells are thought to be activated by endogenous molecules called danger-associated molecular patterns (DAMPs) (Kono and Rock, 2008). These damage signals can activate immune cells through pattern recognition receptors (PPR) of which there are three major families: Toll-like receptors (TLRs), nucleotide-binding domain leucine-rich repeat containing proteins (also known as NOD-like receptors, NLRs), and Rig1-like receptors (RLRs). All of which are found within the brain. The engagement of these receptors results in the induction of specific pathways and release of cytokines that contribute to resolving the injury. One type of DAMP is an intracellular compound released from dying cells as a result of loss of membrane integrity. Another is an extracellular compound modified by the enzymes released from dying cells. Molecules that function as DAMPs include nucleic acids, lipids, proteins that normally are not present to immune cells until released or unmasked with cells death occurring due to tissue injury. Intracellular DAMPs include such things as high mobility group box 1 (HMGB1) and peroxiredoxin (Prx) family proteins. Compounds in the extracellular matrix can also function as DAMPs (Frey et al., 2013). For example, fibrinogen and proteoglycans (hyaluronan, biglycan, and versican) may activate the TLRs and initiate an inflammatory response (Barbelova et al., 2009; Petrey and de la Motte, 2014). However, there are data suggesting that many of the proteoglycans also serve in a neuroprotective fashion.

3. Inflammasome activation

Inflammasomes are important multiprotein complexes that are activated by a variety of PAMPs and DAMPs that cleave and release mature pro-inflammatory cytokines interleukin-1 beta (IL-1β) and interleukin-18 (IL-18). They consist of a cytosolic sensor belonging to the AIM2 (absent in melanoma 2), or NLR, an adaptor protein ASC (apoptosis-associated speck-like protein containing a CARD), and an effector caspase, primarily capspase-1. There are a variety of inflammasomes, many of which fall into the NLR domain. NLRP1 and AIM2 inflammasomes have been characterized in neurons (Adamczak et al., 2014; Kaushal et al., 2015; Yogarajah et al., 2017) and the NLRC4, NLRP2, NLRP3 inflammasomes in astrocytes (Freeman et al., 2017; Minkiewicz et al., 2013). Components for multiple inflammasomes are expressed in microglia (Burm et al., 2015; de Rivero Vaccari et al., 2014). The NLRP1 inflammasome has been implicated in ethanol induced, neuronal-driven inflammation in the CNS (Wang et al., 2015) and the NLRP3 inflammasome responds to a number of activators, including sterile activators such as, asbestos, silica crystals, aluminum salts, and polystyrene nanoparticles (Franchi et al., 2009; Hornung et al., 2008; Lunov et al., 2011). Extracellular Aβ is an example of an aberrant protein that can trigger activation of the NLRP3 inflammasome in microglia due to lysosomal damage and cathepsin B release following fibril uptake (Schnaars et al., 2013). This reaction to multiple sterile and physical activators has served as the basis for much of the research examining pharmacological and environmental agents as triggers for inflammasome activation. Exactly how the activation and regulation of cellular inflammasomes contributes to nervous system injury remains a question that is driving current research activities (de Rivero Vaccari et al., 2014; Lang et al., 2018).

Unlike most inflammasomes, NLRP3 typically requires a two-step activation. Activation of the toll-like receptor (TLR), tumor necrosis factor receptors (TNFR), or interleukin-1 receptors (IL-1R) initiates an intracellular cascade of effects, including activation of NF-κB, leading to the upregulation of NLRP3 and pro-IL-1β within the cell, as well as, immediate facilitation of post-transcriptional changes to NLRP3 to free ubiquitinated binding sites by BRCC3 (BRCA1/2-Containing Complex Subunit 3) (Afonina et al., 2017; Elliott and Sutterwala, 2015; Franchi et al., 2009). Delivery of a secondary "trigger" such as PAMPs, DAMPs, and intact pathogens to the "primed" cell causes the release of the repressed state of

NLRP3. Upon release, NLRP3 activates the inflammasome forming a multiprotein complex comprised of the cytosolic sensor NLRP3, ASC, and caspase 1. Caspase 1 and caspase 18 facilitate the cleavage of pro-interleukin-1β (Zambetti et al., 2012). Cell membrane leakage due to pyroptosis then allows for the release of mature IL-1β into the extracellular space where it can act upon surrounding cells. Critical roles for IL-1β and its receptor IL-1R have been identified in the amplification of cytokine/chemokine networks and in disease development in EAE (Gris et al., 2010; Sutton et al., 2006) with microglia phagocytosis of myelin debris (Williams et al., 1994). Neurotoxin prion infection can lead to NLRP3 inflammasome activation in microglia (Shi et al., 2012), as can West Nile virus infection (Kaushik et al., 2012). Cultured microglia respond to live *Staphylococcus aureus* by producing IL-1β through an NLRP3 inflammasome manner (Hanamsagar et al., 2011) which plays a critical role for bacterial clearance during acute *S. aureus* infection (Kieliam et al., 2004; Xiong et al., 2012). In addition to mature IL-1β, the pyroptosis-related release of the large inflammasome complex, termed "specks," may also represent a stimulus for activating phagocytic cells in the immediate environment and thus contributing to a prolonged propagation of inflammation (Franklin et al., 2014). While inflammasome activation is an efficient producer of mature IL-1β, inflammasome independent mechanisms exist, including cathepsin B or caspase 11 dependent pathways (Lee et al., 2013; Viganò and Mortellaro, 2013; Wang et al., 2015; Wu et al., 2013) as well as bacterial pore-forming toxins and extracellular ATP (Hanamsagar et al., 2011). Thus, an upregulation of mature IL-1β does not necessarily indicate an inflammasome mechanism.

The mechanism by which NLRP3 is activated by the secondary "trigger" is not fully understood. However, there is mounting evidence that mitochondrial damage plays a central role (Mills et al., 2017). As an example, the complex 1 inhibitor, rotenone, can serve as a secondary trigger for NLRP3 inflammasome activation (Lawana et al., 2017). Conditions that have been found to facilitate NLRP3 activation include increased ROS production, calcium influx, potassium efflux, reduction in NAD+, externalization of cardiolipin from mitochondria, and presence of mitochondrial DNA in the cytoplasm; all of which are results of mitochondrial dysfunction and damage (Gurung et al., 2015). Additionally, mitophagy, the process by which cells clear damaged mitochondria, has been implicated in inhibition of NLRP3 signaling. Nakahira et al. (2011) reported that inhibition of mitophagy in macrophages resulted in heightened activation of the NLRP3 inflammasome that paralleled an increase in mitochondrial DNA released into the cytosol from damaged mitochondria and not cleared.

In the peripheral immune cell, activation of the inflammasome serves in a regulatory capacity in the induction of pyroptotic cell death to remove the damaging immune cell. How this translates to central nervous system macrophages remains in question given the long-lived nature of these cells. However, the need to remove severely damaged cells, i.e., mitochondrial damage, could be of even greater importance in an organ that requires such a tight regulation to maintain homeostasis. Given the increasing evidence centralizing mitochondria in innate immune signaling (Mills et al., 2017) and the dynamic nature of microglia, the NLRP3 inflammasome could be crucial in ensuring maintenance of a fully functioning population of microglia.

4. Primary CNS cells for innate immunity

While the CNS does not readily allow infiltration of peripheral immune-related cells and secreted factors, the basic host defense tasks are performed primarily in astrocytes and microglia. The endpoint for the innate immune system is the production of cytokines and the initiation of phagocytosis. The engagement of PRRs converges on the associated transcription factor signaling to induce expression of cytokines, chemokines, reactive oxygen species, and free radicals.

5. The neurovascular unit

The neurovascular unit represents the coordinated contributions of microglia, astrocytes, and endothelial cells and pericytes in the innate immune response and the bioactive molecules that may modulate the function of the BBB. Endothelial cells are located at the luminal side of the BBB and are in continuous contact and cross talk with blood circulation. They are able to link peripheral immune responses to the CNS by acting as mediators and sensors of immune processes in the periphery. Given their location, they are considered the first line of defense and may play an active role in innate immunity in the CNS (Danese et al., 2007). Under physiological conditions, endothelial cells are immunologically quiescent and exhibit a basal expression of adhesion molecules (Alvarez et al., 2011). Upon detection of an immunological threat, endothelial cells produce pro-inflammatory mediators such as IL-1β (Creagh and O'Neill, 2006), intracellular adhesion molecule-1 (ICAM-1), vascular cell adhesion molecule-1 (VCAM-1) (Dinarello, 2009). Through this process, neutrophil infiltration into the CNS across cerebral endothelial cells can be enhanced (Stanimirovic et al., 1997). CNS pericytes are in the lineage of vascular smooth muscle cells and cover up to 30% of endothelial layer circumference. They play an important role in regulating endothelial cell proliferation, survival, migration, and cerebrovascular branching (Hellström et al., 2001). Pericytes are packed at the abluminal side of cerebral endothelial cells facing the basement membrane, controlling endothelial functions, and therefore play a central role in integrating luminal signals generated from cerebral endothelial cells to CNS parenchyma (Hermann and Elali, 2012). As contractile cells, these cells can modify cerebral blood flow including creating blood entrapment in CNS capillary beds (Yemisci et al., 2009). Additionally, pericytes display some macrophage features and antigen presentation capacities (Balabanov et al., 1996; Graeber et al., 1990; Hickey and Kimura, 1988; Kovac et al., 2011). Pericytes downregulate genes associated with vascular permeability (Daneman et al., 2010) and induce activity of ABCB1 transporters in brain endothelial cells (Al Ahmad et al., 2011). Through these mechanisms, they induce BBB formation and help to maintain integrity. The loss of pericytes can lead to a vascular injury allowing extravasation of blood-borne molecules into the brain parenchyma (Winkler et al., 2011). Both pericytes and endothelial cells are ensheathed by the basal lamina constituting the peri-vascular space to provide physical support and backup immune surveillance and scavenging center. This is accomplished by creating a niche for perivascular microglia that patrol CNS vasculature. Although perivascular microglia perform normal microglial functions, they are different due to their interaction and cross talk with cerebral endothelial cells and the fact that they are rapidly replaced by blood circulating bone marrow-derived cells (Bechmann et al., 2001; Gehrmann et al., 1995).

6. Astrocytes

Astrocytes are the most abundant glial cell type in the CNS and are necessary for maintaining brain homeostasis. To accomplish this task, astrocytes provide metabolites and growth factors to neurons, maintain the extracellular balance of ions, fluids, and neurotransmitters, as well as, support for synaptic formation and plasticity (Colombo and Farina, 2016). Their contribution to the BBB can be considered as a first role in innate immunity.

Astrocyte end-feet surround capillary endothelial cells, forming the glia limitans separating the perivascular space from the brain. This structure serves to limit the entry of peripheral immune cells into the brain in the absence of inflammation (Jensen et al., 2013). A wide range of insults can serve to stimulate astrocytes, including innate immune signals in the CNS. The resulting morphological response, astrogliosis, is characterized by the upregulation of glial fibrillary acidic protein (GFAP) and hypertrophy of the cell body and processes. While the normal astrocyte remains within a defined self-identified space and does not infringe on the area of another astrocyte, with astrogliosis, dense hypertrophied processes can overlap and eventually result in the formation of a "glial scar" (Sofroniew and Vinters, 2010). The response of astrocytes can vary as a function of the type of insult/injury, the localized region, and the molecular signaling events (Colombo and Farina, 2016; Farina et al., 2007; Sofroniew and Vinters, 2010).

While there is a limited body of data that lends support to a role of astrocytes in the innate immune response, evidence continues to grow describing astrocytes as immune competent cells showing an ability to detect PAMPS and DAMPS through a repertoire of PRRs. Astrocytes express a number of PRRs NLRC4, NLRP1, NLRP3 (Freeman et al., 2017; Minkiewicz et al., 2013), the mannose receptor as a pathogenic sensor recognizing mammosylated ligands from microbes (Burudi and Regnier-Vigouroux, 2001; Burudi et al., 1999), and scavenger receptors (SR) including, scavenger receptor b (SR-B; Husemann and Silverstein, 2001), macrophage receptor (MARCO) (Alarcon et al., 2005), receptor for advanced glycation end products (RAGE) (Sasaki et al., 2001), and scavenger receptor with C-type lectin (SRCL) (Nakamura et al., 2006) for the recognition and internalization of many ligands including lipids, proteins, collagen and apoptotic cells. The most well-studied mechanism of DAMP and PAMP recognition are the TLRs with astrocytes expressing only a few of the known TLRs. A basal expression of TLR2 is observed in primary mouse astrocytes (Bowman et al., 2003) with induction demonstrated following various PAMPs such as, flagellin, CpG oligodeoxynucleotides (Bowman et al., 2003), peptidoglycan (Bowman et al., 2003; Esen et al., 2004), and lipopolysaccharide (LPS) (Carpentier et al., 2005). Induction of TLR3 in primary rodent astrocytes can be observed by polyinosinic-polycytidylic acid (poly I:C); (Carpentier et al., 2005; Scumpia et al., 2005). Basal expression of TLR4 in primary mouse astrocytes can be upregulated following LPS (Bowman et al., 2003; Carpentier et al., 2005; Gorina et al., 2011); yet, this was not demonstrated in human astrocytes (Farina et al., 2005).

In addition to the astrocytic glial limitans as a structural aspect of the BBB to limit influx of leukocytes into the CNS, the astrocyte may serve a role in recruiting immune cells to the brain, when required (Abbott et al., 2006). This process is linked with the expression of chemokines at the BBB (Carpentier et al., 2005; Farina et al., 2007). Chemokines are a family of small cytokines or signaling proteins secreted by cells. They primarily serve to facilitate directed chemotaxis in nearby cells. These chemotaxis responses can serve in a developmental or homeostatic manner or can contribute to the recruitment of immune cells to a site of injury. Inflammatory chemokines serve as chemoattractants for leukocytes, monocytes, neutrophils, and other effector cells across the BBB or within the brain parenchyma. Astrocytes are capable of modulating chemokine expression in a pathogen-specific manner. For example, the production of the chemokines CCL2, CXCL1, CCL20, and CCL3 by cultured astrocytes can be induced by exposure to LPS and flagellin and CXCL10 and CCL5 with poly I:C (McKimmie and Graham, 2010). This allows the astrocyte a mechanism by which it can contribute to maintaining the health of the CNS by attracting leukocytes from the periphery to clear infection within the CNS.

In addition to the various PPRs, astrocytes possess receptors for pro-inflammatory cytokines (Sofroniew, 2014). Cytokines can be produced and released in autocrine, paracrine and endocrine manner, allowing for astrocytes to be influenced by cytokines from a wide variety of cells. Transcriptome analysis of primary mouse astrocytes demonstrated that astrocytes express receptors for IL-1β, IL-6, interferon gamma (IFN-γ), TGF-β1, and tumor necrosis factor alpha (TNF-α) (Choi et al., 2005; Hamby et al., 2012; Tada et al., 1994). Additional studies have demonstrated that astrocytes stimulated with IFN-γ, IL-1β, or TNF-α directly regulate the expression of genes related to innate immunity, antigen presentation, apoptosis, leukocyte migration, and BBB permeability (John et al., 2002; Rivieccio et al., 2005). While not as robust as microglia, astrocytes have been reported to release a number of pro-inflammatory cytokines in response to danger signals such as, IL-1β, IL-6, and TNF-α (Farina et al., 2005; Jack et al., 2005; Park et al., 2006). TNF-α or IFN-γ stimulated primary rodent astrocytes release IL-6, INF-α, IFN-β, and inducible nitric oxide synthetase (iNOS) (Carpentier et al., 2005), while TGF-β1 stimulated astrocytes showed an upregulation of the chemokine, CXCR6, production (McKimmie and Graham, 2010).

In vivo studies have examined the role of astrocytes in regulating the inflammatory response in the brain. The majority of these studies manipulated critical components of the immune transcription pathways in astrocytes. In mice, in which nuclear factor kappa-B (NF-κB) was selectively inactivated in astrocytes, functional recovery following spinal cord injury showed improvement (Brambilla et al., 2005). The associated diminished expression of CXCL10 and CCL2 was considered to reduced leukocyte recruitment to the lesion. Further studies from the same group showed an associated decreased expression of TNF, IFN, CXCL9, CXCL10, CCL3, CXCR3, and CCL5 with the improved recovery (Brambilla et al., 2009). TGF-β1 is often considered an anti-inflammatory or growth promoting factor. In mice with a targeted deletion of TGF-β1 in astrocytes, an increased expression of inflammatory molecules has been reported in a model of ischemia (Cekanaviciute et al., 2014a) and following infection with *Toxoplasma* (Cekanaviciute et al., 2014b). Thus, the existing data suggest that, while not the primary immune cell of the brain, the astrocyte plays a significant role in the innate immune response.

7. Microglia

Tissue resident macrophages represent distinct classes of macrophages with different transcriptional profiles (del Río Hortega, 1932). Resident macrophages in the CNS, microglia, were described by Virchow as early as the mid-19th century. These cells were classified as morphologically distinct from neurons and astrocytes (Cajal, 1913) then further classified as distinct from oligodendrocytes (del Río Hortega, 1932). By the 1980s, their distinct macrophage-like features were established, including their role in immune defense (Kreutzberg, 1996) and they are now best known as the mononuclear phagocytes of the CNS. In contrast to other CNS-specific cells, microglia originate from a primitive monocyte population derived from the yolk sack during a defined window of time before vascularization or definitive hematopoiesis (Alliot et al., 1999; Ginhoux et al., 2010; Gomez Perdiguero et al., 2015; Kierdorf et al., 2013; Schulz et al., 2012). Microglia are present as early as embryonic day (ED)8, the very early stages of neurogenesis at the time of neuroepithelial cell transformation into cortical progenitor cells (radial glial (RG)) (Alliot et al., 1999). From ED12 they accumulate within the ventricular zone (VZ)/subventricular zone (SVZ) and contact immature RG (Antony et al., 2011; Swinnen et al., 2012). Microglia are temporally and spatially aligned with astrogliogenesis and their physical proximity leads to the speculation that microglia play a role in RG differentiation into astrocytes (Monier et al., 2006). In light of these observations, microglia could have a role in maintaining the progenitor cell pool and regulating progenitor cell migration, or in influencing the timing of neuronal or glial differentiation.

There are approximately 6500 microglia per cubic millimeter of the CNS tissue (Beggs et al., 2012; Nimmerjahn et al., 2005). These morphologically heterogeneous cells (Harry and Kraft, 2012; Lawson et al., 1990; Mittelbronn et al., 2001) perform crucial physiological functions for tissue development, architecture refinement, and remodeling/repairing tissue. Maintenance and expansion of the resident microglia population is the primary route for self-renewal rather than recruitment of blood-borne monocytes (Ajami et al., 2007; Goldmann et al., 2013; Schulz et al., 2012; Sieweke and Allen, 2013). The limited replication and turnover of microglia make them a relatively stable population that is maintained throughout life. Within the brain parenchyma, microglia assume a context-dependent phenotype. In their quiescent surveillance phenotype, microglia display fine mobile processes extending into the surrounding microenvironment following a non-overlapping pattern. This allows the cells to maintain constant surveillance of the brain parenchyma for tissue changes (Davalos et al., 2005; Nimmerjahn et al., 2005).

While microglia are viewed as a single cell category, recent data support the coexistence of different subsets of microglia that have tightly adapted to specific requirements of the brain region. Several highly enriched genes, classified as "sensome" genes, have been identified in microglia that allow the cells to sense and interact with their local environment (Hickman et al., 2013). These sensome genes included those for putative purinergic receptors, *P2ry12* and *P2ry13*, transmembrane protein 119 (*Tmem119*), G protein-coupled receptor 34 (*Gpr34*), the 1-type lectin receptor, sialic acid-binding immunogloubulin-type lectin H (*Siglec-h*), triggering receptor expressed on myeloid cells 2 (*Trem2*), and the fractalkine receptor, *Cx3cr1*. Many of these same genes as well as Sal-like 1 (*Sall1*) and Fc receptor-like S (*Fcrls*) have been reported in quiescent microglia (Butovsky et al., 2014; Chiu et al., 2013). These sensing genes allow microglia cells to efficiently survey their environment to detect any aberrant occurrences that require quick action.

7.1 Phagocytic actions

The process of recognizing, attaching, engulfing, and internalizing extracellular material for clearance and degradation is orchestrated in the CNS primarily by microglia. Unique molecular signals for phagocytosis likely rely on secreted "find-me," "eat-me," and "don't-eat-me" signals also utilized by professional phagocytes (Ravichandran, 2010). The ability of microglia to recognize material for clearance is dependent upon receptor signaling. Soluble forms of lysophosphatidylcholine (LPC), CX3CL1, ATP, and UTP are known signals released by apoptotic cells that serve as "find-me" signals to attract phagocytes. "Eat-me" signals serve to recognize dead cells for engulfment with the best-characterized signal being phosphatidylserine, a cell membrane component that translocates from the inner to the outer leaflet of the plasma membrane of the apoptotic cell (Fadok et al., 2000) and targets them for phagocytosis (Ravichandran, 2003). Phagocytic engulfment is initiated through a direct binding to phagocyte receptors such as T-cell immunoglobulin and mucin-domain-containing molecule-1 (Tim4) (Miyanishi et al., 2007); brain-specific angiogenesis inhibitor-1 (BAI1; Park et al., 2007), stabilin-2 (Park et al., 2008), or the metabotropic P2Y6 receptor that recognizes the nucleotide UDP released from injured neurons (Koizumi et al., 2007). Purified microglia display the capacity to engulf phosphatidylserine-coated cells or particles (Konduru et al., 2009; Liu et al., 2013; Witting et al., 2000). An alternative "eat-me" signal is TREM2, a member of the immunoreceptor tyrosine-based activation motif (ITAM) receptors contributing to phagocytic activity (Neumann and Takahashi, 2007; Takahashi et al., 2005, 2007). TREM2 ligand is upregulated on apoptotic neurons and mediates signal transduction via association with SNAX adaptor protein-12 (DAP12) on microglia for phagocytosis (Hsieh et al., 2009; Takahashi et al., 2005). The UTP/UDP released from apoptotic neurons activates microglial P2Y6 receptors for phagocytosis (Koizumi et al., 2007). Microglia change over the course of the lifespan and such changes may drive pathogenic progression of diseases or injury through a diminution of neuroprotective functions, increase in neurotoxicity, and dysregulation of responses to signals and perturbations (Flanary et al., 2007; Lopes et al., 2008; Luo et al., 2010; Perry et al., 1993; Sheffield and Berman, 1998). With aging, microglia display decreased process motility (Hefendehl et al., 2014; Tremblay et al., 2012; Vaughan and Peters, 1974) that might compromise surveillance features and responsivity (Damani et al., 2011; Nimmerjahn et al., 2005). The cells assume a more reactive/activated phenotype with cytoplasmic structures showing excessive beading and spheroid swellings that may be reflective of dystrophy and senescence (Conde and Streit, 2006; Flanary and Streit, 2005; Streit et al., 2008) as supported by telomere shortening and decreased telomerase activity (Flanary and Streit, 2005; Flanary et al., 2007). Microglial co-localization around degenerating neurons is similar to that seen at younger ages; however, there is a higher incidence of clumping, irregular distribution, and accumulation of phagocytic debris, such as lysosomal lipopigments, cellular elements, vacuoles, and large vesicles (Hart et al., 2012; Hefendehl et al., 2014; Perry et al., 1993; Tremblay et al., 2012) that may represent an impaired dysfunction in microglia catabolic autophagy and the ability of the cell to degrade unnecessary or dysfunctional proteins through the lysosome (Wong et al., 2013). The increase in cell number has been proposed to reflect a compensatory mechanism to provide a sufficient level of cells for protection in the face of diminished functional capability (Streit and Xue, 2010).

7.2 Microglia/neuronal interactions

Under normal conditions, microglia assume a neural-specific phenotype that is tightly regulated by the CNS environment (Schmid et al., 2009). The communication between microglia and neurons is now recognized as a key mechanism in the regulation of microglial function. For example, neurons constitutively express cell-surface and secreted factors that serve as microglia inhibitors. The predominant expression of the fractalkine receptor CX3CR1 on microglia (Cardona et al., 2006; Gautier et al., 2012; Jung et al., 2000) drives much of this interaction as the ligand, CX3CL1 is expressed on neuronal membranes (Bazan et al., 1997; Mizoue et al., 1999; Rossi et al., 1998). The binding of CX3CL1 to CX3CR1 on microglia has been considered neuroprotective due to the inhibition of microglial activation (Bachstetter et al., 2011; Cardona et al., 2006; Rogers et al., 2011). Although other studies suggest an alternative pro-inflammatory interaction in neurodegenerative disease (Fuhrmann et al., 2010; Mattison et al., 2013). These interactions between neurons and microglia in the mature brain suggest a role to repress inflammatory activities with the minimization of inducible nitric oxide synthase, IL-1β, TNFα, and IL-6 production (Cardona et al., 2006). The interaction between the membrane glycoprotein, CD200, on neurons and microglial CD200R is related to the regulation of microglial activation and local inflammation (Frank et al., 2018; Hoek et al., 2000; Lyons et al., 2007) and functions to suppress inflammatory responses (Denieffe et al., 2013; Gorczynski, 2005; Meuth et al., 2008). Interactions with astrocytes have been more recently examined. Microglia express most of the chemokine receptors (CCR3, CCR5, CXCR3, and CX3CR1) depending on the microenvironment (Glabinsk and Ransohoff, 2001; Prinz et al., 2011) which makes them highly responsive to respective ligands produced by astrocytes (Cowell et al., 2006; Flynn et al., 2003; Gautier et al., 2012). Recent work suggests that astrocyte-activated microglia promoted neuronal dendritic growth in culture as compared to neuronal death induced by exposure to endothelial-activated microglia (Xing et al., 2018).

7.3 Microglia changes with injury

While a hallmark of microglia reactions is a change in the morphological phenotype of the cells, actions may or may not require major morphological transitions while more subtle changes in microglia morphology may reflect significant functional alterations. Originally the description of a morphological change as a result of injury was characterized by increased soma size and the retraction of elongated fine processes to shorter, coarser cytoplasmic processes displaying a bushy appearance. This was thought to continue to progress to a fully amoeboid morphology depending on the nature and severity of the signal and the biological process required (Hanisch and Kettenmann, 2007; Ransohoff and Perry, 2009). Mechanisms associated with the swelling and process extension of microglia involve activation of chloride channels (Ducharme et al., 2007; Zierler et al., 2008) and the phosphorylation state of cofilin, a member of the actin depolymerizing factor family (Hadas et al., 2012). While the transcription factor NF-κB regulated the inflammatory response of microglia (Wilms et al., 2003), disruption of the actin cytoskeleton has also been shown to activate this transcription factor and thus, the physical changes may be upstream of NF-κB (Nemeth et al., 2004). The fact that microglia adjust to their environment and that this may often be reflected in observable morphological changes has led to a less than specific use of the terminology of "activation." It is likely that the broad variety of microglia morphologies reflect different functional changes of the cells or simply, a limited difference in actual function but rather a difference in the nature of the initiation signal. With removal of the stimulating factor, microglia can then downregulate and return to the more ramified quiescent phenotype. The question remains, however, as to whether this downregulated cell is representative of the "normal" cells or whether the cell has been altered in some manner that may manifest in subsequent activation scenarios.

The initial concept of a stereotypic response of microglia with limited response variability has been replaced by data showing that microglia respond with a variety of different morphological changes by integrating multifarious inputs (Colton, 2009; Hanisch and Kettenmann, 2007; Kierdorf et al., 2013;

Prinz and Mildner, 2011; Ransohoff and Perry, 2009; Sierra et al., 2013; Tremblay and Majewska, 2011). Microglia are influenced by a plethora of factors including those that activate receptors (Biber et al., 2007; Domercq et al., 2013; Hanisch and Kettenmann, 2007; Kettenmann et al., 2011; Ransohoff and Cardona, 2010) as well as endogenously by transcription factor activation (Kierdorf and Prinz, 2013). It is thought that microglial receptors (Kettenmann et al., 2011; Lee, 2013; Pocock and Kettenmann, 2007; Sierra et al., 2007) can act as molecular switches to control microglial responses (Kettenmann et al., 2011) through alterations in calcium signals (Seifert et al., 2011). Microglia express P1 receptors for nucleosides and monophosphate containing nucleotides (adenosine and AMP) and most of the receptors of the P2 family for di- and tri-phosphate containing nucleotides (e.g., ADP, ATP). The level of expression varies with age, sex, and source of microglia (Crain et al., 2009) and can regulate both pro-inflammatory and anti-inflammatory functions. It is possible that the regular influence of local neurotransmitters may act to suppress microglial reactivity and promote ramification thus, a change in local neurotransmitter release could serve as a signal for microglia reactivity. Microglia express a1, a2, and b2 adrenergic receptors (AR) (Kohm and Sanders, 2001; Mori et al., 2002). Norepinephrine can induce a neuroprotective phenotype in microglia (McName et al., 2010; Mori et al., 2002) and anti-inflammatory properties are conferred by a7 nicotinic receptors (Carnevale et al., 2007). However, b2AR stimulation increases microglial NADPH oxidase (nicotinamide adenine dinucleotide phosphate-oxidase; a membrane bound enzyme complex that produces superoxide) and pro-inflammatory cytokine production (Wang et al., 2003). Activation of group I metabotropic glutamate receptors mGluR5 (Biber et al., 1999) downregulates inflammatory activities (Piers et al., 2011) while mGluRs2 and 3 activation promotes a pro-inflammatory response (Kaushal and Schlichter, 2008; Taylor et al., 2005). In culture, mGluRs 4, 5, and 8 are expressed and coupled to adenylate cyclase activation, promoting a neuroprotective phenotype (Taylor et al., 2003). As a possible mechanism to reduce excitotoxicity, microglia glutamate transporters GLT1/EAAT-2 (SLC1A2) and GLAST/EAAT-1 are upregulated following injury (Kettenmann et al., 2011). Microglia express functional GABAb receptors that reduce interleukin secretion (Kuhn et al., 2004) as well as GABAA receptors to modulate microglial motility and promote an anti-inflammatory or protective phenotype (Kettenmann et al., 2011; Lee, 2013). In addition to neurotransmitter receptors, microglia also express glucocorticoid receptors. This allows for glucocorticoid and mineralocorticoid hormones and glucocorticoids to act upon microglia by serving as priming signals (Frank et al., 2012). It is thought that the signaling is related to a protective mechanism with evidence of a downregulation of microglial glucocorticoid receptor levels by inflammatory challenge (Sierra et al., 2008) or with sustained glucocorticoid levels due to chronic stress (Park et al., 2011). Thus, a wide variety of receptors can serve to induce or modulate the microglia response.

7.4 Cell polarization

Macrophages respond to endogenous stimuli generated following infection or injury. This represents a product of a cell-mediated immune response that is essential for a host defense but also one that requires tight regulation. In the absence of microorganisms, a similar but sterile inflammatory response occurs often as a result of trauma, ischemia-reperfusion injury, or chemical exposure (Chen and Nunez, 2010; McPherson et al., 2014; Shechter and Schwartz, 2013). In general, the macrophage response is characterized based on the nature of the activating stimulus and the resulting production of factors (Mosser and Edwards, 2008). A conceptual framework for such activation identifies two polar extremes of signals, classical (M1) or alternative (M2) (Biswas and Mantovani, 2010). This characterization has been beneficial for describing immune responses occurring during acute infections, allergies, asthma, and obesity (Chinetti-Gbaguidi and Staels, 2011). While this dichotomy has been found to be limited, and representative of extremes of LPS/IFN or IL4/1L13 stimulation as a working hypothesis this concept has been applied, cautiously, to more fully characterize microglial responses (Boche et al., 2013; Colton, 2009).

LPS and IFNγ are classic inducers of M1 activation (Martinez and Gordon, 2014). LPS is recognized by TLR4 upon which activation induces myeloid differentiation primary response gene 88 (MyD88) and MaL/Tirap (Toll-interleukin 1 receptor domain containing adaptor protein)-dependent pathways. These pathways lead to production of pro-inflammatory cytokines (e.g., IFNγ, IL-12, TNF-α, IL-6, and IL-1β), chemokines (CCL2, CXCL10, and CXCL11), and antigen presentation molecules, such as MHC. Regulators include NF-κB, activator protein 1 (AP-1), interferon regulatory factors, signal transducers and activators of transcription 1 (STAT1), and early growth response (EGR) family members (Martinez et al., 2006). IFNγ signals through the IFNγ receptors (IFNγR1 and IFNγR2) for gene expression of cytokine receptors (IL-15R, IL-2RA, and IL-6R) and cell activation factors

(CD36, CD38, CD69, and CD97). IFNγ also regulates inducible SH2-containing protein (CISH), N-myc-interactor (NMI), protein tyrosine phosphatase, receptor type, C (PTPRC), protein tyrosine phosphatase, receptor type, O (PTPRO), and suppressor of cytokine signaling 1 (SOCS1) (Hu and Ivashkiv, 2009; Waddell et al., 2010). Mediators of IFNγ-induced signaling include STAT1, JAK2, and IRF1. There are similarities between gene expression after IFNγ and LPS stimulation of macrophages; however, these M1 responses are not considered homologous (Murray et al., 2014). Gene expression profiles following stimulation with a combination of the cytokines or various other stimuli are different from LPS or IFNγ profiles alone generating a vast spectrum rather than a polarized dichotomy of responses (Martinez et al., 2006).

Upon stimulation with TLR agonists, microglia switch their reliance on oxidative phosphorylation metabolism (Cherry et al., 2014) toward glycolytic metabolism to maintain mitochondrial function and ensure cell survival (Gimeno-Bayon et al., 2014; Orihuela et al., 2015; Voloboueva et al., 2013). It has been proposed that a two-stage process occurs with M1 activation with cells in the initial stage capable of utilizing both oxidative and glycolytic metabolism then shifting to rely on glycolytic metabolism for survival. Under both stages, activation of the pentose phosphate pathway occurs (Orihuela et al., 2015). A key feature of M1 macrophages is their ability to produce reactive oxygen species (ROS) to facilitate killing of phagocytized bacteria (West et al., 2011). Whether or not this occurs with all levels of M1 stimulation in microglia is not as clearly defined given that pro-inflammatory cytokine production can be induced in the absence of nitric oxide (NO) or ROS production. When the response is severe enough to result in ROS production, the associated intracellular damage is limited by increased generation of nicotinamide adenine dinucleotide phosphate (NADPH) required for maintenance of reduced glutathione (Kletzien et al., 1994; Salvemini et al., 1999) and NO production (Bredt and Snyder, 1990; Knowles and Moncada, 1994). However, while attempting to protect the cell of source, these produce damage to surrounding tissue. While these inflammatory factors serve in the host defense process they can also lead to the creation of a toxic microenvironment created with the production and release of pro-inflammatory cytokines, chemokines, redox molecules (NOX, PHOX, iNOS), scavenger receptors MARCO, co-stimulatory proteins (CD40), and MHCII (Boche et al., 2013; Colton and Wilcock, 2010; Gordon and Martinez, 2010; Hanisch and Kettenmann, 2007; Henkel et al., 2009; Martinez and Gordon, 2014;

Mosser and Edwards, 2008; Varnum and Ikezu, 2012). The outcome of a M1 polarizing event is dependent upon a number of features, not the least of which is the severity or length of the response and whether or not it includes production of iNOS, ROS, or activation of NOD-like receptor family pyrin domain-containing 3 (NLRP3) inflammasome complex (Bordt and Polster, 2014; de Rivero Vaccari et al., 2015).

While the initial response of macrophages to injury has been known for some time, positive influences to facilitate resolution of inflammation through anti-inflammatory factors, to deactivate pro-inflammatory cell phenotypes and reestablish tissue homeostasis, have had a shorter time of consideration. In the early 1990s, the concept of alternative activation was developed largely based on work showing a role for IL-4 in the induction of an alternative (M2) activation state (Stein et al., 1992). In M2 macrophages, glucose consumption is significantly lower as compared to M1 (Rodríguez-Prados et al., 2010) and the sedoheptulose kinase carbohydrate kinase-like protein is critical for regulating the pentose phosphate pathway (Galvan-Peña and O'Neill, 2014). Under this state, expression of the anti-inflammatory cytokines (*IL-4, IL-10, IL-13*, and *Tgfβ*) as well as, arginase-1 (*Arg-1*), *CD206*, and Chitinase-3-like-3 (*Ym-1* in rodents) was induced (Colton, 2009; Henkel et al., 2009). Upon further study, subclasses of M2 activation have been identified and, as with the M1 phenotype, they are dependent upon the inducting stimuli. The M2a activation state is induced by parasitic products or associated signals (IL-4 and IL-13) providing a longer-term function for resolution and repair (Gordon, 2003; Lawrence and Natoli, 2011; Mills, 2012; Wynn et al., 2013). In this case, signaling occurs through IL-4 receptor alpha (IL-4Rα) leading to inhibition of NF-κB signaling induced by M1 activation. M2b polarization is observed with triggering of Fc gamma receptors, TLRs, and immune complexes (Martinez and Gordon, 2014). M2c polarization occurs in response to specific anti-inflammatory factors such as, IL-10, TGFβ, and glucocorticoids (Gordon, 2003; Martinez et al., 2008; Mills, 2012; Vodovotz et al., 1993). In addition, cells do not appear to be committed to one phenotype but rather can shift from M2b phenotype to a mixture of M1 and M2a/b (Lisi et al., 2014). M2 polarization of microglia is similar to peripheral macrophages (Chhor et al., 2013; Fenn et al., 2012; Freilich et al., 2013; Liu et al., 2012), displaying different mRNA profiles for IL-4 and IL-10 stimulation including *Arg1*, mannose receptor 1 (*Mrc1*), chitinase-like 3 *YM-1*, found in inflammatory zone 1 (*Fizz1*) and peroxisome proliferator-activated receptor (*PPAR*) (Michelucci et al., 2009). In addition, the specificity of such markers remains in question. For example, the primary "marker" for M2, Arg1, is also induced in M1 macrophages and expressed in some resident and mycobacteria-infected macrophages (El Kasmi et al., 2008). In addition, induction of arginase 1 could be related to neurotoxicity based on arginine deprivation (Vogel et al., 2013). While these associations have been demonstrated in vitro, a number of the M2-related products can be induced in vivo by sterile wounds in the absence IL-4 or IL-13 (Crane et al., 2014) suggesting an alternative stimulus.

Michelucci et al. (2009) published a wide-ranging transcription and functional profile using the M1/M2 differentiation spectrum in microglia characterizing the inflammatory and functional phenotype in the murine microglia cell line, MMGT12, and in murine primary microglia upon stimulation with LPS, IFN-γ, IL-4 or, IL-10. Transcription profile analysis showed that cells exposed to LPS or IFN-γ were characterized by a classical M1 phenotype (*Il-1b*, *Il-6*, *Tnfa*, *Nos-II*, *Cox-2*, *Ccl2*, *Ccl20*, *Ccr2*) and those stimulated with IL-10 or IL-4 were characterized by M2-related mRNA expression (*Arg-1*, *CD206*, *Ym-1*, *Fizz1*, *Ppar*). Based upon this framework, one might consider that the response occurs in a sequential fashion with M1 preceding M2, however, studies have demonstrated that this is not necessarily the case. Exposure of microglia to branched-chain amino acids results in an intermediate M1/M2 phenotype of an enhanced IL-10 expression and phagocytic activity but an increase in free radical production. In vivo, a concurrent expression of M1 and M2 factors has been identified. For example, CNS biopsies from multiple sclerosis patients showed markers for both M1 and M2 macrophage populations including CD40, CD64, CD86, and CD32, mannose receptor, and CD163 were synergistically displayed (Vogel et al., 2013). Other studies have demonstrated concurrent expression of M1 and M2 markers in various disease or injury models (Chiu et al., 2013; McPherson et al., 2014; Morganti et al., 2016; Orihuela et al., 2015; Stirling et al., 2014). However, it is not yet clear whether the mixed expression pattern can occur within individual cells or rather if the region of interest is comprised of cells showing different profiles. While this framework has been put forth to aid in interpretation of experimental data too often a limited dichotomy of M1 and M2 is presented in the literature. This has indeed lead to a potential confusion if not a hindering of the larger body of experimental data trying to organize our understanding of a very complex and dynamic cell within the CNS (Ransohoff, 2016). From work assessing transcriptional regulation during human macrophage activation, an extended version of the current M1 versus M2 polarization model has been proposed. This spectrum model contains nine distinct activation programs (Xue et al., 2014). A distinct axis of response was demonstrated upon stimulation with IFNγ or IL-4, with a similar axis demonstrated between LPS and IL-13 polarization. With stimuli not directly linked to M1 or M2 polarization such as, free fatty acids, high-density lipoprotein, or combinations associated with chronic inflammation, a spectrum of signatures was displayed. In addition to immune-related factors, the spectrum recruited clusters associated with cell death, biosynthetic processes of small molecules, and metabolic and catabolic process as critical aspects of macrophage activation.

8. Summary

It is well accepted that the CNS maintains a resident cell population that participates in the genesis of immune responses during injury/infection and significantly contributes to maintaining homeostasis within the tissue. Microglia have been identified as the primary immune cell in the brain, however, more is becoming known about the role and influence of the astrocyte in these functions. In any case, immune responses in the brain do not occur in isolation but rather involve multiple cell types and structures including those within the parenchyma, in the vasculature system, within the BBB, and now incorporating the possible influence of the CNS lymphatic drainage system. Gaining a better understanding of resident immune cells of the CNS and their functions will contribute to our understanding of the critical and necessary actions taken to maintain a healthy environment. In the area of neurotoxicology, consideration of an innate immune response may involve a direct response to chemical exposure, systemic toxicity, or secondary as a reaction to dysfunctional or damaged neural cells including cells associated with the neurovasculature. Alternatively, chemical exposure may alter the normal function of microglia thus creating an environment in which the normal responses to maintain homeostatic balance are compromised. This could manifest as the inability of the cells to identify a danger, localize to the site of injury, mount a sufficient or appropriate sequence of inflammatory responses, or phagocytize and clear aberrant proteins and cellular debris.

References

Abbott, N.J., Ronnback, L., Hansson, E., 2006. Astrocyte-endothelial interactions at the blood-brain barrier. Nat. Rev. Neurosci. 7, 41–53.

Adamczak, S.E., de Rivero Vaccari, J.P., Dale, G., et al., 2014. Pyroptotic neuronal cell death mediated by the AIM2 inflammasome. J. Cereb. Blood Flow Metab. 34 (4), 621–629.

Afonina, I.S., Zhong, Z., Karin, M., et al., 2017. Limiting inflammation—the negative regulation of NF-κB and the NLRP3 inflammasome. Nat. Immunol. 18 (8), 861–869.
Ajami, B., Bennett, J.L., Krieger, C., et al., 2007. Local self-renewal can sustain CNS microglia maintenance and function throughout adult life. Nat. Neurosci. 10 (12), 1538–1543.
Al Ahmad, A., Taboada, C.B., Gassmann, M., et al., 2011. Astrocytes and pericytes differentially modulate blood-brain barrier characteristics during development and hypoxic insult. J. Cereb. Blood Flow Metab. 31, 693–705.
Alarcon, R., Fuenzalida, C., Santibanez, M., et al., 2005. Expression of scavenger receptors in glial cells. Comparing the adhesion of astrocytes and microglia from neonatal rats to surface-bound beta-amyloid. J. Biol. Chem. 280, 30406–30415.
Alliot, F., Godin, I., Pessac, B., 1999. Microglia derive from progenitors, originating from the yolk sac, and which proliferate in the brain. Brain Res. Dev. Brain Res. 117 (2), 145–152.
Alvarez, J.I., Dodelet-Devillers, A., Kebir, H., et al., 2011. The hedgehog pathway promotes blood-brain barrier integrity and CNS immune quiescence. Science 334, 1727–1731.
Antony, J.M., Paquin, A., Nutt, S.L., 2011. Endogenous microglia regulate development of embryonic cortical precursor cells. J. Neurosci. Res. 89, 286–298.
Asperlund, A., Antila, S., Proulx, S.T., et al., 2015. A dural vascular system that drains brain insterstitial fluid and macromolecules. J. Exp. Med. 212, 991–999.
Bachstetter, A.D., Morganti, J.M., Jernberg, J., et al., 2011. Fractalkine and CX3CR1 regulate hippocampal neurogenesis in adult and aged rats. Neurobiol. Aging 32 (11), 2030–2044.
Balabanov, R., Washington, R., Wagnerova, J., et al., 1996. CNS microvascular pericytes express macrophage-like function, cell surface integrin alpha M, and macrophage marker ED-2. Microvasc. Res. 52, 127–142.
Banerjee, S., Bhat, M.A., 2007. Neuron-glial interactions in blood-brain barrier formation. Annu. Rev. Neurosci. 30, 235–258.
Barbelova, A., Moreth, K., Tsalastra-Greul, W., et al., 2009. Biglycan, a danger signal that activates the NLRP3 inflammasome via Toll and P2X receptors. J. Biol. Chem. 284, 24035–24048.
Bazan, J.F., Bacon, K.B., Hardiman, G., et al., 1997. A new class of membrane-bound chemokine with a XC3C motif. Nature 385, 640–644.
Bechmann, I., Priller, J., Kovac, A., et al., 2001. Immune surveillance of mouse brain perivascular spaces by blood-borne macrophages. Eur. J. Neurosci. 14, 1651–1658.
Beggs, S., Trang, T., Salter, M.W., 2012. P2X4R+ microglia drive neuropathic pain. Nat. Neruosci. 15 (8), 1068–1073.
Biber, K., Laurie, D.J., Berthele, A., et al., 1999. Expression and signaling of group I metabotropic glutamate receptors in astrocytes and microglia. J. Neurochem. 72, 1671–1680.
Biber, K., Neumann, H., Inoue, K., et al., 2007. Neuronal 'on' and 'off' signals control microglia. Trends Neurosci. 30 (11), 596–602.
Biswas, S.K., Mantovani, A., 2010. Macrophage plasticity and interaction with lymphocyte subsets: cancer as a paradigm. Nat. Immunol. 11 (10), 889–896.
Boche, D., Perry, V.H., Nicoll, J.A., 2013. Review: activation patterns of microglia and their identification in the human brain. Neuropathol. Appl. Neurobiol. 39 (1), 3–18.
Bordt, E.A., Polster, B.M., 2014. NADPH oxidase- and mitochondria-derived reactive oxygen species in proinflammatory microglial activation: a bipartisan affair? Free Radic. Biol. Med. 76c, 34–46.
Bowman, C.C., Rasley, A., Tranguch, S.L., et al., 2003. Cultured astrocytes express toll-like receptors for bacterial products. Glia 43, 281–291.
Brambilla, R., Bracchi-Ricard, V., Hu, W.H., et al., 2005. Inhibition of astroglial nuclear factor kappaB reduces inflammation and improves functional recovery after spinal cord injury. J. Exp. Med. 202, 145–156.

Brambilla, R., Hurtado, A., Persaud, T., et al., 2009. Transgenic inhibition of astroglial NF-kappa B leads to increased axonal sparing and sprouting following spinal cord injury. J. Neurochem. 110 (2), 765–778.
Bredt, D.S., Snyder, S.H., 1990. Isolation of nitric oxide synthetase, a calmodulin-requiring enzyme. Proc. Natl. Acad. Sci. U. S. A. 87 (2), 682–685.
Burm, S.M., Zuiderwijk-Sick, E.A., 't Jong, A.E., et al., 2015. Inflammasome-induced IL-1β secretion in microglia is characterized by delayed kinetics and is only partially dependent on inflammatory caspases. J. Neurosci. 35 (2), 678–687.
Burudi, E.M., Regnier-Vigouroux, A., 2001. Regional and cellular expression of the mannose receptor in the post-natal developing mouse brain. Cell Tissue Res. 303, 307–317.
Burudi, E.M., Riese, S., Stahl, P.D., et al., 1999. Identification and functional characterization of the mannose receptor in astrocytes. Glia 25, 44–55.
Butovsky, O., Jedrychowski, M.P., Moore, C.S., et al., 2014. Identification of a unique TGF-beta-dependent molecular and functional signature in microglia. Nat. Neurosci. 17 (1), 131–143.
Cajal, R.Y., 1913. Contribucion al conocimiento de la neuroglia del cerebro humano. Trab. Lab. Invest. Biol. 11, 254.
Cardona, A., Pioro, E.P., Sasse, M.E., et al., 2006. Control of microglial neurotoxicity by the fractalkine receptor. Nat. Neurosci. 9, 917–924.
Carnevale, D., De Simone, R., Minghetti, L., 2007. Microglia-neuron interaction in inflammatory and degenerative disease: role of cholinergic and noradrenergic systems. CNS Neurol. Disord. Drug Targets 6, 388–397.
Carpentier, P.A., Begolka, W.S., Olson, J.K., et al., 2005. Differential activation of astrocytes by innate and adaptive immune stimuli. Glia 49, 360–374.
Cekanaviciute, E., Dietrich, H.K., Axtell, R.C., et al., 2014a. Astrocytic TGF-beta signaling limits inflammation and reduces neuronal damage during central nervous system toxoplasma infection. J. Immunol. 193, 139–149.
Cekanaviciute, E., Fathali, N., Doyle, K.P., et al., 2014b. Astrocytic transforming growth factor-beta signaling reduces subacute neuroinflammation after stroke in mice. Glia 62, 1227–1240.
Chen, G.Y., Nunez, G., 2010. Sterile inflammation: sensing and reacting to damage. Nat. Rev. Immunol. 10 (12), 826–837.
Chen, B.P., Kuziel, W.A., Lane, T.E., 2001. Lack of CCR2 results in increased mortality and impaired leukocyte activation and trafficking following infection of the central nervous system with a neurotropic coronavirus. J. Immunol. 167 (8), 4585–4592.
Cherry, J.D., Olschowka, J.A., O'Banion, M.K., 2014. Are "resting" microglia more "m2"? Front. Immunol. 5, 594. https://doi.org/10.3389/fimmu.2014.00594.
Chhor, V., Le Charpentier, T., Lebon, S., et al., 2013. Characterization of phenotype markers and neuronotoxic potential of polarised primary microglia in vitro. Brain Behav. Immun. 32, 70–85.
Chinetti-Gbaguidi, G., Staels, B., 2011. Macrophage polarization in metabolic disorders: functions and regulation. Curr. Opin. Lipidol. 22 (5), 365–372.
Chiu, I.M., Morimoto, E.T., Goodarzi, H., et al., 2013. A neurodegeneration-specific gene-expression signature of acutely isolated microglia from an amyotrophic lateral sclerosis mouse model. Cell Rep. 4, 385–401.
Choi, W.H., Ji, K.A., Jeon, S.B., et al., 2005. Anti-inflammatory roles of retinoic acid in rat brain astrocytes: suppression of interferon-gamma-induced JAK/STAT phosphorylation. Biochem. Biophys. Res. Commun. 329, 125–131.
Colombo, E., Farina, C., 2016. Astrocytes: key regulators of neuroinflammation. Trends Immunol. 37, 608–620.
Colton, C.A., 2009. Heterogeneity of microglial activation in the innate immune response in the brain. J. Neuroimmune. Pharmacol. 4 (4), 399–418.

Colton, C., Wilcock, D.M., 2010. Assessing activation states in microglia. CNS Neurol. Disord. Drug Targets 9 (2), 174–191.
Conde, J.R., Streit, W.J., 2006. Microglia in the aging brain. J. Neuropathol. Exp. Neurol. 65 (3), 199–203.
Cowell, R.M., Xu, H., Parent, J.M., et al., 2006. Microglial expression of chemokine receptor CCR5 during rat forebrain development and after perinatal hypoxia-ischemic. J. Neuroimmunol. 173, 155–165.
Crain, J., Nikodemova, M., Watters, J., 2009. Expression of P2 nucleotide receptors varies with age and sex in murine brain microglia. J. Neuroinflammation 6, 24. https://doi.org/10.1186/1742-2094-6-24.
Crane, M.J., Daley, J.M., van Houtte, O., 2014. The monocyte to macrophage transition in the murine sterile wound. PLoS One 9 (1), e86660. https://doi.org/10.1371/journal.pone.0086660.
Creagh, E.M., O'Neill, L.A., 2006. TLRs, NLRs and RLRs: a trinity of pathogen sensors that co-operate in innate immunity. Trends Immunol. 27, 352–357.
Damani, M.R., Zhao, L., Fontainhas, A.M., et al., 2011. Age-related alterations in the dynamic behavior of microglia. Aging Cell 10 (2), 263–276.
Daneman, R., Zhou, L., Kebede, A.A., 2010. Pericytes are required for blood-brain barrier integrity during embryogenesis. Nature 468 (7323), 562–566.
Danese, S., Dejana, E., Fiocchi, C., 2007. Immune regulation by microvas- cular endothelial cells: directing innate and adaptive immunity, coagulation, and inflammation. J. Immunol. 178, 6017–6022.
Davalos, D., Grutzendler, J., Yang, G., et al., 2005. ATP mediates rapid microglial response to local brain injury in vivo. Nat. Neurosci. 8 (6), 752–758.
de Rivero Vaccari, J.P., Dietrich, W.D., Keane, R.W., 2014. Activation and regulation of cellular inflammasomes: gaps in our knowledge for central nervous system injury. J. Cereb. Blood Flow Metab. 34 (3), 369–375.
de Rivero Vaccari, J.C., Brand 3rd, F.J., Berti, A.F., et al., 2015. Mincle signaling in the innate immune response after traumatic brain injury. J. Neurotrauma 32 (4), 228–236.
del Río Hortega, P., 1932. Microglia. In: Penfield, W. (Ed.), Cytology and Cellular Pathology of the Nervous System. Hoeber, New York, pp. 481–534.
Denieffe, S., Kelly, R.J., McDonald, C., et al., 2013. Classical activation of microglia in CD200-deficient mice is a consequence of blood brain barrier permeability and infiltration of peripheral cells. Brain Behav. Immun. 34, 86–97.
DeSalvo, M.K., Mayer, N., Mayer, F., et al., 2011. Physiologic and anatomic characterization of the brain surface glia barrier of *Drosophila*. Glia 59, 1322–1340.
Dinarello, C.A., 2009. Immunological and inflammatory functions of the interleukin-1 family. Annu. Rev. Immunol. 27, 519–550.
Domercq, M., Vázquez-Villoldo, N., Matute, C., 2013. Neurotransmitter signaling in the pathophysiology of microglia. Front. Cell. Neurosci. 7, 49.
Ducharme, G., Newell, E.W., Pinto, C., et al., 2007. Small-conductance Cl- channels contribute to volume regulation and phagocytosis in microglia. Eur. J. Neurosci. 26, 2119–2130.
El Kasmi, K.C., Qualls, J.E., Pesce, J.T., et al., 2008. Toll-like receptor-induced arginase 1 in macrophages thwarts effective immunity against intracellular pathogens. Nat. Immunol. 9 (12), 1399–1406.
Elliott, E.I., Sutterwala, F.S., 2015. Initiation and perpetuation of NLRP3 inflammasome activation and assembly. Immunol. Rev. 265 (1), 35–52.
Esen, N., Tanga, F.Y., DeLeo, J.A., et al., 2004. Toll-like receptor 2 (TLR2) mediates astrocyte activation in response to the gram-positive bacterium *Staphylococcus aureus*. J. Neurochem. 88, 746–758.

Fadok, V.A., Bratton, D.G., Rose, D.M., et al., 2000. A receptor for phosphatidylserine-specific clearance of apoptotic cells. Nature 405 (6782), 85–90.
Farina, C., Krumbholz, M., Giese, T., et al., 2005. Preferential expression and function of toll-like receptor 3 in human astrocytes. J. Neuroimmunol. 159, 12–19.
Farina, C., Aloisi, F., Meinl, E., 2007. Astrocytes are active players in cerebral innate immunity. Trends Immunol. 28, 138–145.
Fenn, A.M., Henry, C.J., Huang, Y., et al., 2012. Lipopolysaccharide-induced interleukin (IL)-4 receptor-alpha expression and corresponding sensitivity to the M2 promoting effects of IL-4 are impaired in microglia of aged mice. Brain Behav. Immun. 26 (5), 766–777.
Finsen, B., Owens, T., 2011. Innate immune responses in the central nervous system inflammation. FEBS Lett. 585 (23), 3806–3812.
Flanary, B.E., Streit, W.J., 2005. Effects of axotomy on telomere length, telomerase activity, and protein in activated microglia. J. Neurosci. Res. 82 (2), 160–171.
Flanary, B.E., Sammons, N.W., Nguyen, C., et al., 2007. Evidence that aging and amyloid promote microglial cell senescence. Rejuvenation Res. 10 (1), 61–74.
Flynn, G., Maru, S., Loughlin, J., et al., 2003. Regulation of chemokine receptor expression in human microglia and astrocytes. J. Neuroimmunol. 136, 84–93.
Foldi, M., Csanda, E., Simon, M., et al., 1968. Lymphogenic haemangiopathy. Prelymphatic pathways in the wall of the cerebral and cervical blood vessels. Angiologica 5, 250–262.
Franchi, L., Eigenbrod, T., Núñez, G., 2009. Cutting edge: TNF-alpha mediates sensitization to ATP and silica via the NLRP3 inflammasome in the absence of microbial stimulation. J. Immunol. 183 (2), 792–796.
Frank, M.G., Thompson, B.M., Watkins, L.R., 2012. Glucocorticoid mediate stress-induced priming of microglial pro-inflammatory responses. Brain Behav. Immun. 26, 337–345.
Frank, M.G., Fonken, L.K., Annis, J.L., et al., 2018. Stress disinhibits microglia via down-regulation of CD200R: a mechanism of neuroinflammatory priming. Brain Behav. Immun. 69, 62–73.
Franklin, B.S., Bossaller, L., De Nardo, D., et al., 2014. The adaptor ASC has extracellular and 'prionoid' activities that propagate inflammation. Nat. Immunol. 15 (8), 727–737.
Freeman, L., Guo, H., David, C.N., et al., 2017. NLR members NLRC4 and NLRP3 mediate sterile inflammasome activation in microglia and astrocytes. J. Exp. Med. 214, 1351–1370.
Freilich, R.W., Woodbury, M.E., Ikezu, T., 2013. Integrated expression profiles of mRNA and miRNA in polarized primary murine microglia. PLoS One 8 (11), e79416. https://doi.org/10.1371/journal.pone.0079416.
Frey, H., Schroeder, N., Manon-Jensen, T., et al., 2013. Biological interplay between proteoglycans and their innate immune receptors in inflammation. FEBS J. 280, 433–436.
Fuhrmann, M., Bittner, T., Jung, C.K., et al., 2010. Microglial Cx3cr1 knockout prevents neuron loss in a mouse model of Alzheimer's disease. Nat. Neurosci. 13, 411–413.
Galea, I., Bechmann, I., Perry, V.H., 2007. What is immune privilege (not)? Trends Immunol. 28 (1), 12–18.
Galvan-Peña, S., O'Neill, L.A., 2014. Metabolic reprograming in macrophage polarization. Front. Immunol. 5, 420.
Gautier, E.L., Shay, T., Miller, J., et al., 2012. Gene-expression profiles and transcriptional regulatory pathways that underlie the identity and diversity of mouse tissue macrophages. Nat. Immunol. 13, 1118–1128.
Gehrmann, J., Matsumoto, Y., Kreutzberg, G.W., 1995. Microglia: intrinsic immuneffector cell of the brain. Brain Res. Brain Res. Rev. 20, 269–287.

Gimeno-Bayon, J., Lopez-Lopez, A., Rodriguez, M.J., et al., 2014. Glucose pathways adaptation supports acquisition of activated microglia phenotype. J. Neurosci. 92 (6), 723–731.
Ginhoux, F., Jung, S., 2014. Monocytes and macrophages: developmental pathways and tissue homeostasis. Nat. Rev. Immunol. 14 (6), 392–404.
Ginhoux, F., Greter, M., Leboeuf, M., et al., 2010. Fate mapping analysis reveals that adult microglia derive from primitive macrophages. Science 330, 841–845.
Glabinsk, A.R., Ransohoff, R.M., 2001. Targeting the chemokine system for multiple sclerosis treatment. Curr. Opin. Investig. Drugs 2, 1712–1719.
Goldmann, T., Wieghofer, P., Müller, P.F., et al., 2013. A new type of microglia gene targeting shows TAK1 to be pivotal in CNS autoimmune inflammation. Nat. Neurosci. 16 (11), 1618–1626.
Gomez Perdiguero, E., Klapproth, K., Schulz, C., et al., 2015. Tissue-resident macrophages originate from yolk-sac-derived erythro-myeloid progenitors. Nature 518 (7540), 547–551.
Gorczynski, R.M., 2005. CD200 and its receptors as targets for immunoregulation. Curr. Opin. Investig. Drugs 6, 483–488.
Gordon, S., 2003. Alternative activation of macrophages. Nat. Rev. Immunol. 3 (1), 23–35.
Gordon, S., Martinez, F.O., 2010. Alternative activation of macrophages: mechanism and functions. Immunity 32 (5), 593–604.
Gorina, R., Font-Nieves, M., Marquez-Kisinousky, L., et al., 2011. Astrocyte TLR4 activation induces a proinflammatory environment through the interplay between MyD88-dependent NFkappaB signaling, MAPK, and Jak1/Stat1 pathways. Glia 59, 242–255.
Graeber, M.B., Streit, W.J., Kiefer, R., et al., 1990. New expression of myelomonocytic antigens by microglia and perivascular cells following lethal motor neuron injury. J. Neuroimmunol. 27, 121–132.
Gris, D., Ye, Z., Iocca, H.A., et al., 2010. NLRP3 plays a critical role in the development of experimental autoimmune encephalomyelitis by mediating Th1 and Th17 responses. J. Immunol. 185, 974–981.
Gurung, P., Lukens, J.R., Kanneganti, T.D., 2015. Mitochondria: diversity in the regulation of NLRP3 inflammasome. Trends Mol. Med. 3, 193–201.
Hadas, S., Spira, M., Hanisch, U.K., et al., 2012. Complement receptor-3 negatively regulates the phagocytosis of degenerated myelin through tyrosine kinase Syk and cofilin. J. Neuroinflammation 9, 166.
Hamby, M.E., Coppola, G., Ao, Y., et al., 2012. Inflammatory mediators alter the astrocyte transcriptome and calcium signaling elicited by multiple G-protein-coupled receptors. J. Neurosci. 32, 14489–14510.
Hanamsagar, R., Torres, W., Kielian, T., 2011. Inflammasome activation and IL-1beta/IL-18 processing are influenced by distinct pathways in microglia. J. Neurochem. 119, 736–748.
Hanisch, U.K., Kettenmann, H., 2007. Microglia: active sensor and versatile effector cells in the normal and pathologic brain. Nat. Neurosci. 10 (11), 1387–1394.
Harry, G.J., Kraft, A.D., 2012. Microglia in the developing brain: a potential target with lifetime effects. Neurotoxicology 33 (2), 191–206.
Hart, A.D., Wyttenbach, A., Perry, V.H., et al., 2012. Age related changes in microglial phenotype vary between CNS regions: grey versus white matter differences. Brain Behav. Immun. 26 (5), 754–765.
Hauser, S.L., Oksenberg, J.R., 2006. The neurobiology of multiple sclerosis: genes, inflammation, and neurodegeneration. Neuron 52 (1), 61–76.
Hawkins, B.T., Davis, T.P., 2005. The blood-brain barrier/neurovascular unit in health and disease. Pharmacol. Rev. 57 (2), 173–185.

Hefendehl, J.K., Neher, J.J., Suhs, R.B., et al., 2014. Homeostatic and injury-induced microglia behavior in the aging brain. Aging Cell 13 (1), 60–69.
Hellström, M., Gerhardt, H., Kalén, M., et al., 2001. Lack of pericytes leads to endothelial hyperplasia and abnormal vascular morphogenesis. J. Cell Biol. 153, 543–553.
Hendrix, S., Nitsch, R., 2007. The role of T helper cells in neuroprotection and regeneration. J. Neuroimmunol. 184 (1–2), 100–112.
Henkel, J.S., Beers, D.R., Zhao, W., et al., 2009. Microglia in ALS: the good, the bad, and the resting. J. NeuroImmune Pharmacol. 4 (4), 389–398.
Hermann, D.M., Elali, A., 2012. The abluminal endothelial membrane in neurovascular remodeling in health and disease. Sci. Signal. 5, re4.
Hickey, W.F., Kimura, H., 1988. Perivascular microglial cells of the CNS are bone marrow-derived and present antigen in vivo. Science 239, 290–292.
Hickman, S.E., Kingery, N.D., Ohsumi, T.K., et al., 2013. The microglial sensome revealed by direct RNA sequencing. Nat. Neurosci. 16 (12), 1896–1905.
Hoek, R.M., Ruuls, S.R., Murphy, C.A., et al., 2000. Down-regulation of the macrophage lineage through interaction with OX2 (CD200). Science 290, 1768–1771.
Hornung, V., Bauernfeind, F., Halle, et al., 2008. Silica crystals and aluminum salts activate the NALP3 inflammasome through phagosomal destabilization. Nat. Immunol. 9 (8), 847–856.
Hsieh, C.L., Kokke, M., Sputa, S.C., 2009. A role for TREM2 ligands in the phagocytosis of apoptotic neuronal cells by microglia. J. Neurochem. 109 (4), 1144–1156.
Hu, X., Ivashkiv, L.B., 2009. Cross-regulation of signaling pathways by interferon-gamma: implications for immune responses and autoimmune diseases. Immunity 31, 539–550.
Husemann, J., Silverstein, S.C., 2001. Expression of scavenger receptor class B, type I, astrocytes and vascular smooth muscle cells in normal adult mouse and human brain and in Alzheimer's disease brain. Am. J. Pathol. 158, 825–832.
Ifergan, I., Kebir, H., Bernard, M., 2008. The blood-brain barrier induces differentiation of migrating monocytes into the Th17-polarizing dendritic cells. Brain 131 (Pt 3), 785–799.
Iliff, J.J., Wang, M., Liao, Y., et al., 2012. A paravascular pathway facilitates CSF flow through the brain parenchyma and the clearance of interstitial solutes, including amyloid β. Sci. Transl. Med. 4, 147ra111.
Jack, C.S., Arbour, N., Manusow, J., et al., 2005. TLR signaling tailors innate immune responses in human microglia and astrocytes. J. Immunol. 175, 4320–4330.
Jensen, C.J., Massie, A., De Keyser, J., 2013. Immune players in the CNS: the astrocyte. J. NeuroImmune Pharmacol. 8, 824–839.
Jessen, N.A., Munk, A.S.F., Lundgard, I., et al., 2015. The glymphatic system: a beginner's guide. Neurochem. Res. 40, 2583–2599.
John, G.R., Shankar, S.L., Shafit-Zagardo, B., et al., 2002. Multiple sclerosis: re-expression of a developmental pathway that restricts oligodendrocyte maturation. Nat. Med. 8, 1115–1121.
Jung, S., Aliberti, J., Graemmel, P., et al., 2000. Analysis of fractalkine receptor Cx(3)CR1 fumction by targeted deletion and green fluorescent protein reporter gene insertion. Mol. Cell. Biol. 20, 4106–4114.
Kaushal, V., Schlichter, L.C., 2008. Mechanisms of microglia-mediated neurotoxicity in a new model of the stroke penumbra. J. Neurosci. 28, 2221–2230.
Kaushal, V., Dye, R., Pakavathkumar, P., 2015. Neuronal NLRP1 inflammasome activation of caspase-1 coordinately regulates inflammatory interleukin-1-beta production and axonal degeneration-associated caspase-6 activation. Cell Death Differ. 22 (10), 1676–1686.
Kaushik, D.K., Gupta, M., Kumawat, K.L., et al., 2012. NLRP3 inflammasome: key mediator of neuroinflammation in murine Japanese encephalitis. PLoS One 7, e32270.

Kawai, T., Akira, S., 2010. The role of pattern-recognition receptors in innate immunity: update on toll-like receptors. Nat. Immunol. 11, 373–384.
Kettenmann, H., Hanisch, U.K., Noda, M., et al., 2011. Physiology of microglia. Physiol. Rev. 91 (2), 461–553.
Kida, S., Pantazis, A., Weller, R.O., 1993. CSF drains directly from the subarachnoid space into nasal lymphatics in the rat. Anatomy, histology, and immunological significance. Neuropathol. Appl. Neurobiol. 19, 480–488.
Kieliam, T., Bearden, E.D., Baldwin, A.C., et al., 2004. IL-1 and TNF-alpha play a pivotal role in the host immune response in a mouse model of *Staphylococcue aureus*-induced experimental brain abscess. J. Neuropathol. Exp. Neurol. 63, 381–396.
Kierdorf, K., Prinz, M., 2013. Factors regulating microglia activation. Front. Cell. Neurosci. 7, 44. https://doi.org/10.3389/fncel.2013.00044.
Kierdorf, K., Erny, D., Goldmann, T., et al., 2013. Microglia emerge from erythromyeloid precursors via Pu.1- and Irf8-dependent pathways. Nature Neurosci. 16 (3), 273–280.
King, I.L., Dickendesher, T.L., Segal, B.M., 2009. Circulating Ly-6C+ myeloid precursors migrate to the CNS and play a pathogenic role during autoimmune demyelinating disease. Blood 113 (14), 3190–3197.
Kletzien, R.F., Harris, P.K., Foellmi, L.A., 1994. Glucose-6-phosphate dehydrogenase: a "housekeeping" enzyme subject to tissue-specific regulation by hormones, nutrients, and oxidant stress. FASEB J. 8 (2), 174–181.
Knowles, R.G., Moncada, S., 1994. Nitric oxide synthases in mammals. Biochem. J. 298 (Pt2), 249–258.
Kohm, A.P., Sanders, V.M., 2001. Norepinephrine and b2-adrenergic receptor stimulation regulate CD4+ T and B lymphocyte function in vitro and in vivo. Pharmacol. Rev. 53, 487–525.
Koizumi, S., Shigemoto-Mogami, Y., Nasu-Tada, K., 2007. UDP acting at P2Y6 receptors is a mediator of microglial phagocytosis. Nature 446 (7139), 1091–1095.
Konduru, N.V., Tyurina, Y.Y., Feng, W., et al., 2009. Phosphatidylserine targets single-walled carbon nanotubes to professional phagocytes in vitro and in vivo. PLoS One 4 (2), e4398. https://doi.org/10.1371/journal.pone.0004398.
Kono, H., Rock, K.L., 2008. How dying cells alert the immune system to danger. Nat. Rev. Immunol. 8, 279–289.
Kovac, A., Erickson, M.A., Banks, W.A., 2011. Brain microvascular pericytes are immunoactive in culture: cytokine, chemokine, nitric oxide, and LRP-1 expression in response to lipopolysaccharide. J. Neuroinflammation 8, 139.
Kreutzberg, G.W., 1996. Microglia: a sensor for pathological events in the CNS. Trends Neurosci. 19 (8), 312–318.
Kuhn, S.A., van Landeghem, F.K.H., Zacharias, R., et al., 2004. Microglia express GABAB receptors to modulate interleukin release. Mol. Cell. Neurosci. 25, 312–322.
Laman, J.D., Weller, R.O., 2013. Drainage of cells and souble antigens from the CNS to regional lymph nodes. J. NeuroImmune Pharmacol. 8, 840–856.
Lang, Y., Chu, F., Shen, D., et al., 2018. Role of inflammasomes in neuroimmune and neurodegenerative diseases: a systematic review. Mediators Inflamm. 2018, 1549549. https://doi.org/10.1155/2018/1549549.
Lawana, V., Singh, N., Sarkar, S., et al., 2017. Involvement of c-Abl kinase in microglial activation of NLRP3 inflammasome and impairment in autolysosomal system. J. NeuroImmune Pharmacol. 12, 624–660.
Lawrence, T., Natoli, G., 2011. Transcriptional regulation of macrophage polarization: enabling diversity with identity. Nat. Rev. Immunol. 11 (11), 750–761.
Lawson, L.J., Perry, V.H., Dri, P., et al., 1990. Heterogeneity in the distribution and morphology of microglia in the normal adult mouse brain. Neuroscience 39, 151–170.

Lee, M., 2013. Neurotransmitters and microglial-mediated neuroinflammation. Curr. Protein Pept. Sci. 14 (1), 33–42.
Lee, H.M., Kang, J., Lee, S.J., et al., 2013. Microglial activation of the NLRP3 inflammasome by the priming signals derived from macrophages infected with mycobacteria. Glia 61 (3), 441–452.
Li, J., Gran, B., Zhang, G.X., et al., 2003. Differential expression and regulation of IL-23 and IL-12 subunits and receptors in adult mouse microglia. J. Neurol. Sci. 215 (1–2), 95–103.
Lisi, L., Stigliano, E., Lauriola, L., et al., 2014. Proinflammatory-activated glioma cells induce a switch in microglial polarization and activation status, from a predominant M2b phenotype to a mixture of M1 and M2a/B polarized cells. ASN Neuro 6 (3), 171–183.
Liu, H.C., Zheng, M.H., Du, Y.L., 2012. N9 microglial cells polarized by LPS and IL4 show differential responses to secondary environmental stimuli. Cell. Immunol. 278 (1–2), 84–90.
Liu, Y., Yang, X., Guo, C., et al., 2013. Essential role of MFG-E8 for phagocytic properties of microglial cells. PLoS One 8 (2), e55754. https://doi.org/10.1371/journal.pone.0055754.
Lopes, K.O., Sparks, D.L., Streit, W.J., 2008. Microglial dystrophy in the aged and Alzheimer's disease brain is associated with ferritin immunoreactivity. Glia 56 (10), 1048–1060.
Louveau, A., Harris, T.H., Kipnis, J., 2015a. Revisiting the mechanisms of CNS immune privilege. Trends Immunol. 36 (10), 569–577.
Louveau, A., Smirnov, I., Keyes, T.J., et al., 2015b. Structural and functional features of central nervous system lymphatic vessels. Nature 523 (7560), 337–341.
Lunov, O., Syrovets, T., Loos, C., et al., 2011. Amino-functionalized polystyrene nanoparticles activate the NLRP3 inflammasome in human macrophages. ACS Nano 5 (12), 9648–9657.
Luo, X.G., Ding, J.Q., Chen, S.D., 2010. Microglia in the aging brain: relevance to neurodegeneration. Mol. Neurodegener. 5, 12. https://doi.org/10.1186/1750-1326-5-12.
Lyons, A., Downer, E.J., Crotty, S., et al., 2007. CD200 ligand receptor interactions modulates microglial activation in vivo and in vitro: a role for IL-4. J. Neurosci. 27, 8309–8313.
Martinez, F.O., Gordon, S., 2014. The M1 and M2 paradigm of macrophage activation: time for reassessment. F1000Prime Rep. 6, 13. https://doi.org/10.12703/P6-13.
Martinez, F.O., Gordon, S., Locati, M., et al., 2006. Transcriptional profiling of the human monocyte-to-macrophage differentiation and polarization: new molecules and patterns of gene expression. J. Immunol. 177 (10), 7303–7311.
Martinez, F.O., Sica, A., Mantovani, A., et al., 2008. Macrophage activation and polarization. Front. Biosci. 13, 453–461.
Mattison, H.A., Nie, H., Gao, H., et al., 2013. Suppressed pro-inflammmatory respons of microglia in CX3CR1 knockout mice. J. Neuroimmunol. 257, 110–115.
McKimmie, C.S., Graham, G.J., 2010. Astrocytes modulate the chemokine network in a pathogen-specific manner. Biochem. Biophys. Res. Commun. 394, 1006–1011.
McName, E.N., Ryan, K.M., Kilroy, D., et al., 2010. Noradrenaline induced IL-1ra and IL-1 type II receptor expression in primary glial cells and protects against IL-1β-induced neurotoxicity. Eur. J. Pharmacol. 626, 219–228.
McPherson, C.A., Merrick, B.A., Harry, G.J., 2014. In vivo molecular markers for pro-inflammatory cytokine M1 stage and resident microglia and tremethyltin-induced hippocampal injury. Neurotox. Res. 25 (1), 45–56.
Meuth, S.G., Simon, O.J., Grimm, A., et al., 2008. CNS inflammation and neuronal degeneration is aggravated by impaired CD200-CD200R-mediated macrophage silencing. J. Neuroimmunol. 194, 62–69.

Michelucci, A., Heurtaux, T., Grandbarbe, L., 2009. Characterization of the microglial phenotype under specific pro-inflammatory and anti-inflammatory conditions: effects of oligomeric and fibrillar amyloid-beta. J. Immunol. 210 (1–2), 3–12.
Mildner, A., Schmidt, H., Nitsche, M., et al., 2007. Microglia in the adult brain arise from Ly-6ChiCCR2+ monocytes only under defined host conditions. Nat. Neurosci. 10 (12), 1544–1553.
Mildner, A., Mack, M., Schmidt, H., et al., 2009. CCR2+Ly-6Chi monocytes are crucial for the effector phase of autoimmunity in the central nervous system. Brain 132, 2487–2500.
Mills, C.D., 2012. M1 and M2 macrophages: oracles of health and disease. Crit. Rev. Immunol. 32 (6), 463–488.
Mills, E.L., Kelly, B., O'Neill, L.A.J., 2017. Mitochondria are the powerhouses of immunity. Nat. Immunol. 18 (5), 488–498.
Minkiewicz, J., de Rivero Vaccari, J.P., Keane, R.W., 2013. Human astrocytes express a novel NLRP2 inflammasome. Glia 87, 3655–3667.
Mittelbronn, M., Dietz, K., Schluesener, J.H., et al., 2001. Local distribution of microglia in the normal adult human central nervous system differs by up to one order of magnitude. Acta Neuropathol. 101, 249–255.
Mix, E., Meyer-Rienecker, H., Zettl, U.K., 2008. Animal models of multiple sclerosis for the development and validation of novel therapies—potential and limitations. J. Neurol. 255 (suppl 6), 7–14.
Miyanishi, M., Tada, K., Koike, M., et al., 2007. Identification of Tim4 as a phosphatidylserine receptor. Nature 450 (7168), 435–439.
Mizoue, L.E., Bazan, J.F., Johnson, E.C., et al., 1999. Solution structure and dyanmics of the CX3C chemokine domain of fractalkine and its interaction with an N-terminal fragment of CX3CR1. Biochemistry 38, 1402–1414.
Monier, A., Evrard, P., Gressens, P., et al., 2006. Distribution and differentiation of microglia in the human encephalon during the first two trimesters of gestation. J. Comp. Neurol. 499 (4), 565–582.
Morganti, J.M., Riparip, L.K., Rosi, S., 2016. Call off the dog(ma): M1/M2 polarization is concurrent following traumatic brain injury. PLoS One 11 (1), e0148001.
Mori, K., Ozaki, E., Zhang, B., et al., 2002. Effects of norepinephrine on rat cultured microglial cells that express a1, a2, b1 and b2 adrenergic receptors. Neuropharmacology 43, 1026–1034.
Mosser, D.M., Edwards, J.P., 2008. Exploring the spectrum of macrophage activation. Nat. Rev. Immunol. 8 (12), 958–969.
Muldoon, L.L., Alvarez, J.I., Begley, D.J., 2013. Immunologic privilege in the central nervous system and the blood-brain barrier. J. Cereb. Blood Flow Metab. 33 (1), 13–21.
Murray, P.J., Allen, J.E., Biswas, S.K., et al., 2014. Macrophage activation and polarization: nomenclature and experimental guidelines. Immunity 41 (1), 14–20.
Nakahira, K., Haspel, J.A., Rathinam, V.A.K., et al., 2011. Autophagy proteins regulate innate immune response by inhibiting NALP3 inflammasome-mediated mitochondrial DNA release. Nat. Immunol. 12 (3), 222–230.
Nakamura, K., Ohya, W., Funakoshi, H., et al., 2006. Possible role of scavenger receptor SRCL in the clearance of amyloid-beta in Alzheimer's disease. J. Neurosci. Res. 84, 874–890.
Nemeth, Z.H., Deitch, E.A., Davidson, M.T., et al., 2004. Disruption of the actin cytoskeleton results in nuclear factor-kappaB activation and inflammatory mediator production in cultured human intestinal epithelial cells. J. Cell. Physiol. 200, 71–81.
Neumann, H., Takahashi, K., 2007. Essential role of the microglial triggering receptor expressed on myeloid cells-2 (TREM2) for central nervous tissue immune homeostasis. J. Neuroimmunol. 184 (1–2), 92–99.

Nimmerjahn, A., Kirchhoff, F., Helmchen, F., 2005. Resting microglial cells are highly dynamic surveillants of brain parenchyma in vivo. Science 308, 1314–1318.
Orihuela, R., McPherson, C.A., Harry, G.J., 2015. Microglial M1/M2 polarization and metabolic states. Br. J. Pharmacol. 173 (4), 649–665.
Park, C., Lee, S., Cho, I.H., et al., 2006. TLR3-mediated signal induces proinflammatory cytokine and chemokine gene expression in astrocytes: differential signaling mechanisms of TLR3-induced IP-10 and IL-8 gene expression. Glia 53, 248–256.
Park, D., Tosello-Trampont, A.C., Elliott, M.R., et al., 2007. BAI1 is an engulfment receptor for apoptotic cells upstream of the ELMO/Dock180/Rac module. Nature 450 (7168), 430–434.
Park, S.Y., Jung, M.Y., Kim, H.J., et al., 2008. Rapid cell corpse clearance by stabilin-2, a membrane phosphatidylserine recetor. Cell Death Differ. 15 (1), 192–201.
Park, J., Yoo, K.-Y., Lee, C., et al., 2011. Comparison of glucocorticoid receptor and ionized calcium-binding adapter molecule 1 immunoreactivity in the adult and aged gerbil hippocampus following repeated restraint stress. Neurochem. Res. 36, 1037–1045.
Perry, V.H., Matyszak, M.K., Fearn, S., 1993. Altered antigen expression of microglia in the aged rodent CNS. Glia 7, 60–67.
Petrey, A.C., de la Motte, C.A., 2014. Hyaluronan, a crucial regulator of inflammation. Front. Immunol. 5, 101.
Piers, T.M., Heales, S.J., Pocock, J.M., 2011. Positive allosteric modulation of metabotropic glutamate receptor 5 down-regulates fibrinogen-activated microglia providing neuronal protection. Neurosci. Lett. 505, 140–145.
Pocock, J.M., Kettenmann, H., 2007. Neurotransmitter receptors on microglia. Trends Neurosci. 30 (10), 527–535.
Prinz, M., Mildner, A., 2011. Microglia in the CNS: immigrants from another world. Glia 59 (2), 177–187.
Prinz, M., Priller, J., Sisodia, S.S., et al., 2011. Heterogeneity of CNS myeloid cells and their roles in neurodegeneration. Nat. Neurosci. 14, 1227–1235.
Ransohoff, R.M., 2016. A polarizing question: do M1 and M2 microglia exist? Nat. Neurosci. 19 (8), 987–991.
Ransohoff, R.M., Brown, M.A., 2012. Innate immunity in the central nervous system. J. Clin. Invest. 122 (4), 1164–1171.
Ransohoff, R.M., Cardona, A.E., 2010. The myeloid cells of the central nervous system parenchyma. Nature 468 (7321), 253–262.
Ransohoff, R.M., Perry, V.H., 2009. Microglial physiology: unique stimuli, specialized responses. Annu. Rev. Immunol. 27, 119–145.
Ravichandran, K.S., 2003. "Recruitment signals" from apoptotic cells: invitation to a quiet meal. Cell 113 (7), 817–820.
Ravichandran, K.S., 2010. Find-me and eat-me signals in apoptotic cell clearance: progress and conundrums. J. Exp. Med. 207 (9), 1807–1817.
Rivieccio, M.A., John, G.R., Song, X., et al., 2005. The cytokine IL-1beta activates IFN response factor 3 in human fetal astrocytes in culture. J. Immunol. 174, 3719–3726.
Rodríguez-Prados, J.C., Través, P.G., Cuenca, J., et al., 2010. Substrate fate in activated macrophages: a comparison between innate, classic, and alternative activation. J. Immunol. 185 (1), 605–614.
Rogers, J.T., Morganti, J.M., Bachstetter, A.D., et al., 2011. CX3CR1 deficiency leads to impairment of hippocampal cognitive function and synaptic plasticity. J. Neurosci. 31, 16241–16250.
Rossi, D.L., Hardiman, G., Copeland, N.G., et al., 1998. Cloning and characterization of a new type of mouse chemokine. Genomics 47, 163–170.
Salvemini, F., Franze, A., Iervolino, A., et al., 1999. Enhanced glutathione levels and oxido resistance mediated by increased glucose-6-phosphate dehydrogenase expression. J. Biol. Chem. 274 (5), 2750–2757.

Sasaki, N., Toki, S., Chowei, H., Saito, et al., 2001. Immunohistochemical distribution of the receptor for advanced glycation end products in neurons and astrocytes in Alzheimer's disease. Brain Res. 888, 256–262.

Schmid, C.D., Melchior, B., Masek, K., et al., 2009. Differential gene expression in LPS/IFNgamma activated microglia and macrophages: in vitro versus in vivo. J. Neurochem. 109 (Suppl 1), 117–125.

Schnaars, M., Beckert, H., Halle, A., 2013. Assessing β-amyloid-induced NLRP3 inflammasome activation in primary microglia. Methods Mol. Biol. 1040, 1–8. https://doi.org/10.1007/978-1-62703-523-1_1.

Schulz, C., Gomez Perdiguero, E., Chorro, L., et al., 2012. A lineage of myeloid cells independent of Myb and hematopoietic stem cells. Science 336 (6077), 86–90.

Scumpia, P.O., Kelly, K.M., Reeves, W.H., et al., 2005. Double-stranded RNA signals antiviral and inflammatory programs and dysfunctional glutamate transport in TLR3-expressing astrocytes. Glia 52, 153–162.

Seifert, S., Pannell, M., Uckert, W., et al., 2011. Transmitter- and hormone-activated Ca2+ responses in adult microglia/brain macrophages in situ recorded after viral transduction of a recombinant Ca2+ sensor. Cell Calcium 49, 365–375.

Shechter, R., Schwartz, M., 2013. CNS sterile injury: just another wound healing? Trends Mol. Med. 19 (3), 135–143.

Sheffield, L.G., Berman, N., 1998. Microglial expression of MHC class II increases in normal aging of nonhuman primates. Neurobiol. Aging 19 (1), 47–55.

Shi, R., Yang, L., Kouadir, M., et al., 2012. The NALP3 inflammasome is involved in neurotoxic prion peptide-induced microglial activation. J. Neuroinflammation 9, 73.

Sierra, A., Gottfried-Blackmore, A.C., McEwen, B.S., et al., 2007. Microglia derived from aging mice exhibit an altered inflammatory profile. Glia 55 (4), 412–424.

Sierra, A., Gottfried-Blackmore, A., Milner, T.A., et al., 2008. Steroid hormone receptor expression and function in microglia. Glia 56, 659–674.

Sierra, A., Abiega, O., Shahraz, A., et al., 2013. Janus-faced microglia: beneficial and detrimental consequences of microglial phagocytosis. Front. Cell. Neurosci. 7, 6.

Sieweke, M.H., Allen, J.E., 2013. Beyond stem cells: self-renewal of differentiated macrophages. Science 342 (6161), 1242974.

Sofroniew, M.V., 2014. Multiple roles for astrocytes as effectors of cytokines and inflammatory mediators. Neuroscientist 20, 160–172.

Sofroniew, M.V., Vinters, H.V., 2010. Astrocytes: biology and pathology. Acta Neuropathol. 119, 7–35.

Stanimirovic, D., Shapiro, A., Wong, J., et al., 1997. The induction of ICAM-1 in human cerebromicrovascular endothelial cells (HCEC) by ischemia-like conditions promotes enhanced neutrophil/HCEC adhesion. J. Neuroimmunol. 76, 193–205.

Stein, M., Keshav, S., Harris, N., et al., 1992. Interleukin 4 potently enhances murine macrophage mannose receptor activity: a marker of alternative immunologic macrophage activation. J. Exp. Med. 176 (1), 287–292.

Stirling, D.P., Cummins, K., Mishra, M., Teo, W., Yong, W., Stys, P., 2014. Toll-like receptor 2-mediated alternative activation of microglia is protective after spinal cord injury. Brain 137, 707–723.

Streit, W.J., Xue, Q.-S., 2010. The Brain's aging immune system. Aging Dis. 1 (3), 254–261.

Streit, W.J., Miller, K.R., Lopes, K.O., et al., 2008. Microglial degeneration in the aging brain—bad news for neurons? Front. Biosci. 13, 3423–3438.

Sutton, C., Brereton, C., Keogh, B., 2006. A crucial role for interleukin (IL)-1 in the induction of IL-17-producing T cells that mediate autoimmune encephalomyelitis. J. Exp. Med. 203, 1685–1691.

Swinnen, M., Smolders, S., Avila, A., et al., 2012. Complex invasion pattern of the cerebral cortex by microglial cells during development of the mouse embryo. Glia 61 (2), 150–163.

Tada, M., Diserens, A.C., Desbaillets, I., et al., 1994. Analysis of cytokine receptor messenger RNA expression in human glioblastoma cells and normal astrocytes by reverse-transcription polymerase chain reaction. J. Neurosurg. 80, 1063–1073.
Takahashi, K., Rochford, C.D., Neumann, H., 2005. Clearance of apoptotic neurons without inflammation by microglial triggering receptor expressed on myeloid cells-2. J. Exp. Med. 201 (4), 647–657.
Takahashi, K., Prinz, M., Stagi, M., et al., 2007. TREM2-transduced myeloid precursors mediate nervous tissue debris clearance and facilitate recovery in an animal model of multiple sclerosis. PLoS Med. 4 (4), e124.
Taylor, D.L., Diemel, L.T., Pocock, J.M., 2003. Activation of microglial group III metabotropic glutamate receptors protects neurons against microglial neurotoxicity. J. Neurosci. 23, 2150–2160.
Taylor, D.L., Jones, F., Kubota, E.S.F.C.S., et al., 2005. Stimulation of microglial metabotropic glutamate receptor mGlu2 triggers tumor necrosis factor α-induced neurotoxicity in concert with microglial-derived Fas ligand. J. Neurosci. 25, 2952–2964.
Tremblay, M.E., Majewska, A.K., 2011. A role for microglia in synaptic plasticity? Commun. Integr. Biol. 4 (2), 220–222.
Tremblay, M.E., Zettel, M.L., Ison, J.R., et al., 2012. Effects of aging and sensory loss on glial cells in mouse visual and auditory cortices. Glia 60 (4), 541–558.
Vallieres, L., Sawchenko, P.E., 2003. Bone marrow-derived cells that populate the adult mouse brain preserve their hematopoietic identity. J. Neurosci. 23 (12), 5197–5207.
Varnum, M.M., Ikezu, T., 2012. The classification of microglial activation phenotypes on neurodegeneration and regeneration in Alzheimer's disease brain. Arch. Immunol. Ther. Exp. 60 (4), 251–266.
Vaughan, D.W., Peters, A., 1974. Neuroglial cells in the cerebral cortex of rats from young adulthood to old age: an electron microscope study. J. Neurocytol. 3 (4), 405–429.
Viganò, E., Mortellaro, A., 2013. Caspase-11: the driving factor for noncanonical inflammasomes. Eur. J. Immunol. 43 (9), 2240–2245.
Vodovotz, Y., Bogdan, C., Paik, J., et al., 1993. Mechanisms of suppression of macrophage nitric oxide release by transforming growth factor beta. J. Exp. Med. 178 (2), 605–613.
Vogel, D.Y., Vereyken, E.J., Glim, J.E., 2013. Macrophages in inflammatory multiple sclerosis lesions have an intermediate activation status. J. Neuroinflammation 10, 35.
Voloboueva, L.A., Emery, J.F., Sun, X., et al., 2013. Inflammatory response of microglial BV-2 cells includes a glycolytic shift and is modulated by mitochondrial glucose-regulated protein 75/mortalin. FEBS Lett. 587 (6), 756–762.
Waddell, S.J., Popper, S.J., Rubins, K.H., et al., 2010. Dissecting interferon-induced transcriptional programs in human peripheral blood cells. PLoS One 5, e9753. https://doi.org/10.1371/journal.pone.0009753.
Waisman, A., Liblau, R.S., Becher, B., 2015. Innate and adaptive immune responses in the CNS. Lancet Neurol. 14 (9), 945–955. https://doi.org/10.1016/S1474-4422(15)00141-6.
Wang, H., Yu, M., Ochani, M., et al., 2003. Nicotinic acetylcholine receptor alpha7 subunit is an essential regulator of inflammation. Nature 421 (6921), 384–388.
Wang, X., Chu, G., Yang, Z., et al., 2015. Ethanol directly induced HMGB1 release through NOX2/NLRP1 inflammasome in neuronal cells. Toxicology 334, 104–110.
West, A.P., Brodsky, I.E., Rahner, C., et al., 2011. TLR signalling augments macrophage bactericidal activity through mitochondrial ROS. Nature 472 (7344), 476–480.
Williams, K., Ulvestad, E., Waage, A., et al., 1994. Activation of adult human derived microglia by myelin phagocytosis in vitro. J. Neurosci. Res. 38, 433–443.
Wilms, H., Rosenstiel, P., Sievers, J., et al., 2003. Activation of microglia by human neuromelanin in NF-kappaB dependent and involves p38 mitogen-activated protein kinase: implication for Parkinson's disease. FASEB J. 17 (3), 500–502.
Winkler, E.A., Bell, R.D., Zlokovic, B.V., 2011. Central nervous system pericytes in health and disease. Nat. Neurosci. 14, 1398–1405.

Witting, A., Muller, P., Herrmann, A., et al., 2000. Phagocytic clearance of apoptotic neurons by microglia/brain macrophages in vitro: involvement of lectin-, integrin-, and phosphatidylserine-mediated recognition. J. Neurochem. 75 (3), 1060–1070.
Wong, P.M., Puente, C., Ganley, I.G., et al., 2013. The ULK1 complex: sensing nutrient signals for autophagy activation. Autophagy 9 (2), 124–137.
Wu, Z., Sun, L., Hashioka, S., et al., 2013. Differential pathways for interleukin-1B production activated by chromogranin A and amyloid β in microglia. Neurobiol. Aging 34 (12), 2715–2725.
Wynn, T.A., Chawla, A., Pollard, J.W., 2013. Macrophage biology in development, homeostasis and disease. Nature 496 (7446), 445–455.
Xing, C., Li, W., Deng, W., et al., 2018. A potential gliovascular mechanism for microglial activation: differential phenotypic switching of microglia by endothelium versus astrocytes. J. Neuroinflammation 15 (1), 143.
Xiong, J., Burkovetskaya, M., Karpuk, N., et al., 2012. IL-1R1 (interleukin-1 receptor type I) signaling is essential for host defence and hemichannel activity during acute central nervous system bacterial infection. ASN Neuro 4, e00085.
Xue, J., Schmidt, S.V., Sander, J., 2014. Transcriptome-based network analysis reveals a spectrum model of human macrophage activation. Immunity 40 (2), 274–288.
Yemisci, M., Gursoy-Ozdemir, Y., Vural, A., et al., 2009. Pericyte contraction induced by oxidative-nitrative stress impairs capillary reflow despite successful opening of an occluded cerebral artery. Nat. Med. 15, 1031–1037.
Yogarajah, T., Ong, K.C., Perera, D., Wong, K.T., 2017. AIM2 inflammasome-mediated pyroptosis in enterovirus A71-infected neuronal cells restricts viral replication. Sci. Rep. 7 (1), 5845. https://doi.org/10.1038/s41598-017-05589-2.
Zambetti, L.P., Laudisi, F., Licandro, G., et al., 2012. The rhapsody of NLRPs: master players of inflammation...and a lot more. Immunol. Res. 53 (1–3), 78–90.
Zierler, S., Frei, E., Grissmer, S., et al., 2008. Chloride influx provokes lamellipodium formation in microglial cells. Cell. Physiol. Biochem. 21 (1–3), 55–62.

Further reading

Johnson, J.D., Zimomra, Z.R., Stewart, L.T., 2013. Beta-adrenergic receptor activation primes microglia cytokine production. J. Neuroimmunol. 254, 161–164.
Owens, T., 2009. Toll-like receptors in neurodegeneration. Curr. Top. Microbiol. Immunol. 336, 105–120.
Prinz, M., Priller, J., 2010. Tickets to the brain: role of CCR2 and CX(3)CR1 in myeloid cell entry in the CNS. J. Neuroimmunol. 224, 80–84.

CHAPTER TWO

Neuroinflammation in organophosphate-induced neurotoxicity

Michelle Guignet, Pamela J. Lein*
Department of Molecular Biosciences, University of California-Davis, Davis, CA, United States
*Corresponding author: e-mail address: pjlein@ucdavis.edu

Contents

1. Introduction

The term "organophosphate" (OP) refers to a group of synthetic organic compounds that have in common a pentavalent phosphorus joined to sulfur or oxygen via a covalent double bond (Costa, 2018). OP compounds were first developed in the early 20th century as insecticides. In the 1930s, it was discovered that the principal molecular mechanism of their insecticidal activity was inhibition of acetylcholinesterase (AChE), an enzyme that is conserved across species, including humans. This discovery unfortunately led to the development during World War II of potent OP

Advances in Neurotoxicology, Volume 3
ISSN 2468-7480
https://doi.org/10.1016/bs.ant.2018.10.003

nerve agents with high human toxicity, including tabun, sarin, cyclosarin, soman, VR, and VX (Costa, 2018; Pereira et al., 2014). Since World War II, hundreds of OP compounds have been developed and marketed as insecticides, fire retardants, plasticizers, and fuel additives, but their most predominant use has been as insecticides (Costa, 2018; Voorhees et al., 2016). By the 1970s, OPs largely replaced organochlorine pesticides because they were equally efficacious, but significantly less persistent in the environment, and by the year 2000, >70% of all insecticides used in the United States were OPs (Costa, 2018; Voorhees et al., 2016). However, over the past decade, residential uses of OP pesticides have been largely phased out and commercial applications of OPs are being increasingly restricted in the United States and many European countries due to increasing concerns of adverse public health impacts (Rohlman et al., 2011; Voorhees et al., 2016). Despite the decline in their use, OP insecticides remain among the most commonly used group of insecticides worldwide, with particularly heavy use in developing countries because of their lower cost relative to newer insecticides (Costa, 2018; Voorhees et al., 2016). As a result of their continued widespread use as insecticides, OPs also remain among the most commonly detected anthropogenic contaminants in the global environment, including human tissues (Barr et al., 2011; Clune et al., 2012; Voorhees et al., 2016).

OPs are primarily neurotoxicants that poison insects by inhibiting AChE, an enzyme that catalyzes the hydrolysis of the neurotransmitter acetylcholine to terminate cholinergic neurotransmission. The phosphate form of OPs, which depending on the OP may be the parent compound or an active metabolite, inhibits AChE by phosphorylating the serine residue in the catalytic triad of the enzyme's active site (Costa, 2006; Eddleston et al., 2008). Inhibition of AChE increases the amount and residence time of acetylcholine at nicotinic and muscarinic cholinergic receptors, resulting in cholinergic overstimulation of target tissues, which immobilizes the insect. However, with few exceptions, OPs are not highly specific for targeted insect species, and mammals, including humans, are susceptible to their neurotoxic effects (Costa, 2018; Rohlman et al., 2011; Voorhees et al., 2016). Acute OP intoxication is estimated to cause >3 million life-threatening human poisonings each year, with approximately 1 million of these cases attributed to occupational exposures, and over 250,000 deaths as a result of self-poisoning, which represents approximately about one-third of the world's suicide cases (Gunnell et al., 2007; Mew et al., 2017; Pereira et al., 2014). In humans, acute OP poisoning causes both peripheral and central cholinergic effects that collectively define the clinical toxidrome known as cholinergic crisis (Eddleston et al., 2008; Hulse et al., 2014). This toxidrome is characterized by parasympathomimetic symptoms, muscle fasciculation followed by flaccid paralysis, centrally-mediated seizures that can quickly progress to *status epilepticus* (SE), and respiratory failure. Cholinergic crisis is frequently fatal unless rapidly treated with atropine to block muscarinic receptors, oxime to reactivate AChE, and benzodiazepines to reduce seizure activity (Bird et al., 2003; Eddleston et al., 2008). However, clinical trials have revealed no added benefit of using oximes to treat poisoning by various OP insecticides, despite the fact that oxime treatment effectively reactivated blood AChE (Buckley et al., 2011). Even with treatment, survivors of cholinergic crisis often face significant long-term morbidity, including cognitive dysfunction, affective disorders, and spontaneous recurrent seizures (also known as acquired epilepsy) that typically manifest well after cholinesterase activity has recovered to pre-exposure levels (Chen, 2012; De Araujo Furtado et al., 2012; Pereira et al., 2014). Preclinical studies have replicated the effects observed in acutely intoxicated humans, including cholinergic signs, SE, delayed and persistent cognitive dysfunction (Deshpande et al., 2014; Filliat et al., 1999, 2007; Flannery et al., 2016; Myhrer et al., 2005; Raffaele et al., 1987), anxiety-like behavior (Coubard et al., 2008; Prager et al., 2014a; Wright et al., 2010), and spontaneous recurrent seizures (De Araujo Furtado et al., 2010; Shrot et al., 2014). As in humans, long-term effects typically manifest after recovery of cholinesterase activity. Thus, the clinical and preclinical data suggest that while acute, profound AChE inhibition mediates the acute neurotoxic effects of OPs, including seizure activity (Mcdonough and Shih, 1997), it seems likely that the long-term effects of acute OP intoxication are mediated by mechanisms other than or in addition to AChE inhibition.

Repeated low-level OP exposures that do not elicit symptoms of cholinergic crisis are also associated with neurotoxic outcomes in humans, including cognitive deficits and neuropsychiatric conditions such as depression, anxiety, and suicidal tendencies (reviewed in Voorhees et al., 2016). Recent epidemiologic studies suggest that repeated low-level OP exposures may also increase individual risk for neurodevelopmental disorders (Munoz-Quezada et al., 2013; Sagiv et al., 2018a,b) and neurodegenerative disease (Baltazar et al., 2014; Sanchez-Santed et al., 2016). These associations have been challenged by some due to inconsistent outcomes and a lack of epidemiologic evidence demonstrating a dose-response relationship of the neurobehavioral impacts from repeated low-level OP exposures

(Meyer-Baron et al., 2015; Rohlman et al., 2011). However, recent evaluations of the published literature have concluded that the weight-of-evidence supports an association of subclinical occupational (Meyer-Baron et al., 2015; Rohlman et al., 2011) and early-life (Burke et al., 2017) OP exposures with negative impacts on human neurobehavior. Interestingly, these evaluations also concluded that the epidemiologic data do not support an association between OP effects on blood cholinesterase activity and neurobehavioral outcomes (Burke et al., 2017; Meyer-Baron et al., 2015; Rohlman et al., 2011), suggesting there may be non-cholinergic mechanisms of neurotoxicity following repeated low-level OP exposures. A review of the preclinical literature generally supports this hypothesis (Burke et al., 2017). While the most significant and prolonged motor effects are observed following OP exposures that markedly inhibit brain AChE activity, cognitive deficits are not as clearly correlated with AChE inhibition (Bushnell and Moser, 2006). The hypothesis that OP neurotoxicity is not mediated entirely by cholinergic mechanisms is further supported by observations that different OPs elicit different profiles of neurotoxicity despite causing comparable changes in AChE activity and other cholinergic markers (Bushnell and Moser, 2006; Costa, 2006; Pope, 1999). Additionally, AChE knockout mice exhibit symptoms of neurotoxicity comparable to those observed in wildtype mice following OP exposure (Duysen et al., 2001).

Collectively, clinical, epidemiologic, and experimental evidence suggest that mechanisms in addition to or other than AChE inhibition mediate OP neurotoxicity associated with either acute poisoning or repeated low-level exposures (Costa, 2006; Naughton and Terry Jr., 2018; Pope, 1999). This observation has significant implications for the development of effective medical countermeasures for OP poisoning, interventional strategies for preventing OP neurotoxicity, and the use of AChE inhibition as a predictive or diagnostic biomarker of OP-induced neurotoxicity. Various targets and mechanisms other than AChE inhibition have been proposed to mediate OP neurotoxicity, including neuroinflammation (Burke et al., 2017; Costa, 2018; Naughton and Terry, 2018; Pereira et al., 2014). Experimental evidence demonstrates that OPs trigger neuroinflammatory responses under a broad range of exposure paradigms that vary with respect to the type of OP, level of exposure, and exposure duration. There is great interest in confirming a mechanistic link between neuroinflammation and OP neurotoxicity because of the existence of quantitative biomarkers of inflammation that correlate well with neurobehavioral deficits observed in neurodevelopmental disorders and neurodegenerative disease (Chahine et al., 2014; Dziedzic, 2006; Heneka et al., 2015) and the availability of anti-inflammatory therapeutics approved for use in humans. The remainder of this chapter will provide a brief overview of neuroinflammation, a summary of experimental evidence implicating neuroinflammation as a mechanism of OP neurotoxicity, and a discussion of critical data gaps that need to be addressed to better understand the diagnostic and therapeutic implications of OP-induced neuroinflammation.

2. Overview of neuroinflammation

Inflammation is a cardinal host defense response to pathogen invasion, irritants (including toxic chemicals), and physical trauma. Initially defined in the context of clinical symptoms (pain, redness, heat, swelling), the term "inflammation" is currently used to refer to a complex cascade of cellular processes involving upregulation of proinflammatory cytokines and chemokines, macrophage activation, and leukocyte infiltration of the affected tissue. The initial proinflammatory response functions to eliminate the offending stimulus, and this is typically followed by a resolution phase in which the same stimuli promote different cellular changes to facilitate repair and healing. The inflammatory response can be acute or chronic, and can involve cells and molecules of the innate and/or the adaptive immune system. The term "neuroinflammation" refers specifically to an immune-mediated, glial cell-propagated response organized within the nervous system in response to brain injury or systemic inflammation (Kraft and Harry, 2011; Viviani et al., 2014). Neuroinflammation is critical for neuroprotection and neurorepair following injury to the brain (Konishi and Kiyama, 2018; Kraft and Harry, 2011); however, when dysregulated, neuroinflammation can compromise the function of neural circuits or even cause neuronal cell death, which contribute to the initiation, progression and/or exacerbation of neurologic disease (Herman and Pasinetti, 2018; Shi and Holtzman, 2018). The balance between beneficial and harmful effects of neuroinflammation is determined by the context, magnitude, duration, and timing relative to the primary stimulus of the neuroinflammatory response (Disabato et al., 2016). Whether initiated directly in the nervous system or indirectly in response to a peripheral stimulus, neuroinflammatory responses are orchestrated by multiple resident cell types in the nervous system that signal to each other through the production of second messengers, including cytokines, chemokines, and reactive oxygen and nitrogen species (ROS and RNS, respectively) (reviewed in Disabato et al., 2016; Kraft and Harry, 2011; also see Table 1; Fig. 1).

Table 1 Biomarkers of neuroinflammation.

Inflammatory element	Role in inflammation	Experimental application
Soluble biomarkers		
IL-1α, IL-1β, TNF-α, IL-6, IFN-γ	Pro-inflammatory cytokines produced and secreted by injured neurons and activated immune cells that contributes to the increased inflammatory tone	Secreted biomarkers found in both serum and brain tissue. Expression measured at the mRNA level by qPCR and at the protein level by ELISA or multiplex immunoassays (i.e., Luminex™)
IL-4, IL-10	Anti-inflammatory cytokines secreted by immune cells to dampen inflammation	
CXCL1, MIP-1α, GRO/KC	Chemokines secreted by activated macrophages that recruit neutrophils	
ICAM, VCAM, Selectin	Cell adhesion molecules upregulated on the surface of immune cells in response to cytokine stimulation	
TREM1	Cell surface receptor on myeloid cells that when activated triggers downstream immune response	
PGE2	Bioactive lipid product of arachidonic acid metabolism implicated in inflammatory responses	
SOCS3	Cytokine-inducible negative regulator of cytokine signaling	
Cell surface biomarkers		
Iba-1; IB4 (ILB4), Cd11b, OX-42	Microglia and macrophage specific cell surface markers	Spatiotemporal expression assessed by immunohistochemistry (also enables analysis of cell morphology); expression levels quantified by qPCR or immunoblotting
MAC-1	Complement receptor expressed on neutrophils, natural killer cells and macrophages	
CD68	Lysosomal marker upregulated in phagocytically active immune cells	
GFAP	Glial fibrillary acidic protein, cell surface marker for astrocytes	
In vivo imaging biomarkers		
TSPO	Outer mitochondrial membrane protein whose expression upregulated in activated microglial cells and astrocytes	*In vivo* PET imaging biomarker of neuroinflammation

Microglia are the innate immune cells of the central nervous system (CNS) and are typically the first responders to CNS injury or inflammatory stimuli. Under physiologic conditions, microglia are thought to regulate synaptic activity and to provide immune surveillance (Mosser et al., 2017). Expression of cell-surface purinergic receptors, phosphatidylserine, and toll-like receptors (TLRs) enables microglia to recognize damage associated molecular patterns (DAMPs) and other molecular signals released by injured cells, such as ATP (Saijo and Glass, 2011). Microglia respond to these molecular signals by morphologically transforming from a cell with a small soma and long, thin ramified processes that is typical under physiological conditions, to a less ramified morphology with fewer and thicker processes associated with pathological conditions, to an amoeboid, macrophage-like morphology that is characteristic of a fully activated microglial cell (Fig. 1). These morphological changes are accompanied by increased proliferation and phagocytic activity, changes in gene expression and motility, and release of numerous pro- and anti-inflammatory mediators (Corps et al., 2015; Mosser et al., 2017). Activation also upregulates microglial expression of the lysosomal protein CD68 and the 18kDa mitochondrial translocator protein (TSPO), both of which are used as biomarkers of activated microglia (Table 1). Activated microglia perform numerous macrophage-like functions, including phagocytosis, antigen presentation, migration to sites of injury, and recruitment of additional immune cells via release of soluble mediators (Dantzer et al., 2008; Davalos et al., 2005; Nimmerjahn et al., 2005).

A hallmark characteristic of microglial activation is upregulated transcription and translation of cytokines and chemokines (Becher et al., 2017; Zhou et al., 2006; see also Table 1; Fig. 1). These soluble mediators function as an intercellular communication system to coordinate the neuroinflammatory response. Chemokines released by activated microglia, such as monocyte chemoattractant protein 1 (MCP-1) and macrophage inflammatory protein 1 (MIP-1α), establish a chemoattractant gradient to recruit other immune cells to the site of injury. Activated microglia also release both proinflammatory (IL-1β, IL-1α, IL-6, IFN-γ) and anti-inflammatory (IL-10, IL-4, TGF-β, arginase) cytokines, and the balance between these dictates the inflammatory tone (Frank-Cannon et al., 2009; Loane et al., 2014; Turtzo et al., 2014; Wang et al., 2013). Activation also increases microglial release of ROS and RNS. While physiologic levels of ROS and RNS are important signaling molecules in the brain (Kennedy et al., 2012; Massaad and Klann, 2011), excessive or sustained levels that

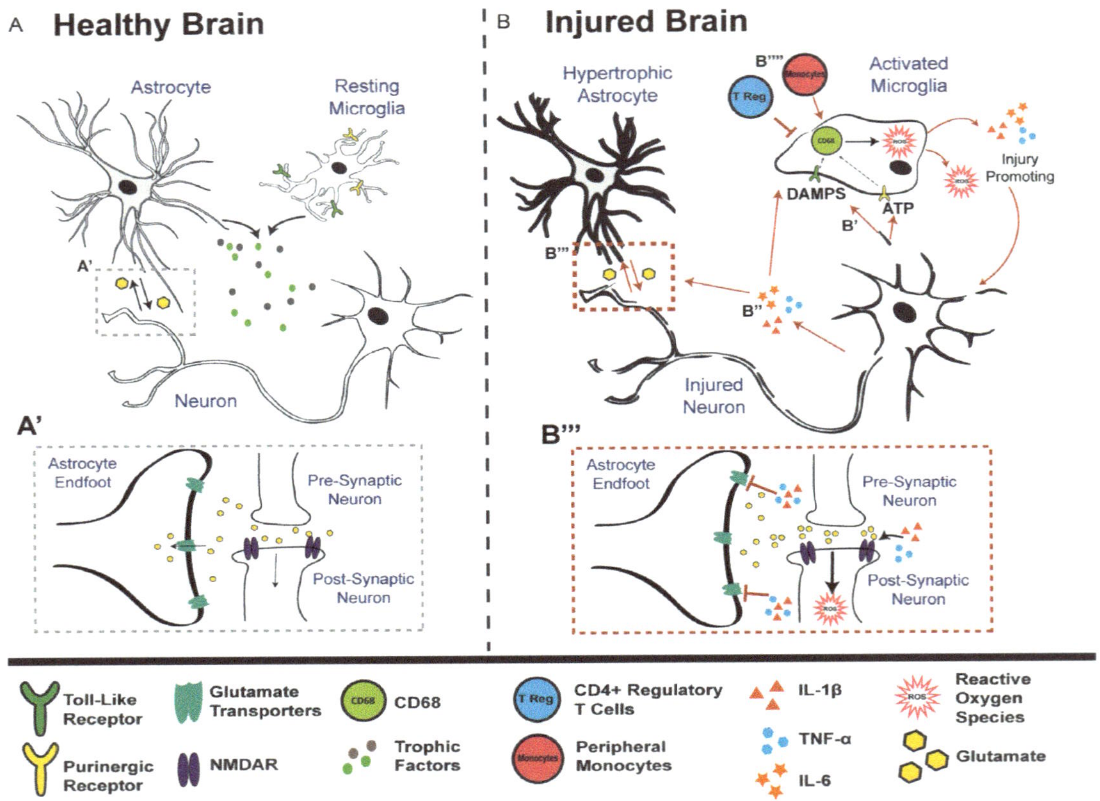

Fig. 1 See legend on opposite page.

overwhelm cellular antioxidant systems can lead to oxidative and nitrative damage to macromolecules, which in turn can cause cellular damage and elicit robust neuroinflammatory responses (Biswas, 2016; Loane et al., 2014).

Astrocytes, which are the predominant cell type in the CNS, are considered to be "pivotal regulators" of neuroinflammation because, depending on context and location, they can either promote neuroinflammation or protect healthy cells from cytotoxic byproducts of neuroinflammation. In response to brain injury, astrocytes undergo a reversible process known as reactive astrogliosis, which is associated with a hypertrophied cellular morphology that coincides with increased expression of the cytoskeletal proteins, glial fibrillary acidic protein (GFAP) and vimentin (Jensen et al., 2013; Kimelberg and Nedergaard, 2010; Liedtke et al., 1996). Activated astrocytes migrate toward and surround the site of injury to form a glial scar, which creates a barrier to prevent damage from spreading into surrounding healthy tissue (Jensen et al., 2013; Pekny et al., 2016). Activated astrocytes within this barrier provide trophic support to promote repair of the damaged tissue (Jensen et al., 2013). Reactive astrocytes can also amplify the neuroinflammatory response by signaling to neighboring cells via release of proinflammatory cytokines and chemokines (Fig. 1).

Oligodendrocytes are the major myelinating glia in the CNS. While damaged myelin has long been recognized as a trigger for microglial

Fig. 1 Schematic of neuroinflammation in the brain. (A) In the healthy brain, both astrocytes and resting microglia provide neurotrophic support to neurons to maintain homeostasis. (A′) Astrocytes modulate glutamatergic neurotransmission in the brain in part through uptake of synaptic glutamate. (B) Acute injury alters signaling between neurons, microglia and astrocytes. (B′) Injured neurons release ATP and damage associated molecular patterns (DAMPs) that bind to purinergic receptors or toll-like receptors (TLRs), respectively, to activate proximal microglia. Activated microglia become amoeboid in shape and increase expression of the lysosomal marker, CD68. Phagocytically active microglia produce elevated levels of ROS and inflammatory cytokines (IL-1β, TNF-α, IL-6) that promote further neurotoxic injury. (B″) Injured neurons also release inflammatory cytokines that trigger reactive astrogliosis as well as microglial activation. (B‴) IL-1β and TNF-α enhance glutamatergic signaling by increasing NMDAR activity and inhibiting glutamate transporters, which collectively promote excitotoxicity-induced ROS production. (B⁗) Breakdown of the blood brain barrier enables peripheral immune cells (T-cells and monocytes) to infiltrate the brain parenchyma and either promote or suppress further immune cell activation depending on the inflammatory context. This image is licensed under a Creative Commons Attribution 3.0 Unported License (Servier Medical Art by Servier and Creative Commons Attribution—Share Alike 4.0 International license).

activation (Clemente et al., 2013; Goldmann and Prinz, 2013; Williams et al., 1994), oligodendrocytes have largely been viewed as passive victims of neuroinflammation. Oligodendrocytes have a high metabolic rate and relatively few endogenous antioxidant defense systems, which makes them particularly vulnerable to increased ROS and RNS produced by sustained or excessive neuroinflammation (Thorburne and Juurlink, 1996). However, emerging evidence challenges this model, arguing instead that oligodendrocytes participate actively in the neuroinflammatory response. In response to injury or inflammatory stimuli, oligodendrocytes upregulate TLR subtypes and produce complement molecules, chemokines, and cytokines that function to coordinate the injury response by signaling to other cells, such as microglia (Balabanov et al., 2007; Bsibsi et al., 2002; Ramesh et al., 2012; Zajicek et al., 1995). Oligodendrocytes have also been demonstrated to produce soluble mediators that modulate microglia activity toward a less inflammatory tone (Peferoen et al., 2014).

Innate and adaptive immune cells of the peripheral immune system, including T cells, B cells, monocytes, and mast cells, play complicated roles in neuroinflammation. In situations of excessive neuroinflammation or compromised blood brain barrier (BBB) integrity, such as is observed in various neurologic disorders, peripheral immune cells that infiltrate the brain parenchyma can perpetuate local neuroinflammatory responses, thereby exacerbating tissue damage (Prinz and Priller, 2017). Conversely, migration of T cells into the brain parenchyma across an intact BBB in response to CNS injury can dampen neuroinflammation by producing neurotrophic factors and cytokines that modulate microglia and astrocyte activity to limit their release of neurotoxic factors, which restricts the spread of neuronal damage (Beers et al., 2008; Schwartz et al., 2003; Shaked et al., 2004).

Given the number of cell types and soluble mediators involved, it is perhaps not surprising that a wide variety of biomarkers are used to detect and quantify neuroinflammation (Table 1). In many cases, microglial and astrocytic activation are assessed by quantifying expression levels or the number of cells expressing biomarkers for microglia and/or reactive astrocytes. However, the pathophysiological relevance of increased expression of these biomarkers is not always clear. For example, upregulation of GFAP can be associated with neuroprotective or neurotoxic activity (Dallerac and Rouach, 2016). Expression of ionized calcium-binding adapter molecule 1 (Iba-1), which is used as a biomarker of microglial cells, does not distinguish between non-inflammatory vs. inflammatory microglia. Early studies distinguished "resting" from "activated" microglia based on cellular morphology (ramified vs. amoeboid, respectively). However, accumulating evidence shows that microglia assume a variety of intermediate morphologies, which are not always linked to an inflammatory role, and the morphology of microglia in the developing brain differs from that observed in the mature brain under both physiologic and pathologic conditions (Mosser et al., 2017). Another common approach for quantifying neuroinflammation is to measure levels of soluble pro- or anti-inflammatory mediators in whole brain. However, this approach cannot distinguish between soluble mediators derived from activated microglia and/or astrocytes vs. those released from oligodendrocytes or peripheral immune cells. Further complicating interpretation, the effect of cytokines is context-dependent such that classic proinflammatory cytokines can have anti-inflammatory effects later in the inflammatory response. Despite these challenges, the currently available body of literature strongly suggests that neuroinflammation may be an important mechanism of OP neurotoxicity, based in large part on clinical and experimental evidence that multiple OPs upregulate one or more mediators of neuroinflammation.

3. Neuroinflammatory responses to neurotoxic OP exposures

Much of the initial data describing neuroinflammatory responses to OPs were generated from preclinical models of acute intoxication with OP nerve agents at doses that induced SE and other signs of cholinergic crisis (Table 2). Subsequently, evidence began to emerge demonstrating that acute intoxication with OP pesticides and even repeated low-level exposures to OP nerve agents or pesticides can also increase neuroinflammation. A general observation emerging from this expanding literature is that the neuroinflammatory response profile varies depending on the OP, the exposure scenario, and the time after exposure when inflammation is assessed (Banks and Lein, 2012; Damodaran and Abou-Donia, 2000; Grauer et al., 2008; Liu et al., 2012).

3.1 Acute OP intoxication increases neuroinflammation

In some of the earliest studies of OP-induced neuroinflammation, acute soman intoxication was observed to significantly increase GFAP at the mRNA and protein level, and to increase the number of amoeboid microglia in the piriform cortex, amygdala, and hippocampus within hours after the onset of SE (Baille et al., 2005; Collombet et al., 2005a; Zimmer et al., 1997).

Table 2 Acute OP intoxication triggers neuroinflammation.

Type of OP	Chemical	Model	Exposure paradigm	Neuroinflammatory response	Brain regions affected	Reference
Nerve Agent	Soman	Rat	77.7 μg/kg (im[a])	↑ GFAP within 1h ↑ Number of OX-42+ cells and morphologic activation within 4h	Piriform Cortex Hippocampus	Zimmer et al. (1997)
	Soman	Rat	99–126 mg/kg (sc[b]) (pre) HI-6[c], 125 mg/kg (ip[d])	↑ *GFAP* mRNA (in situ hybridization) 6–24h	Hippocampus	Baille-Le Crom et al. (1995)
	Soman	Rat	180 μg/kg (sc) (pre) HI-6, 125 mg/kg (ip) (post) AMN[e], 2–4 mg/kg (im)	↑ IB-4+ cells ↑ GFAP at 28 DPE[i]; ↑ *TNF-α* (2h), *VCAM* (2h), *IL-1β* (6h), *E-Selectin* (6h), *IL-6* (6h), *ICAM* (6h) mRNA—resolves within 48h ↑ IL-1α (microglia), IL-1β (microglia), TNF-α, IL-6 (neurons) protein peaks at 12h ↑ CXCL1 (neurons + endothelial cells), MIP-1α (neurons) protein peaks at 12h and precedes neutrophil infiltration into brain	Piriform Cortex Hippocampus Thalamus	Finkelstein et al. (2012), Williams et al. (2003), Dillman et al. (2009), Johnson and Kan (2010), and Johnson et al. (2011)
	Soman	Rat	133 μg/kg (sc) (post) AMN, 2 mg/kg (im)	↑ GFAP at 48h ↑ ILB_4 + cells at 48h	Piriform Cortex Hippocampus Thalamus Amygdala	Angoa-Perez et al. (2010)
	Soman	Rat	100 μg/kg (sc)	↑ *IL-1β* mRNA at 30 min ↑ IL-1β protein peaks at 6h and expressed in astrocytes and endothelial cells	Striatum Hippocampus Cerebellum Frontal Cortex	Svensson et al. (2001)
	Soman	Mouse	110 μg/kg (sc) (post) AMN, 5 mg/kg (ip)	↑ Microglial cell number peaking at 3 DPE ↑ But delayed GFAP immunoreactivity 3–90 DPE; no change in oligodendrocyte health at any time point	Hippocampus Amygdala	Collombet et al. (2005a)

	Soman	Mouse	172 µg/kg (sc) (pre) HI-6, 50 mg/kg (ip)	↑ *IL-1β, TNF-α, IL-6, SOCS3, ICAM* mRNA as early as 1 h in cortex but delayed up to 6 h in hippocampus. Does not resolve for 7 DPE	Cortex Hippocampus Cerebellum	Dhote et al. (2007)
	Sarin	Rat	50 µg/kg/mL (im)	↑ *GFAP* and *vimentin* (northern blot) as early as 1 h in cortex, cerebellum and midbrain but not until 3 days in the brainstem; persisted until 7 days in all four brain regions ↑ *Prostaglandin* (microarray) at 2 h	Cortex Cerebellum Brainstem Midbrain	Damodaran et al. (2002, 2006)
	Sarin	Rat	108 µg/kg (sc) (post) AS[f], 2 mg/kg (im) 2-PAM[g], 25 mg/kg (im) Diazepam,	↑ *TNF-α, IL-1β, IL-10, IL-6, Neurotrophin, TREM1* mRNA within 1 h (peaks by 6 h); does not resolve by 24 h.	Piriform Cortex Hippocampus Thalamus Amygdala	Spradling et al. (2011a,b)
	Sarin	Rat	108 µg/kg (im) (pre) PB[h], 0.1 mg/kg (im); HI-6, 5 mg/kg (im) (post) AS—2 mg/kg (im) Midazolam—1 mg/kg (im)	↑ IL-1β, IL-6, PGE2 protein within 2 h, peaking between 2 and 24 h. Protein levels correlate with seizure duration	Piriform Cortex Hippocampus	Chapman et al. (2006)
Pesticide	Paraoxon	Rat	0.45 mg/kg (im) (post) AS—3 mg/kg (im); Toxogonin—20 mg/kg (im)	↑ IB-4 + cell number at 28 DPE ↑ GFAP at 28 DPE	Piriform Cortex	Finkelstein et al. (2012)
	DFP	Rat	9 mg/kg (ip) (pre) PB, 0.1 mg/kg (im); AMN, 20 mg/kg (im)	↑ *GFAP* mRNA at 24 h post DFP, ↑↑ by 3 DPE ↑ GFAP with distinct spatiotemporal progression beginning as early as 1 h with biphasic response that persists for 21 DPE ↑ CD11b + cell number and morphologic activation at 24 h ↑ IBA-1 and morphological activation persists for 14 DPE, ↑at 21 DPE ↑ TSPO peaks at 7 DPE in whole brain	Piriform Cortex Hippocampus Amygdala Thalamus	Liu et al. (2012), Li et al. (2015), and Flannery et al. (2016)

Continued

Table 2 Acute OP intoxication triggers neuroinflammation.—cont'd

Type of OP	Chemical	Model	Exposure paradigm	Neuroinflammatory response	Brain regions affected	Reference
	DFP	Rat	9 mg/kg (ip) (pre) PB, 0.1 mg/kg (im); AMN, 20 mg/kg (im)	↑ *IL-1β*, *IL-6* mRNA at 24 h No change *TNF-α* mRNA at 24 h	Hippocampus	Li et al. (2015)
	DFP	Rat	3.2 mg/kg (sc) (pre) PB, 0.026 mg/kg (im) (post) AMN, 2 mg/kg (ip); 2-PAM, 25 mg/kg (im)	↑ GFAP at 3 DPE ↑ IBA-1at 3 DPE	Hippocampus Amygdala	Kuruba et al. (2018)
	DFP	Rat	4 mg/kg (sc) (pre) PB, 0.1 mg/kg (im) (post) AS, 2 mg/kg (im); 2-PAM, 25 mg/kg (im)	↑ GFAP and IBA-1 immunoreactivity with varying spatiotemporal progression-persists for 60 DPE	Piriform Cortex Hippocampus Cerebral Cortex Amygdala Thalamus	Siso et al. (2017)
	DFP	Rat	4.5 mg/kg (sc) (pre) PB, 0.1 mg/kg (im) (post) AMN, 2 mg/kg (im); 2-PAM, 25 mg/kg (im)	↑ TNF-α, IL-1β, IL-6, KC/GRO proteinat 24 h	Hippocampus Piriform Cortex	Liang et al. (2018)

[a]im—intramuscular.
[b]sc—subcutaneous.
[c]HI-6—asoxime chloride.
[d]ip—intraperitoneal.
[e]AMN—atropine methyl nitrate.
[f]AS—atropine sulfate.
[g]2-PAM—2-pyridine aldoxime methylchloride.
[h]PB—pyridostigmine bromide.
[i]DPE—day post exposure.

These early observations were subsequently replicated in other preclinical models of acute intoxication with soman (Angoa-Perez et al., 2010) or sarin (Damodaran et al., 2002, 2006). For both OP nerve agents, the temporal patterns of rapid astrogliosis and microglial activation following exposure varied regionally but generally peaked within 24–72h and then subsided. Observations along an extended timeline post-exposure in a mouse model of acute soman intoxication revealed a second wave of increased GFAP expression in the medial and lateral septum and hippocampus of mice beginning 3 days after exposure that persisted for up to 30–90 days depending on the brain region (Collombet et al., 2005a).

Consistent with observations of astrogliosis and microglial activation, transcriptomic analysis of brains from rats intoxicated with sarin identified signaling pathways associated with inflammation among the most significantly altered gene expression changes (Spradling et al., 2011a). Regional analyses of cytokine expression in the brain at the mRNA or protein level indicated that the spatiotemporal patterns varied between cytokines (Table 2). Transcript levels of TNF-α peaked in most brain regions within 2h following soman intoxication, whereas increased mRNA expression of other cytokines, such as IL-1β and IL-6, was delayed, with significant upregulation not observed until 6h post-exposure and even then only in very specific brain regions, such as the piriform cortex (Dhote et al., 2007; Dillman et al., 2009; Svensson et al., 2001; Williams et al., 2003). Analyses of the protein expression of cytokines (TNF-α, IL-1β, IL-6), prostaglandins and chemokines (MCP-1, MIP-1α) similarly indicated a robust but transient response in the brain following soman and sarin exposures (Angoa-Perez et al., 2010; Chapman et al., 2006; Johnson and Kan, 2010; Johnson et al., 2011). More detailed spatial resolution of the expression of these inflammatory mediators by immunohistochemical analyses revealed that neurons, microglia, astrocytes, and components of the BBB vasculature all expressed these mediators to varying degrees (Johnson and Kan, 2010; Johnson et al., 2011).

More recent preclinical research on OP-induced neuroinflammation has focused on acute poisoning with seizurogenic doses of the OP pesticides diisopropylfluorophosphate (DFP) or paraoxon (Table 2). Multiple laboratories have demonstrated that acute intoxication with paraoxon (Finkelstein et al., 2012) or DFP (Ferchmin et al., 2014; Flannery et al., 2016; Kuruba et al., 2018; Li et al., 2015; Liu et al., 2012; Rojas et al., 2015) activates astrocytes and microglia as indicated by significantly increased expression of GFAP and morphological assessment of the activation state of Iba-1 immunopositive cells, respectively. Acute DFP intoxication also elevates mRNA and protein expression of the proinflammatory cytokines, IL-1β, TNF-α, and IL-6 (Li et al., 2015; Liang et al., 2018) and prostaglandins (Rojas et al., 2015) within 1 day after DFP exposure. A contemporary translational tool for assessing neuroinflammation—positron emission tomography (PET) imaging of the 18kDa mitochondrial translocator protein (TSPO)—has also been employed to longitudinally monitor the neuroinflammatory response in the rat model of acute DFP intoxication. TSPO is typically expressed at low levels on the outer mitochondrial membrane of glial cells in the brain; however, upon activation, TSPO expression is significantly upregulated, making it a useful biomarker of microglial activation, and perhaps reactive astrocytes (Guilarte, 2018; Lavisse et al., 2012; Veiga et al., 2007). In general concurrence with immunohistochemical evidence of neuroinflammation, PET imaging of TSPO labeling in the brains of DFP intoxicated rats indicated significantly increased neuroinflammation in the hippocampus and cortex relative to VEH controls at 2, 7 and 21 days, with peak binding of the TSPO ligand, [^{11}C]PK11195 observed at 7 days post-exposure (Flannery et al., 2016).

As observed with OP nerve agents, glial activation was seen within hours to days following DFP or paraoxon exposure (Finkelstein et al., 2012; Flannery et al., 2016; Liu et al., 2012; Siso et al., 2017). But relative to observations with OP nerve agents, this cellular response peaked much later (3–7 days post-exposure). Also in contrast to observations with OP nerve agents, reactive astrogliosis in the DFP model preceded microglial activation, at least in the two brain regions where this relationship was examined, the hippocampus and cortex (Flannery et al., 2016). The neuroinflammatory responses to acute DFP intoxication have been shown to persist for at least 30–60 days post-exposure; however, the temporal profile varied across brain regions (Flannery et al., 2016; Siso et al., 2017). Reactive astrogliosis persisted significantly longer in the cortex relative to the hippocampus, and while a second wave of microglial activation was apparent in both brain regions at 7 days post-exposure, this persisted significantly longer in the hippocampus relative to the cortex. Whether the second wave of apparent microglial cell activation represented microglia and/or monocytes recruited from the periphery was not addressed in this study.

3.2 Evidence that low-level OP exposures also trigger neuroinflammatory responses

There is growing consensus that repeated low-level OP exposures are linked to neurobehavioral deficits in humans following either occupational exposures (Meyer-Baron et al., 2015; Rohlman et al., 2011) or developmental exposures (Bouchard et al., 2011; Burke et al., 2017; Engel et al., 2011). But the question of whether these OP effects are associated with neuroinflammation has only recently been addressed. The data emerging from preclinical models suggest that repeated low-level OP exposures trigger neuroinflammatory responses (Table 3).

The majority of these studies have focused on the OP pesticide chlorpyrifos. In adult mice, repeated administration of chlorpyrifos at levels that did not inhibit red blood cell cholinesterase activity upregulated GFAP protein expression in the hippocampus (Lim et al., 2011). Genes in the IL-6 and TNF-α signaling pathways were upregulated in the forebrain of adult rats exposed to chlorpyrifos at levels that inhibited AChE, but did not cause cholinergic symptoms (Stapleton and Chan, 2009). In a model of developmental chlorpyrifos exposure, GFAP protein was increased in the brain of fetal animals following chlorpyrifos exposure of the pregnant dam on gestational day 17 through 20 at doses that induced systemic toxicity, but not cholinergic symptoms (Garcia et al., 2002). In contrast, postnatal administration of chlorpyrifos to the dam caused an initial sex-specific decrease in GFAP expression in pups followed by a rebound of GFAP expression to levels significantly higher than vehicle controls at 15 days after the last exposure (Garcia et al., 2002). The authors postulated that elevated gestational GFAP reflected chlorpyrifos-induced astrogliosis, whereas postnatal GFAP deficits were caused by depression of cell differentiation during periods of glial proliferation. A similar decrease in GFAP expression was observed in developing rat pups following a single exposure to chlorpyrifos (2 mg/kg, sc) on postnatal day 7 (Ray et al., 2010). Collectively, these observations suggest that astrocytes in the developing brain are susceptible to the toxic effects of chlorpyrifos, but that the outcome, at least in terms of GFAP expression, varies quantitatively and qualitatively as a function of sex and the developmental age at the time of exposure. In other studies of developmental chlorpyrifos exposure, chlorpyrifos in the maternal diet throughout gestation and lactation upregulated the proinflammatory cytokine IL-1β, but downregulated the anti-inflammatory cytokine IL-10 in the hippocampus of weanling rat pups (Gomez-Gimenez et al., 2017). Increased IL-1β levels in the hippocampus correlated negatively with pup performance in a spatial learning task. Direct exposure of pups to chlorpyrifos during the postnatal period was also sufficient to promote neuroinflammation in the amygdala, as evidenced by increased expression of Iba-1 and proinflammatory cytokines (Tian et al., 2015). The potential relevance of these findings to

Table 3 Repeated Low-level OP exposure upregulates biomarkers of neuroinflammation.

Type of OP	Model	Exposure Paradigm	Neuroinflammatory Response	Brain Regions Affected	Reference
Sarin	Adult rat	0.2 or 0.4 mg/m^3, inhalation 1 h/day for 1 or 5 days	Dose-dependent ↑ *IL-1β*, *TNFα*, *IL-6* mRNA determined by RT-PCR of whole brain after 5 days of daily exposure	whole brain	Henderson et al., 2002
Chlorpyrifos	Adult rat	0.5, 1.0, 5, 10, 30, 50 mg/kg single exposure by oral gavage	Microarray analyses indicated disruption of IL-6 and TNFα signaling pathways at 96 h post-exposure	Forebrain	Stapleton and Chan, 2009
Chlorpyrifos	Adult mouse	20 or 40 mg/kg, dermal application to the tail for 7 d	↑ GFAP immunoreactivity determined by IHC[a]	Hippocampus	Lim et al., 2011
Chlorpyrifos	Developing rats (prenatal)	1, 2, 5, 10, 20 or 40 mg/kg/d Dams injected sc daily Gestational days 17-20	↑ GFAP in 20 and 40 mg/kg dose groups as determined by WB[b] at gestational day 21	Forebrain Midbrain and brainstem	Garcia et al., 2002
Chlorpyrifos	Developing rats (postnatal)	1 or 5 mg/kg/d Dams injected sc daily PND[c] 1-4 or PND 11-14	↓ GFAP in cerebellum of male but not female pups at PND 5 following exposure PND 1-4 ↑ GFAP in multiple brain regions of male, but only striatum in female at PND 30 following exposure PND 11-14	Cerebellum (PND 5) Multiple brain regions (PND 30)	Garcia et al., 2002
Chlorpyrifos	Neonatal rat	5 mg/kg, intrahepatic injection Daily, PND 11-14	↑ *TNFα*, *IL-6* mRNA determined by qRT-PCR ↑ HMGB1[d] and Iba-1 immunoreactivity determined by IHC and WB at 12 (Iba-1 only), 24 and 72 h after last injection	Amygdala	Tian et al., 2015

Chlorpyrifos	Developing rats	0.1, 0.3 or 1 mg/kg Daily in maternal diet Gestational day 7 – PND 21	All doses ⬆ IL-β in male but not female pups; 0.3 and 1.0 mg/kg ↓ IL-10 in female but not female pups determined by WB	Hippocampus	Gómez-Giménez et al., 2017
Dichlorvos	Adult rat	2.5 mg/kg, sc[e] Daily for 12 weeks	⬆ MAC expression as determined by IHC, WB of brain homogenates and immunostaining of isolated microglial cells; ⬆ IL-1β, IL-6 , TNFα as determined by IHC and ELISA	Substantia nigra Corpus striatum	Binukumar et al., 2011
Dimethoate	Adult mouse	1.4 mg/kg, ip[f] 3 times a week for 5 weeks	⬆ *TNFα*, *IL-6* mRNA determined by qRT-PCR ⬆ Microglial cell activation determined by morphological assessment of Iba-1 immunopositive cells	Hippocampus (cytokines) DG[g], substantia nigra (microglia)	Astiz et al., 2013
Malathion	Adult mouse	30 or 100 mg/kg, sc Daily for 15 d	Both doses ⬆ GFAP immunoreactivity determined by IHC	Hippocampus (CA1, CA3, DG)	Dos Santos et al., 2016
Malathion	Adult mouse	200 mg/kg, oral gavage Daily for 8 weeks	⬆ Number of Iba-1 immunoreactive cells	Substantia nigra	Ahmed et al., 2017

[a]IHC – immunohistochemistry;
[b]WB – western blotting;\
[c]PND – postnatal day;
[d]HMGB1 – high mobility group box 1;
[e]sc – subcutaneous;
[f]ip – intraperitoneal;
[g]DG – dentate gyrus

the developing human brain is demonstrated in a study showing that *in vitro* exposure of human fetal astrocytes to chlorpyrifos for 1 week increased levels of *GFAP, IL-6*, and other genes associated with proinflammatory cytokine signaling pathways (Mense et al., 2006).

Studies of repeated low-level exposures in adult rodents to OPs other than chlorpyrifos confirm that neuroinflammation is not a response unique to chlorpyrifos (Table 3). For example, in rats exposed to repeated low doses of sarin vapor, brain levels of transcripts for *TNF-α*, *IL-1β* and *IL-6* were significantly increased (Henderson et al., 2002). In another study, rats were administered the OP pesticide dichlorvos for 12 weeks at 2.5 mg/kg/day, a dosing paradigm previously shown to inhibit AChE by 10–55%, but not cause marked signs of cholinergic toxicity (Verma et al., 2009). At the end of the exposure, significant microglial activation was observed in the substantia nigra and corpus striatum (Binukumar et al., 2011). In this model, increased microglial activation preceded neuronal degeneration in the same brain regions and was accompanied by increased levels of proinflammatory cytokines. Two additional OP pesticides, dimethoate and malathion, have been demonstrated to increase biomarkers of neuroinflammation in adult mice exposed to repeated doses that did not cause cholinergic toxicity. Dimethoate increased expression of *TNF-α* and *IL-6* mRNA in the hippocampus and increased the number of activated microglia in the dentate gyrus and substantia nigra, as determined by morphological analyses of Iba-1 immunopositive cells (Astiz et al., 2013). At doses (30 and 100 mg/kg) that inhibited hippocampal AChE by 40% but did not cause cholinergic toxicity, malathion increased GFAP immunoreactivity in the hippocampus coincident with impaired spatial memory and discrimination (Dos Santos et al., 2016). At a higher dose (200 mg/kg), malathion caused significant dopaminergic neurodegeneration in the substantia nigra and deficits in locomotor behavior and significantly increased the number of activated microglia in the substantia nigra (Ahmed et al., 2017).

4. How do OPs trigger neuroinflammatory responses?

4.1 Cholinergic mechanisms of OP-induced neuroinflammation

A key question is whether neuroinflammatory responses to OP intoxication are mechanistically related to AChE inhibition. AChE inhibition is the principal mechanism of OP-induced cholinergic crisis, which is associated with a robust inflammatory response (Table 2). However, results from diverse models of OP intoxication provide a compelling argument against cholinergic mechanisms of OP-induced neuroinflammation. Particularly convincing are data obtained from a preclinical model of Gulf War Illness (O'callaghan et al., 2015) in which adult male mice are administered corticosterone in the drinking water for 4 days at 400 mg/mL, a level associated with high physiological stress to replicate Gulf War theater conditions, prior to injection with DFP (4 mg/kg, ip). This exposure paradigm resulted in brain-wide neuroinflammation as assessed by qPCR of proinflammatory cytokines, and Western blot analyses of phosphorylated and total STAT, a signaling molecule activated by inflammatory mediators. To test the role of AChE inhibition in the neuroinflammatory response, animals were administered corticosterone plus either irreversible (chlorpyrifos oxon or DFP) or reversible (pyridostigmine bromide or physostigmine) inhibitors of AChE (Locker et al., 2017). Chlorpyrifos oxon, and DFP alone caused cortical and hippocampal neuroinflammation and activation of STAT, which were markedly enhanced by pretreatment with corticosterone. In contrast, exposure to the reversible AChE inhibitors alone or with corticosterone pretreatment did not cause neuroinflammation or activate STAT. All of the CNS-acting AChE inhibitors (DFP, chlorpyrifos oxon, and physostigmine) decreased brain AChE activity to an equivalent level in the absence of corticosterone; however, pretreatment with corticosterone abrogated this effect only for the irreversible inhibitors. The lack of correlation between the level of AChE inhibition and the magnitude of neuroinflammation in this model suggests that neuroinflammation was mediated by non-cholinergic mechanisms (Locker et al., 2017). This conclusion was further supported by measurement of acetylcholine levels in specific brain regions, which suggested that the exacerbated neuroinflammatory effects observed in the brains of animals treated with corticosterone and DFP were not driven by accumulation of acetylcholine in the brain (Miller et al., 2018). The second line of evidence are studies demonstrating that repeated low-level OP exposures induce inflammation even in the absence of seizures or significant AChE inhibition (Table 3). For example, repeated low-level doses of sarin elevated cytokine levels in rat brain in the absence of central AChE inhibition, although blood ChE activity was reduced (Henderson et al., 2002). Similarly, chronic chlorpyrifos administration increased GFAP expression in the hippocampus of rats without inhibiting serum cholinesterase (Lim et al., 2011). Observations such as these indicate that prolonged low-level exposure to OPs can trigger neuroinflammatory responses in the absence of AChE inhibition and overt clinical symptoms of OP poisoning.

The mechanism(s) by which the Gulf War Illness exposure paradigm and the repeated low-level OP exposures induce neuroinflammation have yet to be defined, but one possibility is that OPs interact directly with inflammatory cells to trigger the release of inflammatory mediators. There is experimental evidence to support this hypothesis. For example, the OP pesticide parathion stimulated release of TNF-α from primary alveolar macrophages isolated from guinea pig lung and exposed *ex vivo* (Proskocil et al., 2013). Exposure of human whole blood cultures to the OP chlorpyrifos potentiated LPS-induced release of IFN-γ (Duramad et al., 2006). Gene and protein expression profiling of primary cultures of human fetal astrocytes revealed that chlorpyrifos upregulated GFAP and key inflammatory mediators, including IFN-γ and IL-6 (Mense et al., 2006). The molecular mechanisms by which OPs activate immune cells remain largely unknown; however, recent studies demonstrated that stimulation of cholinergic receptors as a consequence of acetylcholine accumulation at immunological synapses was sufficient to activate signaling cascades that evoked inflammatory responses in macrophages (Mitra et al., 2017). While these studies demonstrate that OPs can directly activate peripheral immune cells, it is unknown whether this phenomenon holds true in the brain.

4.2 Is OP-induced neuroinflammation driven by oxidative stress?

Increased brain expression of biomarkers of oxidative stress has been reported in preclinical models of acute OP poisoning (Kim et al., 1999; Liang et al., 2018; Zaja-Milatovic et al., 2009) and repeated low-level OP exposure (reviewed in Soltaninejad and Abdollahi, 2009). Such observations have led many to propose that OP-induced neuroinflammation is directly related to increased oxidative stress (Milatovic et al., 2006; Naughton and Terry, 2018; Pearson and Patel, 2016; Soltaninejad and Abdollahi, 2009). This hypothesis derives in large part from evidence indicating that the excessive cholinergic neurotransmission caused by AChE inhibition following acute OP intoxication increases glutamatergic signaling, which is critical for sustaining seizures initiated by cholinergic mechanisms (Mcdonough and Shih, 1997). This excessive glutamatergic signaling is thought to drive early neuroinflammatory responses via at least two mechanisms (Fig. 2). First, excessive release of neuronal glutamate can cause astroglial dysfunction, which stimulates reactive astrogliosis (reviewed in Viviani et al., 2014). Second, sustained activation of NMDA receptors can cause excitotoxicity as a result of excessive intracellular Ca^{2+} and mitochondrial stress that elevate ROS to levels that are neurotoxic (Arundine and Tymianski, 2003). Neuronal cell death as a result of either necrosis or apoptosis releases signals, such as ATP, that activate

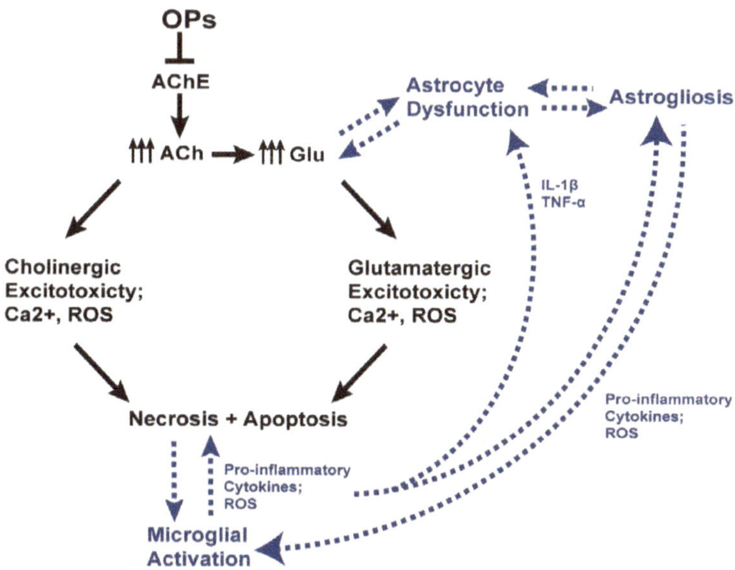

Fig. 2 Mechanisms by which acute OP intoxication triggers neuroinflammation. OPs inhibit acetylcholinesterase (AChE) leading to excessive build-up of acetylcholine (ACh) in cholinergic synapses. During the immediate response (~5 min), cholinergic excitotoxicity is characterized by excessive Ca^{2+} influx and increased ROS production, which ultimately trigger necrotic and apoptotic cell death. Hypercholinergic signaling also activates glutamatergic signaling in the brain, leading to glutamate-induced excitotoxicity. Necrotic cell death releases damage associated molecular patterns (DAMPs) that trigger release of inflammatory cytokines and ROS from activated microglia. These pro-inflammatory mediators can perpetuate the neurotoxic response as well as activate surrounding astrocytes to undergo reactive astrogliosis. Excessive or prolonged neuroinflammation can drive astrocyte dysfunction that promotes glutamate-mediated excitotoxicity. This image is licensed under a Creative Commons Attribution 3.0 Unported License (Servier Medical Art by Servier and Creative Commons Attribution—Share Alike 4.0 International license).

microglia (Chung et al., 2006; Joseph et al., 2005; Potts et al., 2006; Rahman et al., 2006; Wang et al., 2006).

It has long been proposed that the neuroinflammation observed in preclinical models of acute OP intoxication is a response to excitotoxic neurodegeneration. In support of this hypothesis, early neuroinflammatory responses were reported to coincide with neurodegeneration in preclinical models of acute OP intoxication (reviewed in Chen, 2012; Collombet, 2011; De Araujo Furtado et al., 2012; Pereira et al., 2014). Consistent with long-standing observations that the extent of neurodegeneration is directly correlated to the duration of OP-induced seizure activity

(Filliat et al., 1999; Hobson et al., 2017; Lemercier et al., 1983; Mcdonough and Shih, 1997; Siso et al., 2017), a strong positive correlation has recently been reported between seizure activity and various indices of neuroinflammation. Specifically TSPO labeling in the brain as detected by PET imaging of rats acutely intoxicated with the OP pesticide DFP (Flannery et al., 2016) and expression of neuronal cyclooxygenase 2 (COX-2) in the hippocampus, amygdala, piriform cortex, and thalamus of rats acutely intoxicated with the OP nerve agent soman (Angoa-Perez et al., 2010) are positively correlated with seizure duration. However, data from the few studies that have examined spatiotemporal patterns of neurodegeneration and neuroinflammation in the same animal following acute OP intoxication suggest that neuroinflammation precedes neuronal cell death in multiple brain regions. Thus, increased GFAP immunoreactivity was detected in the piriform cortex, amygdala and dentate gyrus within 1 h after acute intoxication with DFP, which preceded any signs of neurodegeneration (Li et al., 2011; Liu et al., 2012). Similarly, in a preclinical model of acute soman intoxication, GFAP immunoreactivity was observed prior to evidence of neuronal injury (Baille-Le Crom et al., 1995; Zimmer et al., 1997). However, these observations do not rule out the possibility that neurodegeneration may trigger other aspects of neuroinflammation, such as microglial activation.

There are significantly fewer data regarding the temporal and mechanistic relationships of either neuroinflammation or neurodegeneration to oxidative stress induced by acute OP intoxication. A study of the spatiotemporal profile of the expression of oxidative stress biomarkers (specifically, the ratio of reduced to oxidized glutathione and levels of nitrosylated tyrosine) in rats acutely intoxicated with DFP reported increased oxidative stress in the brain at 24 and 48 h post-exposure, but not 6 and 12 h post-exposure (Liang et al., 2018). Considered in light of studies that have reported neuroinflammation in preclinical models of acute DFP intoxication at significantly earlier time points (Flannery et al., 2016; Rojas et al., 2015; Siso et al., 2017), these data suggest that neuroinflammation may be driving oxidative stress. However, treatment of rats acutely intoxicated with DFP with a catalytic antioxidant administered at 5 or 15 min post-exposure significantly attenuated both oxidative stress and neuroinflammation, the latter determined by levels of proinflammatory cytokines in the brain, and these effects coincided with decreased neurodegeneration (Liang et al., 2018). Antioxidant treatment has also been reported to attenuate neuroinflammation in preclinical models of repeated low-level OP exposure.

The 6-gingerol-rich fraction from *Zingiber officinale* (ginger) decreased TNF-α levels and neuronal apoptosis in the brain of rats chronically exposed to chlorpyrifos at a level (5 mg/kg daily for 35 days) that caused significant oxidative stress (Abolaji et al., 2017). Crocin, a water-soluble carotenoid isolated from the stigma of *Crocus sativus* (saffron), decreased TNF-α and IL-6 levels in the striatum of rats exposed to malathion (100 mg/kg, ip, daily for 28 days). This exposure paradigm caused neuropathological and behavioral phenotypes relevant to Parkinson's disease, and crocin treatment not only attenuated neuroinflammation but also blocked oxidative stress and improved performance on tests of locomotor behavior (Mohammadzadeh et al., 2018).

In summary, the existing data demonstrate that both acute OP intoxication and repeated low-level OP exposure cause oxidative stress in the brain. However, the data supporting oxidative stress as a principal mechanism driving OP-induced neuroinflammation are equivocal and the functional relationships between oxidative stress, neuroinflammation, and neurodegeneration are not clear. While experiments designed to specifically address this question are required, it seems likely that these are dynamic relationships that change depending on the OP exposure paradigm, the time post-exposure and the brain region(s) involved.

5. What is the pathophysiologic relevance of OP-induced neuroinflammation?

Preclinical studies have provided compelling evidence that both acute high-dose and repeated low-level exposures to OPs induce neuroinflammation. However, the relevance of neuroinflammation to the pathophysiology of neurological effects associated with OP exposure has yet to be determined. Critical questions include: Is neuroinflammation a cause or consequence of the neurotoxic effects of OPs? Is the neuroinflammatory response to OPs neurotoxic or neuroprotective in its actions?

The answers to these questions are emerging primarily from two areas of research: clinical and preclinical studies of neuroinflammation in the context of neurological diseases phenotypically similar to outcomes associated with neurotoxic OP exposures (Table 4), and mechanistic studies in preclinical models of OP neurotoxicity.

Table 4 Neuroinflammatory components implicated in pathogenesis of neurologic disorders.

Disease state	Neuroinflammatory component	Neuroinflammatory response (model)	References
Neurodegenerative disease	IL-1β; TNF-α Complement; Activated microglia	↑ Pro-inflammatory cytokines, complement, microglia colocalized with beta-amyloid plaques in hippocampus (AD[a] human); ↓ Inflammasome activation reduces IL-1β production and protects AD associated deficits in spatial memory (mouse); ↓ microglial activation protects against deficits in spatial memory (TBI[b]—rat); ↑ *TNF-α, IL-1β* polymorphisms associated with greater risk of developing PD[c] (human); ↑ LPS-induced microglia activation and inflammation show selective dopaminergic neurotoxicity (PD—rat)	Reviewed in Simen et al. (2011), Heneka et al. (2015), Liu and Bing (2011), Wahner et al. (2007a,b), Walker and Tesco (2013)
Age-related, non-pathogenic cognitive decline	IL-6 IL-1; Innate immune response genes	↑ IL-6 in serum inversely correlated with attention and working memory (human); ↑ IL-1β in serum associated with decreased executive function (human); ↑ IL-1 and microglia activation associated with impaired spatial learning (mouse; rat); ↑ Inflammation-related gene expression inversely correlated with genes associated with learning and memory (rat)	Reviewed in Simen et al. (2011)

Depression/anxiety	TNF-α; IL-1β; IL-6; Activated microglia; Glial cell activation	↑ TNF-α and IL-1β in CSF predictive of poor performance on psychometric tests for anxiety and depression (human); ↑ microglia activation required for stress-induced anxiety (mouse; rat); ↑ TSPO, IL-1β and IL-6 expression in brains of anxiogenic animals (rat); ↑ *IL-1β* mRNA and protein associated with depressive-like behavior. Reduction in cytokine levels results in reversal of behavior (rat)	Rossi et al. (2017), Wang et al. (2018), Wohleb et al. (2014), Yue et al. (2018)
Epilepsy	IL-1β; COX2; Activated microglia; Astrogliosis	↑ Alteration in phenotype, function and activation of microglia and astrocytes in hippocampus (human); ↑ LPS-induced microglial activation results in downstream astrocytic enhancement of postsynaptic currents (mouse, in vitro); Inhibition of IL-1β signaling slows development of acquired epilepsy (mouse); Pharmacological inhibition of COX2 reduces microglial activation and the development of acquired epilepsy (mouse); ↑ reactive astrogliosis in the absence of BBB[d] pathology triggers the development of acquired epilepsy (mouse); ↓ TSPO expression and GFAP reactivity correlate with reduced incidence of acquired epilepsy (rat)	Review in Vezzani et al. (2013), Jung et al. (2006), Kleen et al. (2012), Pascual et al. (2012), Robel et al. (2015), Vezzani et al. (2000)

[a]AD – Alzheimer's' Disease;
[b]TBI – Traumatic Brain Injury;
[c]PD- Parkinson's Disease;
[d]BBB- Blood Brain Barrier

5.1 Neuroinflammation in neuropsychiatric disorders

Observations from clinical and preclinical studies of neurological diseases with phenotypic outcomes similar to those associated with neurotoxic OP exposures provide indirect support for the hypothesis that neuroinflammation is mechanistically linked to the neurological effects of OPs (Table 4). Acute OP intoxication that causes SE or other signs of cholinergic crisis is associated with delayed and persistent cognitive dysfunction, depression, anxiety, and spontaneous recurrent seizures in both humans and preclinical models (reviewed in Chen, 2012; De Araujo Furtado et al., 2012; Pereira et al., 2014). Similarly, repeated low-level OP exposures have been linked to cognitive deficits and neuropsychiatric conditions such as depression, anxiety, and suicidal tendencies (reviewed in Voorhees et al., 2016). Beyond the field of OP neurotoxicology, there is an extensive literature linking dysregulated neuroinflammation to neurological disease (Skaper et al., 2018), including cognitive deficits in neurodegenerative disease (Guedes et al., 2018; Hong et al., 2016; Mcafoose and Baune, 2009; Wyss-Coray and Rogers, 2012), cognitive impairment in socially defeated mice (Pfau and Russo, 2016), which may be particularly relevant to Gulf War Illness, depression and anxiety (Jeon and Kim, 2018; Kim and Jeon, 2018; Pfau et al., 2018), and electrographic abnormalities, including acquired epilepsy (Devinsky et al., 2013; Rana and Musto, 2018).

Persistent memory impairment is one of the most common neurological consequences observed in humans survivors of acute OP poisoning (Chen, 2012). Years after the 1995 Tokyo subway attack with sarin, exposed rescue personnel and subway workers still exhibited significant deficits on memory tests compared to non-exposed control subjects (Miyaki et al., 2005; Nishiwaki et al., 2001; Yanagisawa et al., 2006). Multiple lines of evidence implicate dysregulated neuroinflammation, especially in the hippocampus, in the pathogenesis of age-related cognitive dysfunction (Simen et al., 2011) and cognitive impairment associated with neurodegenerative disease (Heneka et al., 2015; Liu and Bing, 2011; Wahner et al., 2007a,b). Microglial activation seems particularly important in the pathogenesis of Alzheimer's disease (AD) (Heneka et al., 2015). A network-based integrative analysis of whole-genome expression profiling and genotypic data obtained from brains of individuals with late-onset AD and non-demented control subjects implicated the immune and microglial modules as the molecular systems most strongly associated with the pathophysiology of late-onset AD (Zhang et al., 2013). Consistent with this clinical finding, pharmacologic inhibition of microglial proliferation in a mouse model of

AD shifted microglia to an anti-inflammatory phenotype, prevented synaptic degeneration, and improved memory (Olmos-Alonso et al., 2016). Cognitive impairment following traumatic brain injury (TBI) has similarly been associated with neuroinflammation that persists for years after initial injury. Therapeutic strategies to prevent or mitigate microglial activation have shown efficacy in suppressing production of inflammatory mediators and improving long-term cognitive function in preclinical TBI models (Walker and Tesco, 2013).

Neuropsychiatric disorders, and in particular depression and anxiety, are common morbidities of acute OP intoxication in both humans and preclinical models (Coubard et al., 2008; Delgado et al., 2004; Levin et al., 1976; Okumura et al., 2005; Prager et al., 2014b; Savage et al., 1988). Clinical and experimental evidence implicates neuroinflammation as a significant contributing factor in the etiology of both these neuropsychiatric disorders (Dantzer et al., 2011; Jeon and Kim, 2018; Kim and Jeon, 2018; Pfau et al., 2018). Patients with major depression are reported to have significantly elevated blood levels of proinflammatory cytokines; conversely, non-depressed patients who received chemokine immunotherapy tended to develop clinical signs of depression (Dantzer and Kelley, 1989). In preclinical studies, treatment with anti-inflammatory agents attenuated depressive-like behavior in animals (Wang et al., 2018; Yue et al., 2018). With regard to anxiety disorders, elevated levels of TNF-α and IL-1β in the cerebrospinal fluid of relapsing-remitting multiple sclerosis patients were found to be predictive of poor performance on psychometric tests used to evaluate anxiety levels (Rossi et al., 2017). Blocking the activity of proinflammatory mediators or microglial activation via pharmacologic manipulation and genetic deletion of inflammatory components mitigated the development of stress-induced anxiety-like behavior in preclinical models (Wang et al., 2018; Wohleb et al., 2014; Yue et al., 2018).

Epilepsy is a neurological disease characterized by recurrent seizures. Over the past decade, a prominent role for glial cells in the mechanisms of seizure initiation and recurrence has been defined (Devinsky et al., 2013; Rana and Musto, 2018). Altered phenotype and function of activated astrocytes and microglial cells were detected in experimental and human epileptic tissue, including modifications in the induction of cytokines, chemokines, prostaglandins, complement factors, and cell adhesion molecules involved in both the pro- and anti-inflammatory components of neuroinflammation (Seifert and Steinhauser, 2013; Vezzani et al., 2013). In preclinical models of SE-induced epilepsy, which is thought to be mechanistically similar to acquired epilepsy following OP-induced SE, inflammatory cytokines, particularly IL-1β and TNF-α, promoted the progression of epileptogenesis (Vezzani et al., 2013). Conversely, pharmacological attenuation of inflammation following SE with immunosuppressants or COX 2 inhibitors was efficacious in preventing epileptogenesis (Jung et al., 2006; Vezzani et al., 2013).

5.2 Neuroinflammation in OP neurotoxicity: Neurotoxic or neuroprotective?

The general assumption has been that OP-induced neuroinflammation contributes to the neuropathology and neurological deficits associated with neurotoxic OP exposures. This assumption is based on several lines of evidence. First, histopathological characterization of brains from rats that survived acute OP-induced SE have shown that neuroinflammatory responses typically coincide spatially and temporally with neurodegeneration (Ferchmin et al., 2014; Kim et al., 1999; Li et al., 2015; Liu et al., 2012; Rojas et al., 2015) and either precede or overlap with the onset of behavioral deficits (Coubard et al., 2008; Flannery et al., 2016; Rojas et al., 2015) and recurrent spontaneous seizures (De Araujo Furtado et al., 2010). Second, there is evidence that neuroinflammation exacerbates neuronal damage caused by excitotoxicity (Morimoto et al., 2002), in part via interactions between proinflammatory cytokines and glutamatergic neurotransmission (Fogal and Hewett, 2008). Specifically, IL-1β and TNF-α can hinder astrocytic uptake of synaptic glutamate and increase NMDA receptor activity, actions that promote excitotoxic neuronal cell death (Jander et al., 2000; Lawrence et al., 1998; Vezzani et al., 2013). These interactions are particularly important in the context of acute OP intoxication in which excessive cholinergic stimulation due to AChE inhibition increases glutamatergic signaling (Mcdonough and Shih, 1997). As shown in Fig. 2, excessive activation of NMDA receptors leads to elevated calcium influx into neurons, which perturbs mitochondria to increase ROS production. Excessive ROS ultimately results in apoptotic and necrotic cell death, which triggers neuroinflammatory responses, including increased levels of IL-1β and TNF-α, thereby setting up a feed-forward mechanism. Whether anti-inflammatory agents, particularly those that target IL-1β and TNF-α, can effectively decrease excitotoxic neurodegeneration following acute OP intoxication has yet to be systematically investigated.

The third line of evidence that neuroinflammation contributes to the neurotoxic outcomes of OP exposure are findings from a limited number of preclinical studies suggesting the potential therapeutic benefits of treating

OP intoxication with agents designed to block neuroinflammatory mediators. Most of these studies have been conducted in preclinical models of OP-induced SE. For example, a bifunctional molecule with a non-steroidal anti-inflammatory drug coupled to a pyridostigmine moiety increased survival rates and attenuated brain edema in mice acutely intoxicated with soman (Amitai et al., 2006). Treatment with the anti-inflammatory growth factor neuregulin-1 (NRG-1) inhibited microglial cell activation and reduced delayed neuronal cell death in multiple brain regions in a rat model of acute DFP intoxication (Li et al., 2012). A small molecule inhibitor of the prostaglandin E2 receptor, EP2, significantly decreased microglial activation, but not reactive astrogliosis, during the first few days following acute intoxication with DFP (Rojas et al., 2015). Furthermore, EP2 also attenuated the cognitive deficits in the novel object recognition task often seen following DFP (Rojas et al., 2016). There is at least one study that has tested the effects of anti-inflammatory treatment in a preclinical model of repeated low-level OP exposure. Dexamethasone and acetylsalicylic acid, when administered independently, improved locomotor performance and decreased degeneration of dopaminergic neurons in the substantia nigra of adult male mice exposed to neurotoxic levels of malathion (Ahmed et al., 2017).

While pharmacological studies with anti-inflammatory agents suggest that neuroinflammation contributes to the neurotoxic effects of OPs, there is evidence that neuroinflammatory responses may also function in repairing brain damage following OP-induced SE. In a mouse model of acute soman intoxication, it was observed that glial activation triggered the release of not only the proinflammatory cytokines, IL-6, TNF-α, and IL-1β, but also astrocytic secretion of neurotrophic and other growth factors that promote angiogenesis and neurogenesis (Collombet et al., 2011). These astrocyte-derived growth factors included ciliary neurotrophic factor (CNTF), brain-derived neurotrophic factor (BDNF), fibroblast growth factor (FGF-2), nerve growth factor (NGF), and vascular endothelial growth factor (VEGF) (Collombet et al., 2011). Following acute intoxication with soman, neural progenitor cell proliferation in the subventricular zone and subgranular zone of the dentate gyrus was initially decreased (Collombet et al., 2005a). However, this transient reduction was followed by enhanced neural progenitor cell proliferation above normal levels a month after exposure. The timing of neurogenesis coincided with the onset of astroglial activation, while delayed neuronal death occurred when growth factor levels decreased, approximately 2–3 months after the initial OP exposure (Collombet et al., 2005a, 2007).

These observations suggest that reactive astrocytes may play an important role in neurorepair via secretion of neurotrophic and angiogenic growth factors, and that increasing and/or prolonging the elevated expression of these factors may be of therapeutic benefit in treating acute OP intoxication. In support of this hypothesis, administration of FGF-2 and epidermal growth factor (EGF) to mice acutely intoxicated with soman significantly increased progenitor cells in the CA1 region of the hippocampus and amygdala 1 month after soman exposure (Collombet et al., 2005b). Subsequent studies demonstrated that FGF-2 and EGF co-administration accelerated the rate of neuronal regeneration in the hippocampus, and improved anxiety-like neurobehavioral profiles in soman-exposed mice, as measured by elevated plus maze and fear conditioning (Collombet et al., 2011). This co-treatment did not, however, alter neuronal regeneration in the amygdala, nor did it enhance the restoration of hippocampal-dependent memory-related tasks. Interestingly, administration of these growth factors did not alter the severity or duration of soman-induced seizures, nor did it decrease neuronal cell loss during the first 9 days after soman exposure. What is not known is whether endogenous growth factors released by activated astrocytes *in vivo* influence neuroregeneration and behavioral recovery following acute OP intoxication. Nonetheless, these data suggest that neuroinflammation may be functionally important in the regenerative processes that occur following acute OP intoxication. However, we are still far from a clear understanding of the spatiotemporal profile of inflammatory mediators that contribute to promoting damage as opposed to repair.

6. Critical data gaps and conclusions

Collectively, there is clear and compelling evidence that acute intoxication with OPs at levels that cause an acute and profound inhibition of AChE is associated with robust neuroinflammatory responses in preclinical rodent models (Table 2). While not as extensively studied, emerging data indicate that neuroinflammation is also present in preclinical models of repeated low-level OP exposures (Table 3). However, a key question that has yet to be addressed is whether neurotoxic exposures to OPs similarly elicit neuroinflammatory responses in humans. This critical data gap likely reflects, in part, the technical challenges of assessing neuroinflammation in the living human brain. *In vivo* imaging offers one solution to this challenge. PET imaging of TSPO, a molecular biomarker of neuroinflammation, is already used in the clinical setting (Guilarte, 2018), and has been validated in a preclinical model of acute OP intoxication (Flannery et al., 2016). The recent demonstration that high-order diffusion MRI was effective in quantifying neuroinflammation in a rat model of OP-induced Gulf War Illness (Koo et al., 2018) suggests an alternate *in vivo* imaging approach for measuring neuroinflammation in OP-intoxicated humans. These *in vivo* imaging modalities will also increase the feasibility of performing OP intoxication studies in non-human primates because the same individual can be imaged longitudinally over time, and the ability to capture pre-exposure images enables each individual to serve as its own control.

The mechanism(s) by which OPs trigger neuroinflammation is another outstanding question. While there is compelling evidence from preclinical studies of Gulf War Illness to suggest that cholinergic mechanisms are not the principal mechanism of OP-induced neuroinflammation, data indicating that OPs may directly stimulate peripheral immune cells via cholinergic receptor activation at the immunological synapse suggest that the mechanism(s) by which OPs cause neuroinflammation may be context-specific. Answering this question is critical for gaining insight on novel therapeutic approaches for mitigating neurological damage post-exposure and for preventing brain injury pre-exposure. Elucidating cholinergic involvement in OP-induced neuroinflammation, particularly in the context of repeated low-level OPs exposures, also has important implications regarding the use of AChE inhibition as a point of departure for human health risk assessments.

Another generalization emerging from the literature is that the neuroinflammatory response varies depending on the OP exposure paradigm. Across all OP exposure paradigms, neuroinflammation is a dynamic response that changes over time, and the temporal pattern of the response varies across brain regions. These observations present significant challenges in terms of using neuroinflammatory biomarkers as diagnostic tools because the interpretation of the data may vary depending on the nature and timing of assessment. The available evidence also suggests that the role of neuroinflammation changes over time post-exposure. While the data on this topic are limited, it seems that the early, rapid neuroinflammatory response during the first hours to days post-exposure may be neurotoxic, while the delayed but persistent neuroinflammation that occurs weeks to months post-exposure may be important in neurorepair. A remaining critical data gap is the identification of specific cellular and molecular mechanisms that mediate the harmful vs. beneficial effects of neuroinflammation following OP exposure. Reports suggest that microglial activation, but not reactive astrocytosis, is linked to memory impairment observed in a rat model of acute DFP intoxication (Rojas et al., 2015, 2016), whereas reactive astrocytes have been linked to enhanced neurogenesis and accelerated functional recovery of affective, but not cognitive, behavior in a mouse model of acute soman intoxication (Collombet et al., 2011). Whether neuroinflammation similarly plays a dual role in the context of repeated low-level OP exposure is not known. Regardless, the therapeutic challenge moving forward will be to determine how to modulate neuroinflammatory responses to attenuate harmful effects while sparing beneficial effects, as this type of modulatory strategy is likely to be more effective than simply suppressing neuroinflammation altogether.

References

Abolaji, A.O., Ojo, M., Afolabi, T.T., Arowoogun, M.D., Nwawolor, D., Farombi, E.O., 2017. Protective properties of 6-gingerol-rich fraction from Zingiber officinale (ginger) on chlorpyrifos-induced oxidative damage and inflammation in the brain, ovary and uterus of rats. Chem. Biol. Interact. 270, 15–23.

Ahmed, D., Abdel-Rahman, R.H., Salama, M., El-Zalabany, L.M., Mona Ahmed, E.-H., 2017. Malathion neurotoxic effects on dopaminergic system in mice: role of inflammation. J. Biomed. Sci. 6, 30–36.

Amitai, G., Adani, R., Fishbein, E., Meshulam, H., Laish, I., Dachir, S., 2006. Bifunctional compounds eliciting anti-inflammatory and anti-cholinesterase activity as potential treatment of nerve and blister chemical agents poisoning. J. Appl. Toxicol. 26, 81–87.

Angoa-Perez, M., Kreipke, C.W., Thomas, D.M., Van Shura, K.E., Lyman, M., Mcdonough, J.H., Kuhn, D.M., 2010. Soman increases neuronal COX-2 levels: possible link between seizures and protracted neuronal damage. Neurotoxicology 31, 738–746.

Arundine, M., Tymianski, M., 2003. Molecular mechanisms of calcium-dependent neurodegeneration in excitotoxicity. Cell Calcium 34, 325–337.

Astiz, M., Diz-Chaves, Y., Garcia-Segura, L.M., 2013. Sub-chronic exposure to the insecticide dimethoate induces a proinflammatory status and enhances the neuroinflammatory response to bacterial lypopolysaccharide in the hippocampus and striatum of male mice. Toxicol. Appl. Pharmacol. 272, 263–271.

Baille, V., Clarke, P.G., Brochier, G., Dorandeu, F., Verna, J.M., Four, E., Lallement, G., Carpentier, P., 2005. Soman-induced convulsions: the neuropathology revisited. Toxicology 215, 1–24.

Baille-Le Crom, V., Collombet, J.M., Carpentier, P., Brochier, G., Burckhart, M.F., Foquin, A., Pernot-Marino, I., Rondouin, G., Lallement, G., 1995. Early regional changes of GFAP mRNA in rat hippocampus and dentate gyrus during soman-induced seizures. Neuroreport 7, 365–369.

Balabanov, R., Strand, K., Goswami, R., Mcmahon, E., Begolka, W., Miller, S.D., Popko, B., 2007. Interferon-gamma-oligodendrocyte interactions in the regulation of experimental autoimmune encephalomyelitis. J. Neurosci. 27, 2013–2024.

Baltazar, M.T., Dinis-Oliveira, R.J., De Lourdes Bastos, M., Tsatsakis, A.M., Duarte, J.A., Carvalho, F., 2014. Pesticides exposure as etiological factors of Parkinson's disease and other neurodegenerative diseases—a mechanistic approach. Toxicol. Lett. 230, 85–103.

Banks, C.N., Lein, P.J., 2012. A review of experimental evidence linking neurotoxic organophosphorus compounds and inflammation. Neurotoxicology 33, 575–584.

Barr, D.B., Wong, L.Y., Bravo, R., Weerasekera, G., Odetokun, M., Restrepo, P., Kim, D.G., Fernandez, C., Whitehead Jr., R.D., Perez, J., Gallegos, M., Williams, B.L., Needham, L.L., 2011. Urinary concentrations of dialkylphosphate metabolites of organophosphorus pesticides: National health and nutrition examination survey 1999-2004. Int. J. Environ. Res. Public Health 8, 3063–3098.

Becher, B., Spath, S., Goverman, J., 2017. Cytokine networks in neuroinflammation. Nat. Rev. Immunol. 17, 49–59.

Beers, D.R., Henkel, J.S., Zhao, W., Wang, J., Appel, S.H., 2008. Cd4 + T cells support glial neuroprotection, slow disease progression, and modify glial morphology in an animal model of inherited ALS. Proc. Natl. Acad. Sci. U.S.A. 105, 15558–15563.

Binukumar, B.K., Bal, A., Gill, K.D., 2011. Chronic dichlorvos exposure: microglial activation, proinflammatory cytokines and damage to nigrostriatal dopaminergic system. NeuroMolecular Med. 13, 251–265.

Bird, S.B., Gaspari, R.J., Dickson, E.W., 2003. Early death due to severe organophosphate poisoning is a centrally mediated process. Acad. Emerg. Med. 10, 295–298.

Biswas, S.K., 2016. Does the interdependence between oxidative stress and inflammation explain the antioxidant paradox? Oxidative Med. Cell. Longev. 2016, 5698931.

Bouchard, M.F., Chevrier, J., Harley, K.G., Kogut, K., Vedar, M., Calderon, N., Trujillo, C., Johnson, C., Bradman, A., Barr, D.B., Eskenazi, B., 2011. Prenatal exposure to organophosphate pesticides and IQ in 7-year-old children. Environ. Health Perspect. 119, 1189–1195.

Bsibsi, M., Ravid, R., Gveric, D., Van Noort, J.M., 2002. Broad expression of toll-like receptors in the human central nervous system. J. Neuropathol. Exp. Neurol. 61, 1013–1021.

Buckley, N.A., Eddleston, M., Li, Y., Bevan, M., Robertson, J., 2011. Oximes for acute organophosphate pesticide poisoning. Cochrane Database Syst. Rev. https://doi.org/10.1002/14651858.CD005085.pub2. CD005085.

Burke, R.D., Todd, S.W., Lumsden, E., Mullins, R.J., Mamczarz, J., Fawcett, W.P., Gullapalli, R.P., Randall, W.R., Pereira, E.F.R., Albuquerque, E.X., 2017. Developmental neurotoxicity of the organophosphorus insecticide chlorpyrifos: from clinical findings to preclinical models and potential mechanisms. J. Neurochem. 142 (Suppl. 2), 162–177.

Bushnell, P.J., Moser, V.C., 2006. Behavioral toxicity of cholinesterase inhibitors. In: Gupta, R.C. (Ed.), Toxicology of Organophosphate & Carbamate Compounds. Elsevier, San Diego, CA.

Chahine, L.M., Stern, M.B., Chen-Plotkin, A., 2014. Blood-based biomarkers for Parkinson's disease. Parkinsonism Relat. Disord. 20 (Suppl. 1), S99–103.

Chapman, S., Kadar, T., Gilat, E., 2006. Seizure duration following sarin exposure affects neuro-inflammatory markers in the rat brain. Neurotoxicology 27, 277–283.

Chen, Y., 2012. Organophosphate-induced brain damage: mechanisms, neuropsychiatric and neurological consequences, and potential therapeutic strategies. Neurotoxicology 33, 391–400.

Chung, H.Y., Sung, B., Jung, K.J., Zou, Y., Yu, B.P., 2006. The molecular inflammatory process in aging. Antioxid. Redox Signal. 8, 572–581.

Clemente, D., Ortega, M.C., Melero-Jerez, C., De Castro, F., 2013. The effect of glia-glia interactions on oligodendrocyte precursor cell biology during development and in demyelinating diseases. Front. Cell. Neurosci. 7, 268.

Clune, A.L., Ryan, P.B., Barr, D.B., 2012. Have regulatory efforts to reduce organophosphorus insecticide exposures been effective? Environ. Health Perspect. 120, 521–525.

Collombet, J.M., 2011. Nerve agent intoxication: recent neuropathophysiological findings and subsequent impact on medical management prospects. Toxicol. Appl. Pharmacol. 255, 229–241.

Collombet, J.M., Four, E., Bernabe, D., Masqueliez, C., Burckhart, M.F., Baille, V., Baubichon, D., Lallement, G., 2005a. Soman poisoning increases neural progenitor proliferation and induces long-term glial activation in mouse brain. Toxicology 208, 319–334.

Collombet, J.M., Four, E., Burckhart, M.F., Masqueliez, C., Bernabe, D., Baubichon, D., Herodin, F., Lallement, G., 2005b. Effect of cytokine treatment on the neurogenesis process in the brain of soman-poisoned mice. Toxicology 210, 9–23.
Collombet, J.M., Four, E., Fauquette, W., Burckhart, M.F., Masqueliez, C., Bernabe, D., Baubichon, D., Lallement, G., 2007. Soman poisoning induces delayed astrogliotic scar and angiogenesis in damaged mouse brain areas. Neurotoxicology 28, 38–48.
Collombet, J.M., Beracochea, D., Liscia, P., Pierard, C., Lallement, G., Filliat, P., 2011. Long-term effects of cytokine treatment on cognitive behavioral recovery and neuronal regeneration in soman-poisoned mice. Behav. Brain Res. 221, 261–270.
Corps, K.N., Roth, T.L., Mcgavern, D.B., 2015. Inflammation and neuroprotection in traumatic brain injury. JAMA Neurol. 72, 355–362.
Costa, L.G., 2006. Current issues in organophosphate toxicology. Clin. Chim. Acta 366, 1–13.
Costa, L.G., 2018. Organophosphorus compounds at 80: some old and new issues. Toxicol. Sci. 162, 24–35.
Coubard, S., Beracochea, D., Collombet, J.M., Philippin, J.N., Krazem, A., Liscia, P., Lallement, G., Pierard, C., 2008. Long-term consequences of soman poisoning in mice: part 2. Emotional behavior. Behav. Brain Res. 191, 95–103.
Dallerac, G., Rouach, N., 2016. Astrocytes as new targets to improve cognitive functions. Prog. Neurobiol. 144, 48–67.
Damodaran, T.V., Abou-Donia, M.B., 2000. Alterations in levels of mRNAs coding for glial fibrillary acidic protein (GFAP) and vimentin genes in the central nervous system of hens treated with diisopropyl phosphorofluoridate (DFP). Neurochem. Res. 25, 809–816.
Damodaran, T.V., Bilska, M.A., Rahman, A.A., Abou-Doni, M.B., 2002. Sarin causes early differential alteration and persistent overexpression in mRNAs coding for glial fibrillary acidic protein (GFAP) and vimentin genes in the central nervous system of rats. Neurochem. Res. 27, 407–415.
Damodaran, T.V., Patel, A.G., Greenfield, S.T., Dressman, H.K., Lin, S.M., Abou-Donia, M.B., 2006. Gene expression profiles of the rat brain both immediately and 3 months following acute sarin exposure. Biochem. Pharmacol. 71, 497–520.
Dantzer, R., Kelley, K.W., 1989. Stress and immunity: an integrated view of relationships between the brain and the immune system. Life Sci. 44, 1995–2008.
Dantzer, R., O'connor, J.C., Freund, G.G., Johnson, R.W., Kelley, K.W., 2008. From inflammation to sickness and depression: when the immune system subjugates the brain. Nat. Rev. Neurosci. 9, 46–56.
Dantzer, R., O'connor, J.C., Lawson, M.A., Kelley, K.W., 2011. Inflammation-associated depression: from serotonin to kynurenine. Psychoneuroendocrinology 36, 426–436.
Davalos, D., Grutzendler, J., Yang, G., Kim, J.V., Zuo, Y., Jung, S., Littman, D.R., Dustin, M.L., Gan, W.B., 2005. ATP mediates rapid microglial response to local brain injury in vivo. Nat. Neurosci. 8, 752–758.
De Araujo Furtado, M., Lumley, L.A., Robison, C., Tong, L.C., Lichtenstein, S., Yourick, D.L., 2010. Spontaneous recurrent seizures after status epilepticus induced by soman in Sprague-Dawley rats. Epilepsia 51, 1503–1510.
De Araujo Furtado, M., Rossetti, F., Chanda, S., Yourick, D., 2012. Exposure to nerve agents: from status epilepticus to neuroinflammation, brain damage, neurogenesis and epilepsy. Neurotoxicology 33, 1476–1490.
Delgado, E., Mcconnell, R., Miranda, J., Keifer, M., Lundberg, I., Partanen, T., Wesseling, C., 2004. Central nervous system effects of acute organophosphate poisoning in a two-year follow-up. Scand. J. Work Environ. Health 30, 362–370.
Deshpande, L.S., Phillips, K., Huang, B., Delorenzo, R.J., 2014. Chronic behavioral and cognitive deficits in a rat survival model of paraoxon toxicity. Neurotoxicology 44, 352–357.

Devinsky, O., Vezzani, A., Najjar, S., De Lanerolle, N.C., Rogawski, M.A., 2013. Glia and epilepsy: excitability and inflammation. Trends Neurosci. 36, 174–184.
Dhote, F., Peinnequin, A., Carpentier, P., Baille, V., Delacour, C., Foquin, A., Lallement, G., Dorandeu, F., 2007. Prolonged inflammatory gene response following soman-induced seizures in mice. Toxicology 238, 166–176.
Dillman 3rd, J.F., Phillips, C.S., Kniffin, D.M., Tompkins, C.P., Hamilton, T.A., Kan, R.K., 2009. Gene expression profiling of rat hippocampus following exposure to the acetylcholinesterase inhibitor soman. Chem. Res. Toxicol. 22, 633–638.
Disabato, D.J., Quan, N., Godbout, J.P., 2016. Neuroinflammation: the devil is in the details. J. Neurochem. 139 (Suppl. 2), 136–153.
Dos Santos, A.A., Naime, A.A., De Oliveira, J., Colle, D., Dos Santos, D.B., Hort, M.A., Moreira, E.L., Sunol, C., De Bem, A.F., Farina, M., 2016. Long-term and low-dose malathion exposure causes cognitive impairment in adult mice: evidence of hippocampal mitochondrial dysfunction, astrogliosis and apoptotic events. Arch. Toxicol. 90, 647–660.
Duramad, P., Tager, I.B., Leikauf, J., Eskenazi, B., Holland, N.T., 2006. Expression of Th1/Th2 cytokines in human blood after in vitro treatment with chlorpyrifos, and its metabolites, in combination with endotoxin LPS and allergen Der p1. J. Appl. Toxicol. 26, 458–465.
Duysen, E.G., Li, B., Xie, W., Schopfer, L.M., Anderson, R.S., Broomfield, C.A., Lockridge, O., 2001. Evidence for nonacetylcholinesterase targets of organophosphorus nerve agent: supersensitivity of acetylcholinesterase knockout mouse to VX lethality. J. Pharmacol. Exp. Ther. 299, 528–535.
Dziedzic, T., 2006. Systemic inflammatory markers and risk of dementia. Am. J. Alzheimers Dis. Other Demen. 21, 258–262.
Eddleston, M., Buckley, N.A., Eyer, P., Dawson, A.H., 2008. Management of acute organophosphorus pesticide poisoning. Lancet 371, 597–607.
Engel, S.M., Wetmur, J., Chen, J., Zhu, C., Barr, D.B., Canfield, R.L., Wolff, M.S., 2011. Prenatal exposure to organophosphates, paraoxonase 1, and cognitive development in childhood. Environ. Health Perspect. 119, 1182–1188.
Ferchmin, P.A., Andino, M., Reyes Salaman, R., Alves, J., Velez-Roman, J., Cuadrado, B., Carrasco, M., Torres-Rivera, W., Segarra, A., Martins, A.H., Lee, J.E., Eterovic, V.A., 2014. 4R-cembranoid protects against diisopropylfluorophosphate-mediated neurodegeneration. Neurotoxicology 44, 80–90.
Filliat, P., Baubichon, D., Burckhart, M.F., Pernot-Marino, I., Foquin, A., Masqueliez, C., Perrichon, C., Carpentier, P., Lallement, G., 1999. Memory impairment after soman intoxication in rat: correlation with central neuropathology. Improvement with anticholinergic and antiglutamatergic therapeutics. Neurotoxicology 20, 535–549.
Filliat, P., Coubard, S., Pierard, C., Liscia, P., Beracochea, D., Four, E., Baubichon, D., Masqueliez, C., Lallement, G., Collombet, J.M., 2007. Long-term behavioral consequences of soman poisoning in mice. Neurotoxicology 28, 508–519.
Finkelstein, A., Kunis, G., Berkutzki, T., Ronen, A., Krivoy, A., Yoles, E., Last, D., Mardor, Y., Van Shura, K., Mcfarland, E., Capacio, B.A., Eisner, C., Gonzales, M., Gregorowicz, D., Eisenkraft, A., Mcdonough, J.H., Schwartz, M., 2012. Immunomodulation by poly-ye reduces organophosphate-induced brain damage. Brain Behav. Immun. 26, 159–169.
Flannery, B.M., Bruun, D.A., Rowland, D.J., Banks, C.N., Austin, A.T., Kukis, D.L., Li, Y., Ford, B.D., Tancredi, D.J., Silverman, J.L., Cherry, S.R., Lein, P.J., 2016. Persistent neuroinflammation and cognitive impairment in a rat model of acute diisopropylfluorophosphate intoxication. J. Neuroinflammation 13, 267.
Fogal, B., Hewett, S.J., 2008. Interleukin-1beta: a bridge between inflammation and excitotoxicity? J. Neurochem. 106, 1–23.

Frank-Cannon, T.C., Alto, L.T., Mcalpine, F.E., Tansey, M.G., 2009. Does neuroinflammation fan the flame in neurodegenerative diseases? Mol. Neurodegener. 4, 47.
Garcia, S.J., Seidler, F.J., Qiao, D., Slotkin, T.A., 2002. Chlorpyrifos targets developing glia: effects on glial fibrillary acidic protein. Brain Res. Dev. Brain Res. 133, 151–161.
Goldmann, T., Prinz, M., 2013. Role of microglia in CNS autoimmunity. Clin. Dev. Immunol. 2013, 208093.
Gomez-Gimenez, B., Llansola, M., Hernandez-Rabaza, V., Cabrera-Pastor, A., Malaguarnera, M., Agusti, A., Felipo, V., 2017. Sex-dependent effects of developmental exposure to different pesticides on spatial learning. The role of induced neuroinflammation in the hippocampus. Food Chem. Toxicol. 99, 135–148.
Grauer, E., Chapman, S., Rabinovitz, I., Raveh, L., Weissman, B.A., Kadar, T., Allon, N., 2008. Single whole-body exposure to sarin vapor in rats: long-term neuronal and behavioral deficits. Toxicol. Appl. Pharmacol. 227, 265–274.
Guedes, J.R., Lao, T., Cardoso, A.L., El Khoury, J., 2018. Roles of microglial and monocyte chemokines and their receptors in regulating Alzheimer's disease-associated amyloid-beta and tau pathologies. Front. Neurol. 9, 549.
Guilarte, T.R., 2018. TSPO in diverse CNS pathologies and psychiatric disease: a critical review and a way forward. Pharmacol. Ther. https://doi.org/10.1016/j.pharmthera.2018.09.003. S0163-7258(18)30157-8 [pii].
Gunnell, D., Eddleston, M., Phillips, M.R., Konradsen, F., 2007. The global distribution of fatal pesticide self-poisoning: systematic review. BMC Public Health 7, 357.
Henderson, R.F., Barr, E.B., Blackwell, W.B., Clark, C.R., Conn, C.A., Kalra, R., March, T.H., Sopori, M.L., Tesfaigzi, Y., Menache, M.G., Mash, D.C., 2002. Response of rats to low levels of sarin. Toxicol. Appl. Pharmacol. 184, 67–76.
Heneka, M.T., Carson, M.J., El Khoury, J., Landreth, G.E., Brosseron, F., Feinstein, D.L., Jacobs, A.H., Wyss-Coray, T., Vitorica, J., Ransohoff, R.M., Herrup, K., Frautschy, S.A., Finsen, B., Brown, G.C., Verkhratsky, A., Yamanaka, K., Koistinaho, J., Latz, E., Halle, A., Petzold, G.C., Town, T., Morgan, D., Shinohara, M.L., Perry, V.H., Holmes, C., Bazan, N.G., Brooks, D.J., Hunot, S., Joseph, B., Deigendesch, N., Garaschuk, O., Boddeke, E., Dinarello, C.A., Breitner, J.C., Cole, G.M., Golenbock, D.T., Kummer, M.P., 2015. Neuroinflammation in Alzheimer's disease. Lancet Neurol. 14, 388–405.
Herman, F.J., Pasinetti, G.M., 2018. Principles of inflammasome priming and inhibition: implications for psychiatric disorders. Brain Behav. Immun. 73, 66–84.
Hobson, B.A., Siso, S., Rowland, D.J., Harvey, D.J., Bruun, D.A., Garbow, J.R., Lein, P.J., 2017. From the cover: magneticresonance imaging reveals progressive brain injury in rats acutely intoxicated with diisopropylfluorophosphate. Toxicol. Sci. 157, 342–353.
Hong, S., Dissing-Olesen, L., Stevens, B., 2016. New insights on the role of microglia in synaptic pruning in health and disease. Curr. Opin. Neurobiol. 36, 128–134.
Hulse, E.J., Davies, J.O., Simpson, A.J., Sciuto, A.M., Eddleston, M., 2014. Respiratory complications of organophosphorus nerve agent and insecticide poisoning. Implications for respiratory and critical care. Am. J. Respir. Crit. Care Med. 190, 1342–1354.
Jander, S., Schroeter, M., Stoll, G., 2000. Role of NMDA receptor signaling in the regulation of inflammatory gene expression after focal brain ischemia. J. Neuroimmunol. 109, 181–187.
Jensen, C.J., Massie, A., De Keyser, J., 2013. Immune players in the CNS: the astrocyte. J. Neuroimmune Pharmacol. 8, 824–839.
Jeon, S.W., Kim, Y.K., 2018. The role of neuroinflammation and neurovascular dysfunction in major depressive disorder. J. Inflamm. Res. 11, 179–192.
Johnson, E.A., Kan, R.K., 2010. The acute phase response and soman-induced status epilepticus: temporal, regional and cellular changes in rat brain cytokine concentrations. J. Neuroinflammation 7, 40.

Johnson, E.A., Dao, T.L., Guignet, M.A., Geddes, C.E., Koemeter-Cox, A.I., Kan, R.K., 2011. Increased expression of the chemokines CXCL1 and MIP-1alpha by resident brain cells precedes neutrophil infiltration in the brain following prolonged soman-induced status epilepticus in rats. J. Neuroinflammation 8, 41.

Joseph, J.A., Shukitt-Hale, B., Casadesus, G., 2005. Reversing the deleterious effects of aging on neuronal communication and behavior: beneficial properties of fruit polyphenolic compounds. Am. J. Clin. Nutr. 81, 313S–316S.

Jung, K.H., Chu, K., Lee, S.T., Kim, J., Sinn, D.I., Kim, J.M., Park, D.K., Lee, J.J., Kim, S.U., Kim, M., Lee, S.K., Roh, J.K., 2006. Cyclooxygenase-2 inhibitor, celecoxib, inhibits the altered hippocampal neurogenesis with attenuation of spontaneous recurrent seizures following pilocarpine-induced status epilepticus. Neurobiol. Dis. 23, 237–246.

Kennedy, K.A., Sandiford, S.D., Skerjanc, I.S., Li, S.S., 2012. Reactive oxygen species and the neuronal fate. Cell. Mol. Life Sci. 69, 215–221.

Kim, Y.K., Jeon, S.W., 2018. Neuroinflammation and the immune-kynurenine pathway in anxiety disorders. Curr. Neuropharmacol. 16, 574–582.

Kim, Y.B., Hur, G.H., Shin, S., Sok, D.E., Kang, J.K., Lee, Y.S., 1999. Organophosphate-induced brain injuries: delayed apoptosis mediated by nitric oxide. Environ. Toxicol. Pharmacol. 7, 147–152.

Kimelberg, H.K., Nedergaard, M., 2010. Functions of astrocytes and their potential as therapeutic targets. Neurotherapeutics 7, 338–353.

Kleen, J.K., Scott, R.C., Lenck-Santini, P.P., Holmes, G.L., 2012. Cognitive and behavioral co-morbidities of epilepsy. In: Noebels, J.L., Avoli, M., Rogawski, M.A., Olsen, R.W., Delgado-Escueta, A.V. (Eds.), Jasper's Basic Mechanisms of the Epilepsies, fourth ed, Bethesda, MD.

Konishi, H., Kiyama, H., 2018. Microglial TREM2/DAP12 signaling: a double-edged sword in neural diseases. Front. Cell. Neurosci. 12, 206.

Koo, B.B., Michalovicz, L.T., Calderazzo, S., Kelly, K.A., Sullivan, K., Killiany, R.J., O'callaghan, J.P., 2018. Corticosterone potentiates DFP-induced neuroinflammation and affects high-order diffusion imaging in a rat model of gulf war illness. Brain Behav. Immun. 67, 42–46.

Kraft, A.D., Harry, G.J., 2011. Features of microglia and neuroinflammation relevant to environmental exposure and neurotoxicity. Int. J. Environ. Res. Public Health 8, 2980–3018.

Kuruba, R., Wu, X., Reddy, D.S., 2018. Benzodiazepine-refractory status epilepticus, neuroinflammation, and interneuron neurodegeneration after acute organophosphate intoxication. Biochim. Biophys. Acta 1864, 2845–2858.

Lavisse, S., Guillermier, M., Herard, A.S., Petit, F., Delahaye, M., Van Camp, N., Ben Haim, L., Lebon, V., Remy, P., Dolle, F., Delzescaux, T., Bonvento, G., Hantraye, P., Escartin, C., 2012. Reactive astrocytes overexpress TSPO and are detected by TSPO positron emission tomography imaging. J. Neurosci. 32, 10809–10818.

Lawrence, C.B., Allan, S.M., Rothwell, N.J., 1998. Interleukin-1beta and the interleukin-1 receptor antagonist act in the striatum to modify excitotoxic brain damage in the rat. Eur. J. Neurosci. 10, 1188–1195.

Lemercier, G., Carpentier, P., Sentenac-Roumanou, H., Morelis, P., 1983. Histological and histochemical changes in the central nervous system of the rat poisoned by an irreversible anticholinesterase organophosphorus compound. Acta Neuropathol. 61, 123–129.

Levin, H.S., Rodnitzky, R.L., Mick, D.L., 1976. Anxiety associated with exposure to organophosphate compounds. Arch. Gen. Psychiatry 33, 225–228.

Li, Y., Lein, P.J., Liu, C., Bruun, D.A., Tewolde, T., Ford, G., Ford, B.D., 2011. Spatiotemporal pattern of neuronal injury induced by DFP in rats: a model for delayed neuronal cell death following acute op intoxication. Toxicol. Appl. Pharmacol. 253, 261–269.

Li, Y., Lein, P.J., Liu, C., Bruun, D.A., Giulivi, C., Ford, G.D., Tewolde, T., Ross-Inta, C., Ford, B.D., 2012. Neuregulin-1 is neuroprotective in a rat model of organophosphate-induced delayed neuronal injury. Toxicol. Appl. Pharmacol. 262, 194–204.

Li, Y., Lein, P.J., Ford, G.D., Liu, C., Stovall, K.C., White, T.E., Bruun, D.A., Tewolde, T., Gates, A.S., Distel, T.J., Surles-Zeigler, M.C., Ford, B.D., 2015. Neuregulin-1 inhibits neuroinflammatory responses in a rat model of organophosphate-nerve agent-induced delayed neuronal injury. J. Neuroinflammation 12, 64.

Liang, L.P., Pearson-Smith, J.N., Huang, J., Mcelroy, P., Day, B.J., Patel, M., 2018. Neuroprotective effects of AEOL10150 in a rat organophosphate model. Toxicol. Sci. 162, 611–621.

Liedtke, W., Edelmann, W., Bieri, P.L., Chiu, F.C., Cowan, N.J., Kucherlapati, R., Raine, C.S., 1996. GFAP is necessary for the integrity of CNS white matter architecture and long-term maintenance of myelination. Neuron 17, 607–615.

Lim, K.L., Tay, A., Nadarajah, V.D., Mitra, N.K., 2011. The effect of consequent exposure of stress and dermal application of low doses of chlorpyrifos on the expression of glial fibrillary acidic protein in the hippocampus of adult mice. J. Occup. Med. Toxicol. 6, 4.

Liu, M., Bing, G., 2011. Lipopolysaccharide animal models for Parkinson's disease. Parkinsons Dis. 2011, 327089.

Liu, C., Li, Y., Lein, P.J., Ford, B.D., 2012. Spatiotemporal patterns of GFAP upregulation in rat brain following acute intoxication with diisopropylfluorophosphate (DFP). Curr. Neurobiol. 3, 90–97.

Loane, D.J., Kumar, A., Stoica, B.A., Cabatbat, R., Faden, A.I., 2014. Progressive neurodegeneration after experimental brain trauma: association with chronic microglial activation. J. Neuropathol. Exp. Neurol. 73, 14–29.

Locker, A.R., Michalovicz, L.T., Kelly, K.A., Miller, J.V., Miller, D.B., O'callaghan, J.P., 2017. Corticosterone primes the neuroinflammatory response to gulf war illness-relevant organophosphates independently of acetylcholinesterase inhibition. J. Neurochem. 142, 444–455.

Massaad, C.A., Klann, E., 2011. Reactive oxygen species in the regulation of synaptic plasticity and memory. Antioxid. Redox Signal. 14, 2013–2054.

Mcafoose, J., Baune, B.T., 2009. Evidence for a cytokine model of cognitive function. Neurosci. Biobehav. Rev. 33, 355–366.

Mcdonough Jr., J.H., Shih, T.M., 1997. Neuropharmacological mechanisms of nerve agent-induced seizure and neuropathology. Neurosci. Biobehav. Rev. 21, 559–579.

Mense, S.M., Sengupta, A., Lan, C., Zhou, M., Bentsman, G., Volsky, D.J., Whyatt, R.M., Perera, F.P., Zhang, L., 2006. The common insecticides cyfluthrin and chlorpyrifos alter the expression of a subset of genes with diverse functions in primary human astrocytes. Toxicol. Sci. 93, 125–135.

Mew, E.J., Padmanathan, P., Konradsen, F., Eddleston, M., Chang, S.S., Phillips, M.R., Gunnell, D., 2017. The global burden of fatal self-poisoning with pesticides 2006-15: systematic review. J. Affect. Disord. 219, 93–104.

Meyer-Baron, M., Knapp, G., Schaper, M., Van Thriel, C., 2015. Meta-analysis on occupational exposure to pesticides—neurobehavioral impact and dose-response relationships. Environ. Res. 136, 234–245.

Milatovic, D., Gupta, R.C., Aschner, M., 2006. Anticholinesterase toxicity and oxidative stress. ScientificWorldJournal 6, 295–310.

Miller, J.V., Lebouf, R.F., Kelly, K.A., Michalovicz, L.T., Ranpara, A., Locker, A.R., Miller, D.B., O'callaghan, J.P., 2018. The neuroinflammatory phenotype in a mouse model of gulf war illness is unrelated to brain regional levels of acetylcholine as measured by quantitative hilic-uplc-ms/ms. Toxicol. Sci. 165, 302–313. https://doi.org/10.1093/toxsci/kfy130.

Mitra, A., Sarkar, M., Chatterjee, C., 2017. Modulation of immune response by organophosphate pesticides: mammals as potential model. Proc. Zool. Soc. https://doi.org/10.1007/s12595-017-0256-5.
Miyaki, K., Nishiwaki, Y., Maekawa, K., Ogawa, Y., Asukai, N., Yoshimura, K., Etoh, N., Matsumoto, Y., Kikuchi, Y., Kumagai, N., Omae, K., 2005. Effects of sarin on the nervous system of subway workers seven years after the Tokyo subway sarin attack. J. Occup. Health 47, 299–304.
Mohammadzadeh, L., Hosseinzadeh, H., Abnous, K., Razavi, B.M., 2018. Neuroprotective potential of crocin against malathion-induced motor deficit and neurochemical alterations in rats. Environ. Sci. Pollut. Res. Int. 25, 4904–4914.
Morimoto, K., Murasugi, T., Oda, T., 2002. Acute neuroinflammation exacerbates excitotoxicity in rat hippocampus in vivo. Exp. Neurol. 177, 95–104.
Mosser, C.A., Baptista, S., Arnoux, I., Audinat, E., 2017. Microglia in CNS development: shaping the brain for the future. Prog. Neurobiol. 149–150, 1–20.
Munoz-Quezada, M.T., Lucero, B.A., Barr, D.B., Steenland, K., Levy, K., Ryan, P.B., Iglesias, V., Alvarado, S., Concha, C., Rojas, E., Vega, C., 2013. Neurodevelopmental effects in children associated with exposure to organophosphate pesticides: a systematic review. Neurotoxicology 39, 158–168.
Myhrer, T., Andersen, J.M., Nguyen, N.H., Aas, P., 2005. Soman-induced convulsions in rats terminated with pharmacological agents after 45 min: neuropathology and cognitive performance. Neurotoxicology 26, 39–48.
Naughton, S.X., Terry Jr., A.V., 2018. Neurotoxicity in acute and repeated organophosphate exposure. Toxicology 408, 101–112. https://doi.org/10.1016/j.tox.2018.08.011.
Nimmerjahn, A., Kirchhoff, F., Helmchen, F., 2005. Resting microglial cells are highly dynamic surveillants of brain parenchyma in vivo. Science 308, 1314–1318.
Nishiwaki, Y., Maekawa, K., Ogawa, Y., Asukai, N., Minami, M., Omae, K., Sarin Health Effects Study Group, 2001. Effects of sarin on the nervous system in rescue team staff members and police officers 3 years after the Tokyo subway sarin attack. Environ. Health Perspect. 109, 1169–1173.
O'callaghan, J.P., Kelly, K.A., Locker, A.R., Miller, D.B., Lasley, S.M., 2015. Corticosterone primes the neuroinflammatory response to DFP in mice: potential animal model of gulf war illness. J. Neurochem. 133, 708–721.
Okumura, T., Hisaoka, T., Naito, T., Isonuma, H., Okumura, S., Miura, K., Maekawa, H., Ishimatsu, S., Takasu, N., Suzuki, K., 2005. Acute and chronic effects of sarin exposure from the Tokyo subway incident. Environ. Toxicol. Pharmacol. 19, 447–450.
Olmos-Alonso, A., Schetters, S.T.T., Sri, S., Askew, K., Mancuso, R., Vargas-Caballero, M., Holscher, C., Perry, V.H., Gomez-Nicola, D., 2016. Pharmacological targeting of CSF1R inhibits microglial proliferation and prevents the progression of Alzheimer's-like pathology. Brain 139, 1–17. https://doi.org/10.1093/brain/awv379.
Pascual, O., Ben, Achour S., Rostaing, P., Triller, A., Bessis, A., 2012. Microglia activation triggers astrocyte-mediated modulation of excitatory neurotransmission. Proc. Natl. Acad. Sci. U.S.A. 109, E197–E205.
Pearson, J.N., Patel, M., 2016. The role of oxidative stress in organophosphate and nerve agent toxicity. Ann. N. Y. Acad. Sci. 1378, 17–24.
Peferoen, L., Kipp, M., Van Der Valk, P., Van Noort, J.M., Amor, S., 2014. Oligodendrocyte-microglia cross-talk in the central nervous system. Immunology 141, 302–313.
Pekny, M., Pekna, M., Messing, A., Steinhauser, C., Lee, J.M., Parpura, V., Hol, E.M., Sofroniew, M.V., Verkhratsky, A., 2016. Astrocytes: a central element in neurological diseases. Acta Neuropathol. 131, 323–345.
Pereira, E.F., Aracava, Y., Detolla Jr., L.J., Beecham, E.J., Basinger Jr., G.W., Wakayama, E.J., Albuquerque, E.X., 2014. Animal models that best reproduce the

clinical manifestations of human intoxication with organophosphorus compounds. J. Pharmacol. Exp. Ther. 350, 313–321.
Pfau, M.L., Russo, S.J., 2016. Neuroinflammation regulates cognitive impairment in socially defeated mice. Trends Neurosci. 39, 353–355.
Pfau, M.L., Menard, C., Russo, S.J., 2018. Inflammatory mediators in mood disorders: therapeutic opportunities. Annu. Rev. Pharmacol. Toxicol. 58, 411–428.
Pope, C.N., 1999. Organophosphorus pesticides: do they all have the same mechanism of toxicity? J. Toxicol. Environ. Health B Crit. Rev. 2, 161–181.
Potts, M.B., Koh, S.E., Whetstone, W.D., Walker, B.A., Yoneyama, T., Claus, C.P., Manvelyan, H.M., Noble-Haeusslein, L.J., 2006. Traumatic injury to the immature brain: inflammation, oxidative injury, and iron-mediated damage as potential therapeutic targets. NeuroRx 3, 143–153.
Prager, E.M., Aroniadou-Anderjaska, V., Almeida-Suhett, C.P., Figueiredo, T.H., Apland, J.P., Rossetti, F., Olsen, C.H., Braga, M.F., 2014a. The recovery of acetylcholinesterase activity and the progression of neuropathological and pathophysiological alterations in the rat basolateral amygdala after soman-induced status epilepticus: relation to anxiety-like behavior. Neuropharmacology 81, 64–74.
Prager, E.M., Pidoplichko, V.I., Aroniadou-Anderjaska, V., Apland, J.P., Braga, M.F., 2014b. Pathophysiological mechanisms underlying increased anxiety after soman exposure: reduced gabaergic inhibition in the basolateral amygdala. Neurotoxicology 44, 335–343.
Prinz, M., Priller, J., 2017. The role of peripheral immune cells in the CNS in steady state and disease. Nat. Neurosci. 20, 136–144.
Proskocil, B.J., Bruun, D.A., Jacoby, D.B., Van Rooijen, N., Lein, P.J., Fryer, A.D., 2013. Macrophage TNF-alpha mediates parathion-induced airway hyperreactivity in Guinea pigs. Am. J. Physiol. Lung Cell. Mol. Physiol. 304, L519–L529.
Raffaele, K., Hughey, D., Wenk, G., Olton, D., Modrow, H., Mcdonough, J., 1987. Long-term behavioral changes in rats following organophosphonate exposure. Pharmacol. Biochem. Behav. 27, 407–412.
Rahman, I., Biswas, S.K., Kirkham, P.A., 2006. Regulation of inflammation and redox signaling by dietary polyphenols. Biochem. Pharmacol. 72, 1439–1452.
Ramesh, G., Benge, S., Pahar, B., Philipp, M.T., 2012. A possible role for inflammation in mediating apoptosis of oligodendrocytes as induced by the Lyme disease spirochete Borrelia burgdorferi. J. Neuroinflammation 9, 72.
Rana, A., Musto, A.E., 2018. The role of inflammation in the development of epilepsy. J. Neuroinflammation 15, 144.
Ray, A., Liu, J., Ayoubi, P., Pope, C., 2010. Dose-related gene expression changes in forebrain following acute, low-level chlorpyrifos exposure in neonatal rats. Toxicol. Appl. Pharmacol. 248, 144–155.
Robel, S., Buckingham, S.C., Boni, J.L., Campbell, S.L., Danbolt, N.C., Riedemann, T., Sutor, B., Sontheimer, H., 2015. Reactive astrogliosis causes the development of spontaneous seizures. J. Neurosci. 35, 3330–3345.
Rohlman, D.S., Anger, W.K., Lein, P.J., 2011. Correlating neurobehavioral performance with biomarkers of organophosphorous pesticide exposure. Neurotoxicology 32, 268–276.
Rojas, A., Ganesh, T., Lelutiu, N., Gueorguieva, P., Dingledine, R., 2015. Inhibition of the prostaglandin ep2 receptor is neuroprotective and accelerates functional recovery in a rat model of organophosphorus induced status epilepticus. Neuropharmacology 93, 15–27.
Rojas, A., Ganesh, T., Manji, Z., O'neill, T., Dingledine, R., 2016. Inhibition of the prostaglandin E2 receptor EP2 prevents status epilepticus-induced deficits in the novel object recognition task in rats. Neuropharmacology 110, 419–430.

Rossi, S., Studer, V., Motta, C., Polidoro, S., Perugini, J., Macchiarulo, G., Giovannetti, A.M., Pareja-Gutierrez, L., Calo, A., Colonna, I., Furlan, R., Martino, G., Centonze, D., 2017. Neuroinflammation drives anxiety and depression in relapsing-remitting multiple sclerosis. Neurology 89, 1338–1347.
Sagiv, S.K., Harris, M.H., Gunier, R.B., Kogut, K.R., Harley, K.G., Deardorff, J., Bradman, A., Holland, N., Eskenazi, B., 2018a. Prenatal organophosphate pesticide exposure and traits related to autism spectrum disorders in a population living in proximity to agriculture. Environ. Health Perspect. 126, 047012.
Sagiv, S.K., Harris, M.H., Gunier, R.B., Kogut, K.R., Harley, K.G., Deardorff, J., Bradman, A., Holland, N., Eskenazi, B., 2018b. Erratum: "prenatal organophosphate pesticide exposure and traits related to autism spectrum disorders in a population living in proximity to agriculture". Environ. Health Perspect. 126, 079001.
Saijo, K., Glass, C.K., 2011. Microglial cell origin and phenotypes in health and disease. Nat. Rev. Immunol. 11, 775–787.
Sanchez-Santed, F., Colomina, M.T., Herrero Hernandez, E., 2016. Organophosphate pesticide exposure and neurodegeneration. Cortex 74, 417–426.
Savage, E.P., Keefe, T.J., Mounce, L.M., Heaton, R.K., Lewis, J.A., Burcar, P.J., 1988. Chronic neurological sequelae of acute organophosphate pesticide poisoning. Arch. Environ. Health 43, 38–45.
Schwartz, M., Shaked, I., Fisher, J., Mizrahi, T., Schori, H., 2003. Protective autoimmunity against the enemy within: fighting glutamate toxicity. Trends Neurosci. 26, 297–302.
Seifert, G., Steinhauser, C., 2013. Neuron-astrocyte signaling and epilepsy. Exp. Neurol. 244, 4–10.
Shaked, I., Porat, Z., Gersner, R., Kipnis, J., Schwartz, M., 2004. Early activation of microglia as antigen-presenting cells correlates with t cell-mediated protection and repair of the injured central nervous system. J. Neuroimmunol. 146, 84–93.
Shi, Y., Holtzman, D.M., 2018. Interplay between innate immunity and Alzheimer disease: APOE and TREM2 in the spotlight. Nat. Rev. Immunol. https://doi.org/10.1038/s41577-018-0051-1.
Shrot, S., Ramaty, E., Biala, Y., Bar-Klein, G., Daninos, M., Kamintsky, L., Makarovsky, I., Statlender, L., Rosman, Y., Krivoy, A., Lavon, O., Kassirer, M., Friedman, A., Yaari, Y., 2014. Prevention of organophosphate-induced chronic epilepsy by early benzodiazepine treatment. Toxicology 323, 19–25.
Simen, A.A., Bordner, K.A., Martin, M.P., Moy, L.A., Barry, L.C., 2011. Cognitive dysfunction with aging and the role of inflammation. Ther. Adv. Chronic Dis. 2, 175–195.
Siso, S., Hobson, B.A., Harvey, D.J., Bruun, D.A., Rowland, D.J., Garbow, J.R., Lein, P.J., 2017. Editor's highlight: spatiotemporal progression and remission of lesions in the rat brain following acute intoxication with diisopropylfluorophosphate. Toxicol. Sci. 157, 330–341.
Skaper, S.D., Facci, L., Zusso, M., Giusti, P., 2018. An inflammation-centric view of neurological disease: beyond the neuron. Front. Cell. Neurosci. 12, 72.
Soltaninejad, K., Abdollahi, M., 2009. Current opinion on the science of organophosphate pesticides and toxic stress: a systematic review. Med. Sci. Monit. 15, RA75–90.
Spradling, K.D., Lumley, L.A., Robison, C.L., Meyerhoff, J.L., Dillman 3rd, J.F., 2011a. Transcriptional analysis of rat piriform cortex following exposure to the organophosphonate anticholinesterase sarin and induction of seizures. J. Neuroinflammation 8, 83.
Spradling, K.D., Lumley, L.A., Robison, C.L., Meyerhoff, J.L., Dillman 3rd, J.F., 2011b. Transcriptional responses of the nerve agent-sensitive brain regions amygdala, hippocampus, piriform cortex, septum, and thalamus following exposure to the organophosphonate anticholinesterase sarin. J. Neuroinflammation 8, 84.

Stapleton, A.R., Chan, V.T., 2009. Subtoxic chlorpyrifos treatment resulted in differential expression of genes implicated in neurological functions and development. Arch. Toxicol. 83, 319–333.
Svensson, I., Waara, L., Johansson, L., Bucht, A., Cassel, G., 2001. Soman-induced interleukin-1 beta mRNA and protein in rat brain. Neurotoxicology 22, 355–362.
Thorburne, S.K., Juurlink, B.H., 1996. Low glutathione and high iron govern the susceptibility of oligodendroglial precursors to oxidative stress. J. Neurochem. 67, 1014–1022.
Tian, J., Dai, H., Deng, Y., Zhang, J., Li, Y., Zhou, J., Zhao, M., Zhao, M., Zhang, C., Zhang, Y., Wang, P., Bing, G., Zhao, L., 2015. The effect of HMGB1 on sub-toxic chlorpyrifos exposure-induced neuroinflammation in amygdala of neonatal rats. Toxicology 338, 95–103.
Turtzo, L.C., Lescher, J., Janes, L., Dean, D.D., Budde, M.D., Frank, J.A., 2014. Macrophagic and microglial responses after focal traumatic brain injury in the female rat. J. Neuroinflammation 11, 82.
Veiga, S., Carrero, P., Pernia, O., Azcoitia, I., Garcia-Segura, L.M., 2007. Translocator protein 18 kDa is involved in the regulation of reactive gliosis. Glia 55, 1426–1436.
Verma, S.K., Raheja, G., Gill, K.D., 2009. Role of muscarinic signal transduction and CREB phosphorylation in dichlorvos-induced memory deficits in rats: an acetylcholine independent mechanism. Toxicology 256, 175–182.
Vezzani, A., Moneta, D., Conti, M., Richichi, C., Ravizza, T., De Luigi, A., de Simoni, M.G., Sperk, G., Andell-Jonsson, S., Lundkvist, J., Iverfeldt, K., Bartfai, T., 2000. Powerful anticonvulsant action of IL-1 receptor antagonist on intracerebral injection and astrocytic overexpression in mice. Proc. Natl. Acad. Sci. U. S. A. 97, 11534–11539.
Vezzani, A., Aronica, E., Mazarati, A., Pittman, Q.J., 2013. Epilepsy and brain inflammation. Exp. Neurol. 244, 11–21.
Viviani, B., Boraso, M., Marchetti, N., Marinovich, M., 2014. Perspectives on neuroinflammation and excitotoxicity: a neurotoxic conspiracy? Neurotoxicology 43, 10–20.
Voorhees, J.R., Rohlman, D.S., Lein, P.J., Pieper, A.A., 2016. Neurotoxicity in preclinical models of occupational exposure to organophosphorus compounds. Front. Neurosci. 10, 590.
Wahner, A.D., Bronstein, J.M., Bordelon, Y.M., Ritz, B., 2007a. Nonsteroidal anti-inflammatory drugs may protect against Parkinson disease. Neurology 69, 1836–1842.
Wahner, A.D., Sinsheimer, J.S., Bronstein, J.M., Ritz, B., 2007b. Inflammatory cytokine gene polymorphisms and increased risk of Parkinson disease. Arch. Neurol. 64, 836–840.
Walker, K.R., Tesco, G., 2013. Molecular mechanisms of cognitive dysfunction following traumatic brain injury. Front. Aging Neurosci. 5, 29.
Wang, J.Y., Wen, L.L., Huang, Y.N., Chen, Y.T., Ku, M.C., 2006. Dual effects of antioxidants in neurodegeneration: direct neuroprotection against oxidative stress and indirect protection via suppression of glia-mediated inflammation. Curr. Pharm. Des. 12, 3521–3533.
Wang, G., Zhang, J., Hu, X., Zhang, L., Mao, L., Jiang, X., Liou, A.K., Leak, R.K., Gao, Y., Chen, J., 2013. Microglia/macrophage polarization dynamics in white matter after traumatic brain injury. J. Cereb. Blood Flow Metab. 33, 1864–1874.
Wang, Y.L., Han, Q.Q., Gong, W.Q., Pan, D.H., Wang, L.Z., Hu, W., Yang, M., Li, B., Yu, J., Liu, Q., 2018. Microglial activation mediates chronic mild stress-induced depressive- and anxiety-like behavior in adult rats. J. Neuroinflammation 15, 21.
Williams, K., Ulvestad, E., Waage, A., Antel, J.P., Mclaurin, J., 1994. Activation of adult human derived microglia by myelin phagocytosis in vitro. J. Neurosci. Res. 38, 433–443.
Williams, A.J., Berti, R., Yao, C., Price, R.A., Velarde, L.C., Koplovitz, I., Schultz, S.M., Tortella, F.C., Dave, J.R., 2003. Central neuro-inflammatory gene response following soman exposure in the rat. Neurosci. Lett. 349, 147–150.

Wohleb, E.S., Patterson, J.M., Sharma, V., Quan, N., Godbout, J.P., Sheridan, J.F., 2014. Knockdown of interleukin-1 receptor type-1 on endothelial cells attenuated stress-induced neuroinflammation and prevented anxiety-like behavior. J. Neurosci. 34, 2583–2591.
Wright, L.K., Liu, J., Nallapaneni, A., Pope, C.N., 2010. Behavioral sequelae following acute diisopropylfluorophosphate intoxication in rats: comparative effects of atropine and cannabinomimetics. Neurotoxicol. Teratol. 32, 329–335.
Wyss-Coray, T., Rogers, J., 2012. Inflammation in Alzheimer disease—a brief review of the basic science and clinical literature. Cold Spring Harb. Perspect. Med. 2, a006346.
Yanagisawa, N., Morita, H., Nakajima, T., 2006. Sarin experiences in Japan: acute toxicity and long-term effects. J. Neurol. Sci. 249, 76–85.
Yue, N., Li, B., Yang, L., Han, Q.Q., Huang, H.J., Wang, Y.L., Wang, J., Yu, R., Wu, G.C., Liu, Q., Yu, J., 2018. Electro-acupuncture alleviates chronic unpredictable stress-induced depressive- and anxiety-like behavior and hippocampal neuroinflammation in rat model of depression. Front. Mol. Neurosci. 11, 149.
Zaja-Milatovic, S., Gupta, R.C., Aschner, M., Milatovic, D., 2009. Protection of DFP-induced oxidative damage and neurodegeneration by antioxidants and NMDA receptor antagonist. Toxicol. Appl. Pharmacol. 240, 124–131.
Zajicek, J., Wing, M., Skepper, J., Compston, A., 1995. Human oligodendrocytes are not sensitive to complement. A study of CD59 expression in the human central nervous system. Lab. Investig. 73, 128–138.
Zhang, B., Gaiteri, C., Bodea, L.G., Wang, Z., Mcelwee, J., Podtelezhnikov, A.A., Zhang, C., Xie, T., Tran, L., Dobrin, R., Fluder, E., Clurman, B., Melquist, S., Narayanan, M., Suver, C., Shah, H., Mahajan, M., Gillis, T., Mysore, J., Macdonald, M.E., Lamb, J.R., Bennett, D.A., Molony, C., Stone, D.J., Gudnason, V., Myers, A.J., Schadt, E.E., Neumann, H., Zhu, J., Emilsson, V., 2013. Integrated systems approach identifies genetic nodes and networks in late-onset Alzheimer's disease. Cell 153, 707–720.
Zhou, H., Lapointe, B.M., Clark, S.R., Zbytnuik, L., Kubes, P., 2006. A requirement for microglial TLR4 in leukocyte recruitment into brain in response to lipopolysaccharide. J. Immunol. 177, 8103–8110.
Zimmer, L.A., Ennis, M., Shipley, M.T., 1997. Soman-induced seizures rapidly activate astrocytes and microglia in discrete brain regions. J. Comp. Neurol. 378, 482–492.

CHAPTER THREE

Neuroimmunotoxicology of the heavy metal toxicant lead

Jane Kasten-Jolly[a,*], David A. Lawrence[a,b,*]
[a]Wadsworth Center, New York State Department of Health, Albany, NY, United States
[b]Department of Environmental Health Sciences, University at Albany School of Public Health, Rensselaer, NY, United States
*Corresponding authors: e-mail address: jane.kasten-jolly@health.ny.gov; david.lawrence@health.ny.gov

Contents

1. Introduction

The nervous system is often cited as the system most sensitive to lead (Pb) with blood lead levels below 10 μg/dL (Bellinger, 2008); however, other organ systems including the endocrine, gastrointestinal, hematopoietic, immune, and renal systems also are modified by Pb, and their alterations can influence the central and peripheral nervous systems. It is now evident that there exits an enormous amount of interaction and cross-talk between the nervous, endocrine, and immune systems (Manley et al., 2018). Neuropeptides thyrotropin-releasing hormone (TRH), corticotropin-releasing hormone (CRH), gonadotropin-releasing hormone (GnRH), and growth hormone-releasing hormone (GHRH) produced by the hypothalamus of

Advances in Neurotoxicology, Volume 3
ISSN 2468-7480
https://doi.org/10.1016/bs.ant.2018.10.004

the central nervous system (CNS) affect immune system development and functions via regulation of endocrine activities. Hormones impact immune function, and gonadal function influences the cellularity of the thymus (Calzolari, 1898) and the brain (McCarthy, 2016). Lesions within the CNS (Spector, 1996) and emotions affecting the CNS (Sternberg, 1997) also can modify immune defenses against pathogens. Conversely, cytokines, chemokines, hormones, and other products generated by the endocrine and immune systems affect CNS development, degeneration, and functions. In fact, innate immune cells, microglia, play a critical role in CNS development and neuronal connectivity. Furthermore, neuronal connections are differentially modulated by sex hormones and genes (McCarthy et al., 2008), which involve microglial activation (VanRyzin et al., 2018). Beyond the bidirectional regulation between the nervous and immune systems, these systems duplicate each other's functions, in that some immune cells can generate neurotransmitters, such as epinephrine and norepinephrine, and neurons and glia can synthesize cytokines. An illustration of immune-neuroendocrine interactions is presented (Fig. 1). An example of the commonalities between the immune and nervous systems is exemplified by hematopoietic multivariant mast cells (Gurish and Austen, 2001), which exist in the CNS and throughout the body, express many neurotransmitters, neuropeptides, and cytokines, modulate immune (da Silva et al., 2014) and nervous system functions (Skaper et al., 2014; Traina, 2017), and are modulated by peripheral nerves (Forsythe, 2015). Pb can enhance mast cell deposition of calcium referred to as mastocalcergy (Selye et al., 1964), and calcifications in the CNS may lead to various pathologies (Deng et al., 2015).

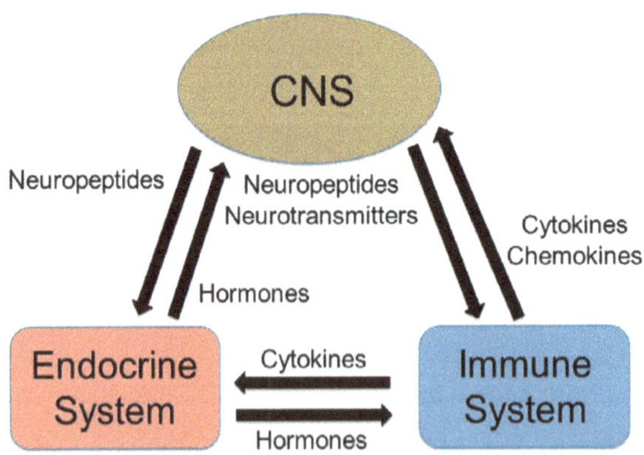

Fig. 1 Diagram of the inter-connections between the Central Nervous System, Immune System, and Endocrine System.

Clearly, molecules from the immune and nervous systems as well as the endocrine system crossover from one system to another creating a different and not necessarily beneficial effect. Such cross-talk also is portrayed by amyloid-beta protein, pro- and anti-inflammatory cytokines, and the neuropeptide serotonin, which affect microglial subsets in the brain and Alzheimer's disease and depression (Ledo et al., 2016; Michelucci et al., 2009; Santos et al., 2016; Skaper et al., 2014). Amyloid peptides can induce mast cell degranulation (Niederhoffer et al., 2009), and mast cells exist near amyloid plaques of patients with Alzheimer's disease (AD) (Maslinska et al., 2007). Dependent on where and amount of a product is expressed a different outcome may occur including induction of a disorder unrelated to its beneficial role for immunosurveillance and protection or neuronal functions affecting behaviors and cognition. The environmental toxicant Pb may further potentiate immunity and behavioral changes as reported for mice with lateralized brain influences on host bacterial defenses (Kim and Lawrence, 2000). Behavioral modifications affected by toxicants such as Pb also are known to differ based on sex (Abazyan et al., 2014; Weston et al., 2014).

Numerous environmental toxins and toxicants are known to have deleterious effects on the INEN (Lawrence and Harry, 2000; Pabello and Lawrence, 2006). The heavy metal toxicant Pb is among these toxicants (Lawrence, 1985; Lawrence and McCabe, 2002). The present review will discuss how Pb affects cross-talk among different organ systems as well as among intracellular organelles, including endoplasmic reticulum (ER), lysosomes, mitochondria, and peroxisomes. Pb-induced stress mechanisms affecting immunity, neuronal apoptosis, autophagy, and health are reviewed. Although Pb has been known to be a neurotoxicant for more than a century; the immunotoxic effects of Pb and immune cell involvement in alteration of neurobehaviors have been described more recently. The central and peripheral effects of Pb on neuroimmune interactions and detrimental outcomes are discussed regarding posited Pb modulatory influences on microglia, the immune cell of the CNS, and modulation of sympathetic nerve effects on immunity.

2. Developmental neuroimmunologic events

Under normal circumstances myeloid progenitors leave the embryonic yolk sac and migrate into the CNS to differentiate into the resident microglia. Microglia like all counterparts in the periphery (macrophages) take on different phenotypes, which can be polarized toward classically activated macrophages (M1) or alternatively activated macrophages (M2) (Porta et al., 2015). Most studies suggest that the M1, which are producers of pro-inflammatory cytokines and nitric oxide (NO), and M2, which are producers of wound-healing cytokines, have plastic phenotypes dependent on the environment. The fluctuations between M1 and M2 are posited to exist to aid balancing of immune responses. For example, M1 involvement in Systemic Inflammatory Response Syndrome (SIRS) is needed to control in vivo expansion of infectious agents, but excessive SIRS is limited with a Compensatory Anti-inflammatory Response Syndrome (CARS), which is driven, in part, by M2 cells (Tsuda et al., 2004). Interestingly, cholinergic (acetylcholine producing) and sympathetic (epinephrine and norepinephrine producing) neurons help to bring on CARS to dampen the inflammatory response. Acetylcholine and epinephrine contribute with IL-10 and other mediators to the anti-inflammatory compensatory response initiated to dampen the inflammatory process (Cavaillon and Annane, 2006). Pb may skew the balance early in development creating aberrant neuronal connections since microglia play a role in neuronal connectivity as well as dendrite pruning. Microglial subsets also have been suggested to differentially influence sex hormone effects on masculinization of brain and behavior (Lenz et al., 2013; VanRyzin et al., 2018). Like microenvironmental effects on macrophages/microglia, which can promote inflammatory or anti-inflammatory responses, beneficial or detrimental mast cell activities (Conti et al., 2017) can be induced by endocrine, immune, or nerve products as well as Pb (Bent et al., 1992; Karaca and Simşek, 2007).

3. Pb effect on the immune system

Although the deleterious effects of Pb on the CNS were evident as far back as the time of the Roman Empire, the effects of Pb on the immune system are more subtle and were not documented until relatively recently (Lanphear et al., 2000; Lawrence and McCabe, 1995; Needleman, 2009). It was noted by Lawrence and McCabe (1995) that mouse litters born on the same day, raised under specified pathogen free (SPF) conditions, and given drinking water containing Pb or regular tap water for 1.5 years were apparently identical with respect to health. Moreover, the mice treated with the Pb-containing water in general had a lower mortality rate than the mice receiving tap water alone. However, this mortality rate was reversed in the presence of an infection. Controlled animal studies undertaken during the

1970s indicated that Pb could reduce host resistance against assaults by various pathogens (Gainer, 1974; Hemphill et al., 1971). Some of Pb's effects on the nervous system are suggested to relate to heightened inflammation due to less immune ability to control infections. Additionally, Pb's suppression of immunity may be like Pb's detrimental neural mechanisms, including upsetting intracellular organelle communications and functions with alteration of calcium distribution leading to altered autophagy and increased apoptosis (Corsetti et al., 2017; Elbaz and Schuldiner, 2011; Filippi-Chiela et al., 2016). Common cell and molecular disruption of immune and nervous system functions are discussed later.

Early studies into the effect of Pb exposure on antibody production were conflicting and varied with time of exposure and Pb concentration (Dietert and Piepenbrink, 2006; Dietert et al., 2000). A breakthrough into how Pb was affecting the immune system came when studies performed with mice indicated that Pb could directly activate B cells causing them to differentiate and increase the expression of major histocompatibility complex (MHC) class II molecules on their surface (McCabe and Lawrence, 1990). Subsequent experiments with T-cell clones demonstrated that Pb inhibited antigen presentation by B cells to the T_{H1} clone, HDK-1, but enhanced presentation to the T_{H2} clone, D10.G4.1. These results were duplicated using other T_{H1} and T_{H2} clones and splenocytes (McCabe and Lawrence, 1991). Further studies employing BALB/c mice to assess Pb effects on $CD4^+$ helper T cell (T_H) subtypes indicated that subcutaneous $PbCl_2$ exposure enhanced the serum concentration of interleukin-4 (IL-4), a cytokine produced by T_{H2} cells and cells that support the preferential development of T_{H2} cells, and decreased the serum concentration of interferon-gamma (IFNγ), a cytokine secreted by T_{H1} cells (Heo et al., 1997). Closer examination into the effect of Pb on IFNγ expression by means of RT-PCR, ribonuclease protection, and qPCR analysis revealed that decreased serum IFNγ concentration was not due to down-regulation of *IFNγ* gene transcription (Heo et al., 2007). Serum concentration of an antibody associated with a T_{H2} response, IgE, was also found to be elevated in the mice exposed to Pb (Dietert et al., 2004; Heo et al., 1996; Snyder et al., 2000). Studies performed with IL-4 deficient mice have demonstrated that IgE production could be completely suppressed in the absence of IL-4 (Finkelman et al., 1991; Kopf et al., 1995). It therefore seemed likely that Pb was increasing the expression of IL-4 and evidence of this came from gene expression studies in the spleens of mice exposed from gestational day (gd)8 to postnatal day (pnd)21 to 0.1 mm Pb via the dam's drinking water (Kasten-Jolly et al., 2010). The results indicated that Pb was increasing expression of IL-4 in the spleen, as well as, increasing expression of genes associated with B-cell activation and differentiation, such as *IL-7*. Elevated levels of serum IgE have been associated with disease of the lungs, such as asthma, and neuropeptides have been reported to modulate synthesis of IgE (Aebisher et al., 1994). $PbCl_2$ was also capable of increasing IL-4 expression in vitro when added to T_{H2} clones in culture (Heo et al., 1996). Preliminary results indicated that increases in IL-4 expression could be due to an increase in cAMP levels after Pb exposure (Heo et al., 1998). It has now been revealed that up-regulation of IL-4 expression is associated with activation of the cAMP-dependent protein kinase A (cAMP/PKA) pathway in anti-CD3/CD28-stimulated naïve T cells (Tokoyoda et al., 2004). Skewing toward T_{H2} immunity might be a contributing factor in the lowering of host resistance to pathogens by Pb. This is supported by the finding that IL-12, a promoter of T_{H1} development, could reverse the inhibition of Pb on host defenses against *Listeria monocytogenes*, which is enhanced by IFNγ activation of macrophages (Kishikawa et al., 1997). Immune cells producing IFNγ are T_{H1} cells and innate lymphoid cells, which include ILC1 and natural killer (NK) cells; IL-12 enhances IFNγ production by these cells. Biosynthesis of IL-12 is inhibited by IL-10 (Aste-Amezaga et al., 1998), and IL-10 production usually occurs concomitantly with that of IL-4. Reagents that polarize the antigen-presenting cells (APCs), which are mainly B cells, macrophages, and dendritic cells support naïve T cell differentiation into T_{H1}, T_{H2}, or other subsets affecting the synthesis of cytokines that promote subset skewing. Thus, the predominance of T_{H2} cells after Pb exposure likely is dependent on the cytokines expressed by the APCs. IL-12 is a heterodimeric cytokine consisting of p35 (IL-12α) disulfide bonded to p40 (IL-12β). IL-4 (Cao et al., 2005) and prostaglandin E2 (PGE_2) (MacKenzie et al., 2013), which are both promoted by Pb, are agonists for IL-10 production. IL-4 seems to play a critical role in IL-10 vs. IL-12 production by APCs (Cao et al., 2002; Homma et al., 2007). The suppression of T_{H1} activity by Pb is like that induced by cold restraint stress with sympathetic nervous system release of norepinephrine signaling β-adrenoceptors on APC, which causes oxidative stress (Mondal et al., 2015). Thus, similar mechanisms may affect immunity due to oxidative stress with involvement of cAMP and PKA leading to mitochondrial damage (Brunyanszki et al., 2014).

Pb exposure dampens the delayed type hypersensitivity response (DTH) (Dietert and Piepenbrink, 2006). This reaction by the immune system is promoted by the T_{H1} cells, which are activated by antigenically modified skin components. These activated T_{H1} cells will promote an inflammatory reaction after re-exposure to the antigen, such as occurs upon exposure to poison ivy. Pb has been found to suppress the DTH response regardless of the time of antigen sensitization or time of Pb exposure (Faith et al., 1979; Laschi-Loquerie et al., 1984; Muller et al., 1977). Pb exposure of rats and chickens during gestation indicated that the effect of Pb on DTH occurs between days 9 and 15 of gestation (Bunn et al., 2001; Lee et al., 2001); this is the time period that differential microglial effects in the brain may be occurring. It was suggested that thymic development and decreased expression of IFNγ both played a role in Pb's suppression of the DTH response (Lee and Dietert, 2003; McCabe et al., 1999). However, these studies did not take into consideration Pb's ability to disrupt the organization of primary or secondary lymphoid organs, or any immunomodulatory effects by CNS or hormones. A 45-day exposure of 1-month-old male C57BL/6 to 200 ppm Pb acetate in their drinking water caused a disruption of splenic architecture, in part, due to increased phagocytosis of erythrocytes and apoptosis (Corsetti et al., 2017).

T cells are activated through engagement of their T cell receptor (TCR) with a peptide presented by a histocompatibility complex (peptide-MHC) located on an APC. However, any cells induced to express MHC class II molecules with modified peptides can present to the $CD4^+$ T_H cells; whereas peptide bound to MHC class I molecules, which exist on all host cells except erythrocytes, stimulate $CD8^+$ T cells. If the T_H cell receives positive signals through MHC class II/peptide activation of its TCR, which is dependent on the affinity of the interaction and receives stimulatory secondary signals through checkpoint regulators, the T_H cell is activated and will differentiate into effector cells and undergo clonal expansion. This process will dramatically raise the bioenergy needs of these cells, which will cause them to increase glucose and glutamine consumption. The activated T-cell's generation of reactive oxygen species (ROS) is buffered by reduced glutathione (GSH). Another function of GSH in activated T cells is regulation of metabolic activity through activation of the mTOR pathway and consequent expression of transcription factors NFAT and Myc, which metabolically reprogram the activated T cells to switch to glycolysis and glutaminolysis to meet their increased energy needs (Mak et al., 2017). It has been demonstrated that Pb decreases the concentration of GSH within cells and tissues (de Souza et al., 2018; Hunaiti and Soud, 2000; Lu et al., 2013). By lowering the cellular concentration of GSH, Pb is decreasing the activated T cell's ability to cope with the build-up of ROS in the cell and with the need for increased energy sources. Mak et al. (2017) found that T cells, which were engineered to not synthesize GSH, had markedly reduced glycolysis when stimulated either in vivo or in vitro. Moreover, T cells that lacked GSH were not able to mount an immune response against a viral pathogen (Mak et al., 2017). Since Pb lowers cellular GSH concentrations, activated T cells would have increased concentrations of the damaging ROS and lowered levels of energy sources, such as ATP, necessary to maintain cell life (Fig. 2). Therefore, Pb could be decreasing viability of the activated T-cells and lowering the number of T-cells in the peripheral blood. Reduced T cell numbers have been reported in human adults and children exposed to Pb (Fischbein et al., 1993; Karmaus et al., 2005; Pinkerton et al., 1998; Sarasua et al., 2000; Sata et al., 1998; Zhao et al., 2004). Decreased T cell numbers and compromised T cell ability to fight off pathogens constitute further ways Pb affects the immune system and its function, which may indirectly affect the endocrine and/or nervous systems.

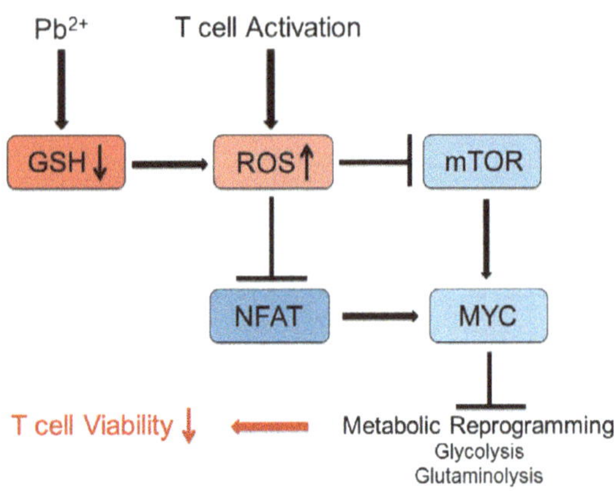

Fig. 2 Scheme for the effect of Pb exposure on metabolic reprogramming in activated T-cells. Activation of T-cells increases the presence of reactive oxygen species (ROS) in the cells, as well as, the need for new energy sources. Pb decreases the level of glutathione, which increases oxidative stress in the cells. Elevated levels of ROS in the cell will block activation of the mTOR pathway leading to a block of metabolic reprogramming. Since the activated cells are undergoing greater damage from the increased intracellular ROS level and decreased energy in the form of ATP, cell viability might be lowered by the Pb exposure.

4. Pb influence on cytokines in the CNS and immune system

Infections first prompt innate cells of the immune system to secret factors, such as cytokines, to serve as "danger" signals to the brain (Steinman, 2004). Most prominent among these factors are the pro-inflammatory cytokines IL-1, TNF, and IL-6 which enter the brain via the circumventricular organs which direct cytokine stimulation into precisely located sites within the hypothalamus (Anisman, 2004; Dantzer et al., 2007, 2008; Pollmächer et al., 2002; Quan et al., 1999a,b). Exactly how these cytokines breach the blood, brain, barrier (BBB) has not been completely established, but various mechanisms have been suggested, including BBB transport receptors (Banks, 2005). It has also been suggested that the pyrogenic cytokines, IL-1, IL-6, and TNF, may induce production of PGE_2 by cerebral endothelial cells or microglia. PGE_2 can cross the BBB and induce fever by stimulating activity of preoptic neurons. Another proposed mechanism has the pyrogenic cytokines transported to the thermoregulatory centers in the hypothalamus via afferent fibers within the vagal nerve (Guyon et al., 2008). Another recent study suggests that central β-adrenergic receptors may increase IL-1 production in the brain, but peripheral β-adrenergic receptors will inhibit cytokine responses to bacterial infections (Garcia-Bueno et al., 2008; Johnson et al., 2008). Once in the brain these cytokines produce what has been termed "sickness behavior," which encompasses malaise and decreased mobility, lack of appetite, and increased body temperature (fever). There is evidence that entry into the brain of pro-inflammatory cytokines from the periphery will result in pro-inflammatory cytokine production by microglial cells—the brain's immune cell (Becher et al., 2000; Calvo et al., 2005; Dantzer et al., 2007; Eikelenboom and Veerhuis, 1996; Hull et al., 1996). It was determined by means of trans-endothelial electrical resistance measurements across the BBB with a microelectrode that Pb in the presence of pyrogenic cytokines induced by administration of lipopolysaccharide (LPS) could increase the permeability of the BBB. Moreover, topical application of glutamate to the blood side of the BBB enhanced the permeability of the BBB (Dyatlov et al., 1998). However, this effect of Pb was most prominent during development of the BBB in the BALB/cByJ suckling mice; after pnd20, the effect was much less pronounced.

As noted earlier, Pb has been found to decrease host resistance to a number of pathogens (Lawrence, 1985). Infection by these pathogens will increase the production of pro-inflammatory cytokines by immune cells in the periphery and a small concentration of these cytokines will then gain access to the brain to promote sickness behavior. Since it was found that Pb could increase the permeability of the BBB a study was undertaken to determine if Pb would increase the concentration of these cytokines in the brain upon infection of Pb-exposed (0.5 mM Pb from pnd0 to pnd21) BALB/c mice with *L. monocytogenes*. The results of this experiment indicated that the Pb exposure prolonged the sickness behavior as measured by food and water intake and body weight loss during the course of the infection out to 10 days post-infection (Dyatlov and Lawrence, 2002). Moreover, the Pb exposure increased serum concentrations of IL-1β at days 1–3 post-infection and IL-6 at days 3–5 post-infection. Treatment of the Pb-exposed mice with IL-1β and IL-6 in the absence of bacterial infection significantly decreases water intake by these mice compared to untreated controls. It was suggested that Pb enhancement of the sickness behavior could promote irreversible damage and result in higher morbidity of the Pb-exposed animals.

In addition to IL-6's contribution to promoting fever during sickness behavior, IL-6 has been linked with cognitive impairment associated with the brain hippocampus (Smith et al., 2007). In transgenic mouse models with over-expression of IL-6, it has been found that increased IL-6 production causes astrocytosis and neurodegeneration (Campbell et al., 1993; Jafarian-Tehrani and Sternberg, 1999; Steffensen et al., 1994). Within the CNS, the main source of IL-6 production is the activated astrocyte. A study of sickness behavior in aged mice (20–22 months old) administered intracerebroventricular LPS indicated that pro-inflammatory cytokine levels, in particular IL-6, were associated with a brain region dependent increase in reactive glial markers (MHC class II, TLR2, and TLR4). Brain regions that exhibited extended pro-inflammatory expression were the cerebellum and the hippocampus (Huang et al., 2008). Elevation of IL-6 has also been associated with stress and depression in otherwise healthy human subjects (Miller et al., 2009). Further, IL-6 has been implicated in alteration of the fetal brain during maternal immune activation (MIA), such as during an infection. Elevation of IL-6 during MIA has been demonstrated to be associated with behavioral and transcriptional changes in the offspring of these pregnancies (Smith et al., 2007). Human illnesses connected with MIA are schizophrenia and autism. Examination of cytokine gene expression in the brain of female and male mice treated with 0.1 mM Pb from gd8 to pnd21 showed that this Pb exposure significantly increased IL-6 expression in the whole brain in both male and female mice. This increase in IL-6 was more pronounced when the Pb concentration was increased to 0.5 mM (Kasten-Jolly et al., 2011). Pb's effect on IL-6 concentration in the brain could, therefore, be creating a situation of astrocytosis resulting in neurodegeneration; the damaging effect of Pb exposure on astrocytes has been reported (Struzynska, 2009).

Another cytokine in the brain affected by the 0.1 mM Pb developmental exposure was TGF-β1 (Kasten-Jolly et al., 2011). Reports of TGF-β1 activity in the brain have been conflicting in that some investigators describe the cytokine as having anti-inflammatory activity and others suggesting that it can have pro-inflammatory activity as well (Boche et al., 2006; Grammas and Ovase, 2002; Makwana et al., 2007). It was found that TGF-β1 when added to astrocyte cultures could enhance astrocyte proliferation and increase the expression of NOS-2. A marker protein for astrocyte cells is GFAP, and it was found that the 0.1 mM Pb exposure was also increasing the transcription of the *gfap* gene in the brain. IL-6 and TGF-β1 through their respective signaling cascades and the transcription factor, p300, can promote transcription of the *gfap* gene (Kasten-Jolly et al., 2011). Increases in astrocyte numbers could upset the delicate balance between astrocytes and neurons during development and interfere with the generation of neuronal connections.

5. Function of immune system cells and factors in the brain and Pb modulation

Interactions and similarities between the brain and the immune system are much stronger than once thought (Steinman, 2004). Communication between neurons through neuronal synapses has been long known; similar communication between APC and T cells has been termed "immune synapse" (Grakoui et al., 1999), and it is now known that peripheral nerves form neuroimmune synapses that can affect immune functions (Dustin, 2012). Immune cells and factors thought to be specific to the immune system have normal and pathophysiologic functions that influence neural development and impact memory, cognition, and the senses (Brombacher et al., 2017; Steinman, 2004; Ziv et al., 2006). It has been found that expression of immune system molecules associated with inflammation, such as MHC, antibodies, and complement components, is increased in the brains of patients suffering from neuropsychiatric lupus and neurodegenerative diseases, such as Alzheimer and Parkinson diseases (Hong et al., 2016; Kuehn et al., 2018; Presumey et al., 2017). Moreover, MHC and TCR molecules have been found to be important for proper development of brain regions whose functions relate to sight, smell, and memory. In the olfactory cortex of mice, MHC class I molecules are co-expressed in vomeronasal neurons and are found in association with pheromone receptors. Male mice exposed to 0.1 mM Pb from gd8 to pnd21 through the dam's drinking water were found to have abnormal expression of the vomeronasal receptor V2r12. Since the vomeronasal system in mice influences odor detection and social behavior, abnormal development due to the Pb exposure of this system in the male mice was proposed to be a possible explanation for the aggressive behavior of these mice toward their litter mates, but not toward a resident-intruder (Kasten-Jolly et al., 2012). Olfactory recognition deficits in male mice exposed to Pb during lactation have been demonstrated (Flores-Montoya et al., 2015).

Low-level Pb exposure has been found to cause damage to the frontal cortex of the brain in females. Brain regions of female of the same litters as the males mentioned above were examined for gene expression and protein concentration of factors associated with inflammation (Kasten-Jolly et al., 2011, 2012). Gene expression analysis of markers of inflammation at pnd21 in brain regions of the females indicated that all inflammation markers, IL-1β, NOS2, IL-33, Casp1, TRAIL, and IL-18 were elevated only in the frontal cortex. Moreover, protein expression of an anti-inflammatory cytokine, IL-10, was significantly decreased in the frontal cortex of these female mice. The frontal cortex of the brain is responsible for control of executive function, and a recent study of women occupationally exposed to Pb suggested that elevated blood Pb levels were associated with altered executive function as measured with the Wisconsin Card Sorting Task (WCST) (Seo et al., 2015). Therefore, in females, low-levels of Pb might generate inflammation and resultant damage to the frontal cortex of the brain, which may culminate in deficits in executive function.

Cells of the immune system within the brain have been found to be important for maintaining proper CNS function. In a study of neurogenesis in the adult brain, it was reported that an "enriched" mouse cage environment resulted in recruitment of T-cells and activation of microglia cells (Ziv et al., 2006). The enriched environment created increased hippocampal neurogenesis and spatial learning abilities, which were impaired in immune-deficient mice. Hippocampal neurogenesis could be restored by administration of T-cells specific for CNS antigen, and these T-cells, along with restoring spatial learning and memory, were able to induce the expression of brain-derived neurotropic factor (BDNF) in the dentate gyrus. This result implied that an immune associated function would underlie aspects of hippocampal plasticity and cell renewal in the adult mice (Ziv et al., 2006). Recently, it has been found that neuroimmune interactions associated with Morris water maze spatial learning and memory tasks increased meningeal T-cells numbers (Derecki et al., 2010). These meningeal T-cells produced IL-4, a cytokine associated with T_{H2} cells. The IL-4 stimulated brain astrocytes to produce BDNF, which is a factor important for visuo-spatial and memory proficiency. IL-4 and IL-13, another T_{H2} cytokine, share the IL-4Rα-chain in their receptors. However, the IL-4 cytokine can signal through the IL-4Rα/γ_χ and the IL-4Rα/IL-13Rα1 receptors, but the IL-13 cytokine can only signal through the IL-4Rα/IL-13Rα1 receptor. Examination of the ability of IL-13 to affect learning and memory revealed that IL-13 deficiency will cause deficits in working and reference memory (Brombacher et al., 2017). In vitro, addition of IL-13 to astrocytes in culture stimulated them to increase production of BDNF. Developmental Pb exposure of mice has been found to skew T-cell help toward T_{H2}, based on expression of IL-4 by these cells (Heo et al., 1996). However, the production of IL-13 by these T_{H2} cells has not been closely studied. Gene expression analysis of the spleens of developmentally Pb-exposed male and female mice at pnd21 has indicated that IL-13 expression was decreased in both male and female mice, but IL-4 expression was higher than controls in the female mice (Kasten-Jolly and Lawrence, 2017). Since IL-13 is expressed by cells other than T_{H2} type T-cells, such as cells of the innate immune system, mast cells and eosinophils, the decrease in IL-13 expression might be due to the Pb effect on these cells. This aspect of Pb's effect on the immune system is worthy of further study.

6. Posited common Pb effects on macrophages/microglia and neurons

Macrophages/microglia are members of the innate immune system and have a key role in the early defense against pathogens. The impact of Pb exposure on macrophages has been the subject of a recent review (Kasten-Jolly and Lawrence, 2014). Macrophages are derived from monocytes in the blood, which after about 8 h in the circulation enter other tissues of the body and differentiate further into sessile or migrating macrophages;

whereas, myeloid progenitors enter the developing brain from the yolk sac and become microglia before hematopoietic stem cells begin differentiation in the fetal liver. Thus, microglia precede the presence of macrophages in any tissue. A primary function of macrophages or microglia is to phagocytize extracellular microorganisms and insoluble particles (injured or dead cells, cellular debris, or activated clotting factors). Once inside the cell these materials are broken down in lysosomes through the actions of hydrolytic enzymes and lysozyme. In the case of live microorganisms, macrophages kill the organisms through oxygen/nitrogen-dependent and independent mechanisms. Activated macrophages secrete inflammatory cytokines and a variety of hydrolytic enzymes, including elastase, and the secretion of these factors can have good or bad influences on health depending on where the macrophages are located. In the case of the brain's microglia, activation has been found to enhance neuronal damage (Cunningham et al., 2002). Microglia activation raises the concentration of harmful ROS within the cell and in extracellular spaces. Prolongation of microglia activation and resulting ROS exposure will culminate in events harmful to the cell, such as lipid peroxidation and protein modification, which will lead to microglia cell death through apoptosis. Interestingly, elimination of the microglia has also been shown to be harmful to neurons. In a model for cerebral ischemia, ablation of the resident microglia was found to be deleterious for neuron survival (Lalancette-Hebert et al., 2007). Decreased numbers of microglia were found in the dentate gyrus of mice exposed to 230 ppm Pb acetate from pnd0 to pnd28 through the dam's drinking water (Sobin et al., 2013). The possibility exists that this decrease in microglial number compared to the no Pb controls and low Pb, 30 ppm, groups was due to prolonged activation of the microglia and inability to overcome the damage created by the ROS exposure.

Many studies have linked Pb exposure to increased oxidative stress in the brain (Liu et al., 2012; Lu et al., 2013; Soltaninejad et al., 2003). Activation of CNS resident microglia cells was determined to be the source of the increased ROS and other factors, such as pro-inflammatory cytokines and hydrolytic enzymes that resulted in neuronal injury and apoptosis. In a study performed with rats exposed to Pb for 30 days via the drinking water, it was reported that 0.1% Pb acetate affected markers of oxidative stress, catalase (CAT), superoxide dismutase (SOD), and malondialdehyde (MDA) in the blood and brain of the exposed animals (Soltaninejad et al., 2003). Histological examination of the hippocampus of these rats revealed demyelinization, collagenous scar formation, and neuronal atrophy.

A further study performed in rats demonstrated that activation of microglia by Pb resulted in long-term potentiation impairment (Liu et al., 2012). Long-term potentiation of hippocampal excitatory synaptic transmission is considered to be important for synaptic plasticity and underlies mechanisms for hippocampus-dependent learning and memory. Another study performed in rats exposed during development, prenatally and lactationally through Pb acetate (0%, 0.05%, 0.1%, and 0.2%) in the dam's drinking water again revealed evidence of oxidative stress in the brain (Lu et al., 2013). The Pb exposure increased MDA concentration and lowered GSH and SOD. Evidence of increased apoptosis was observed in the cortex and hippocampus of these Pb-exposed rats. A study performed in vitro employing the BV-2 microglia cell line indicated that addition of Pb acetate to the cell culture medium induced the cells to secrete pro-inflammatory cytokines and other factors associated with microglial activation (Kumawat et al., 2014). High concentrations of Pb acetate, 25–100 μM, were toxic to the cells and decreased viability as measured at 24 h after Pb exposure. Filtered medium from microglial cells exposed to 25 and 50 μM Pb acetate promoted neuronal death, via apoptosis, when added to the medium of the N2a neuronal cell line. It was concluded from this study that, in the brain, activated microglia could induce death of bystander neuronal cells. Although Pb may enhance neuronal cell death due to effects on microglia, common intracellular modulations also may occur in neurons resulting in neuronal damage. The common mechanisms may include disruption in Ca^{2+} fluxes (Jia et al., 2018; Raffaello et al., 2016) due to oxidants causing ER stress (Corsetti et al., 2017). Pb modulates ryanodine receptors (Jia et al., 2018), which disrupts Ca^{2+} fluxes that modulate autophagy and apoptosis (Filippi-Chiela et al., 2016). Pb impairs lysosomal development (Song et al., 2017); lysosomes are another organelle linked to the ER. Oxidative stress, Ca^{2+} dysregulation and ER loss of homeostasis are tied to Pb toxicity (Flora et al., 2012). ER stress also may generate detrimental mechanisms between mitochondria and peroxisomes, which have mutual regulatory interactions (Demarquoy and Le Borgne, 2015; Fransen et al., 2017). Mitochondria and peroxisomes need to work together to control the levels of oxidants (Schrader et al., 2015), and together they affect lipid metabolism (Lodhi and Semenkovich, 2014). Genistein lessens the neurotoxicity of Pb by promoting the increased expression of antioxidant enzymes ant the antioxidant transcription factor, nuclear factor erythroid 2 p45-related factor 2 (Nrf2) (Su et al., 2016). The Genistein also may attenuate Pb's effects on lipid metabolism and lysosomal activity (Moskot et al., 2014).

In the spleen, Pb can increase oxidative stress within macrophages through many mechanisms as described by Kasten-Jolly and Lawrence (2014). In the blood, Pb promotes damage to erythrocyte membranes and renders these cells vulnerable to phagocytosis by macrophages located in the red pulp of the spleen. Destruction of these senescent red cells increases the amount of ferrous ion (Fe^{2+}) within the macrophages through the activity of heme oxygenase (HO-1), which converts heme to Fe^{2+}, carbon monoxide, and biliverdin. Pb exposure has been found to up-regulate the expression of HO-1 (Shinkai and Kaji, 2012). Excess Fe^{2+} will increase the conversion of hydrogen peroxide (H_2O_2) to highly reactive species, $^{\bullet}HO$ and OH^-, via the Fenton reaction (Jomova and Valko, 2011). These highly reactive species will promote damage to cellular lipids, proteins, and DNA. Many of the enzymes whose function is to alleviate cellular oxidative stress, such as nitric oxide synthase (NOS), peroxidases, and cytochrome *c* oxidases, have heme as a cofactor. Pb has been found to inhibit heme synthesis at several points during the synthesis process by inhibiting the function of δ-aminolevulinic acid dehydratase, protoporphyrinogen oxidase, and ferrochelatase enzymes (de Souza et al., 2018; Valle and Ulmer, 1972). Therefore, by decreasing heme synthesis, Pb is interfering with the cells ability to alleviate oxidative stress. Although Pb exposure elevates concentrations of ROS, it has been found to decrease generation of nitric oxide (NO). Induction of iNOS in splenic macrophages occurs with help from activated T-cells through stimulation by IFNγ and TNF-α and interaction with surface molecules present on the activated T-cells, LFA-1 and CD40L (Tian et al., 1995). Pb was found to inhibit NO production by ConA-stimulated murine splenocytes and further study demonstrated that Pb inhibited NO production in a concentration-dependent manner in cultures of isolated macrophages stimulated with 500 U/mL IFNγ and 1000 U/mL TNF-α (Tian and Lawrence, 1995). The Pb effect on production of NO was found not to be due to decreased mRNA or protein for iNOS. One possibility for the decrease in NO production in the presence of Pb might be by "NO uncoupling" where NOS enzymatic activity would create more super oxide (O_2^-) than NO (Chen et al., 2010). Details of this process have been presented by Kasten-Jolly and Lawrence (2014) and will be summarized here. It was discovered by Crabtree et al. (2013) that S-glutathionylation of specific residues critical for maintaining NOS activity affected "uncoupling" of the NOS activity, and this process was sensitive to the GSH:GSSG ratio within the cell. Pb decreases cellular concentrations of GSH and, therefore, decreases the

GSH:GSSG ratio (Hunaiti and Soud, 2000). Increased S-glutathionylation of NOS in the presence of a lowered GSH:GSSG ratio results in decreased NO production and increased production of O_2^-. By affecting the function of NOS both by reduction of the cofactor, heme, and "NO uncoupling," Pb decreases concentrations of the valuable molecule, NO, and increases oxidative stress by production of O_2^-. In macrophages, NO is a valuable component of the oxygen/nitrogen-dependent mechanism for killing intracellular pathogens, and by decreasing NO production Pb is decreasing the host's ability to ward off pathogens. In the brain decreased production of NO in the presence of Pb has been associated with deficits in synaptic plasticity (Nava-Ruiz et al., 2012). NO from iNOS of macrophages mainly functions in defense against pathogens, but NO from endothelial (eNOS) and neuronal (nNOS) plays vital roles in numerous physiological processes, such as vascular tone, peristalsis and angiogenesis. A cofactor for each NOS is tetrahydrobiopterin (BH4), and ascorbate (vitamin C) can protect from oxidation (Mortensen and Lykkesfeldt, 2014). BH4 is needed for the heme catalytic site of NOS, and production of both BH4 and heme may be inhibited by Pb.

Macrophages also use oxygen/nitrogen-independent mechanisms to kill intracellular pathogens. These include hydrolytic enzymes and lysozyme, defensins, IFNγ-induced GTPases (p47-IRGs and p65-GBPs), and pro-inflammatory cytokines (IL-1, TNF-α, and IL-6). In the spleen, Pb was found to increase expression of several hydrolytic enzymes, including proteases, lipases, and RNase (Kasten-Jolly et al., 2010). Although Pb does not affect the expression of lysozyme, it has been demonstrated to modulate the activity of lysozyme (Olmo et al., 2012). Pb doesn't directly change the expression of IFNγ, but it does skew T_H toward T_{H2} and production of IL-4. Functional activity of IFNγ has been found to be lower upon Pb exposure (Heo et al., 2007). It is for this reason, perhaps, that expression, as measured by microarray, of the IFNγ inducible GTPases was decreased in the brains of mice treated with 0.5 mM Pb from birth to pnd21 through the dam's drinking water and infected on pnd22 with *L. monocytogenes* compared to mice infected only (Kasten-Jolly and Lawrence, 2014). Analysis of the expression of one of these GTPases, Gbp1, by qPCR in the spleen and brain of mice exposed to 0.1 mM Pb from gd8 to pnd21 showed that Pb alone could lower the gene expression of this GTPase (Kasten-Jolly and Lawrence, 2014). IFNγ inducible GTPases have a role in transport of antimicrobial peptides, autophagic machinery, and components of the NADPH oxidase to the phagosomal compartments that contain pathogens

(Kim et al., 2011; MacMicking, 2004; Martens and Howard, 2006; Taylor et al., 2004). By decreasing levels of functional IFNγ and consequently down-regulating the expression of the GTPases needed for intracellular pathogen killing, Pb would be compromising host resistance to pathogens.

CNS autophagy is an important central regulator of cellular health, and disruption of this homeostatic process by Pb exposure might have deleterious effects on learning and memory. Autophagy involves the sequestration, degradation and digestion of intracellular components by lysosomes (Alirezaei et al., 2011). Optimally functioning autophagy machinery will maintain cellular integrity, by removing misfolded proteins and damaged organelles, such as mitochondria (mitophagy) which, when damaged, will release harmful ROS into the cytosol. Additionally, autophagy functions as a critical defense against pathogens and degenerative states. In the CNS, perturbation of the carefully regulated autophagy mechanism will result in neurodegeneration (Cappelletti et al., 2014). By decreasing the expression of the IFNγ inducible GTPases, lactational Pb exposure of the mouse pups appeared to be compromising the efficiency of autophagy with respect to pathogen killing in the brain (Kasten-Jolly and Lawrence, 2014). However, two studies involving Pb exposure in the absence of infection have indicated that the Pb exposure increased autophagy in the hippocampus of the brain (Meng et al., 2016; Zhang et al., 2012). In a study employing Sprague-Dawley rats, Pb acetate was administered through the drinking water at concentrations of 0, 100, 200, and 300 ppm for 8 weeks, and it was shown that Pb exposure increased the accumulation of α-synuclein and tau phosphorylation in the hippocampus, and the accumulation of these factors was directly related to learning and memory as measured by performance in the Morris Water Maze (MWM) test (Zhang et al., 2012). Further analysis by means of western blotting indicated that the Pb exposure was increasing levels of proteins associated with autophagy, Beclin 1 and microtubule-associated protein 1 light chain 3 (LC3), which suggested that the Pb exposure was increasing autophagy in the hippocampus. The Pb exposure also increased expression of caspase-3 which indicated that Pb was increasing apoptosis, and it was concluded that increased autophagy and apoptosis in the hippocampus was the reason for the impairment of learning and memory in the Pb-exposed rats. Examination of Grp78 expression indicated that Pb was increasing endoplasmic stress, and further study indicated the Pb exposure was down-regulating the Akt/mTOR signaling pathway. It was concluded that these factors were the reason for increased autophagy and apoptosis in the hippocampus of the Pb-exposed animals.

In another study with Sprague-Dawley rats exposed during lactation to 300 ppm lead acetate through the dam's drinking water, Pb exposure also increased autophagy as indicated by increases in autophagy marker proteins, Atg5, Beclin-1, and LC3 (Meng et al., 2016). Again, the Pb exposure increased expression of caspase-3 and decreased signaling through the Akt/mTOR pathway. Increased autophagy and apoptosis were proposed to be the cause of poor performance in the MWM test by the lactationally Pb-exposed rats. In this study, it was determined that administration of gangliosides immediately after the Pb exposure could reverse the effects of Pb on autophagy and hippocampal neurodegeneration.

As mentioned earlier, Pb skews T-cell help toward T_{H2}; therefore, Pb elevates IL-4 concentrations in tissues, suggesting that Pb would promote development of the M2 subtype in multiple tissues including the brain. Recently, it has been demonstrated that macrophages skewed toward the M2 subtype are high expressers of heme oxygenase-1 (HO-1) (Naito et al., 2014; Sierra-Filardi et al., 2010). Data obtained by western-blot analysis have shown that Pb exposure will increase expression of HO-1, as well as other factors that contain an antioxidant response element (ARE) within the promoter region of their gene (Shinkai and Kaji, 2012). The transcription factor Nrf2 binds to the ARE region of the gene promoter, and it was found that Pb has a direct effect on increasing the activity of the Nrf2 signaling pathway. BALB/c mice treated with 0.5 mM Pb acetate from pnd0 to pnd21 through the dam's drinking water were found to have increased gene expression of *STAT6* in the brain at pnd21 as determined by Affymetrix gene chip analysis (Kasten-Jolly and Lawrence, 2014). Increased expression of STAT6 has been associated with macrophages polarized toward M2 (Lawrence and Natoli, 2011). Gene expression analysis by means of cytokine gene probe blots gave preliminary indication that IL-10 expression was increased in the non-perfused brain at pnd21 of mice lactationally exposed to 0.5 mM Pb acetate (Fig. 3). Since increased IL-10 gene expression due to Pb exposure was not detected by microarray analysis or qPCR in the brains of perfused mice, it is possible that the IL-10 gene expression found in the non-perfused brain was due to the presence of blood within the brain (Kasten-Jolly et al., 2011). In a study of Pb exposure on serum cytokine concentrations in occupationally exposed workers, it was found that IL-10 was increased in workers with blood Pb levels 9.1–46.0 μg/dL (Valentino et al., 2007). These results lend support to the likelihood of Pb polarizing activated macrophages toward the M2 subtype. Human peripheral blood mononuclear cells (PBMCs) can be polarized to the M2

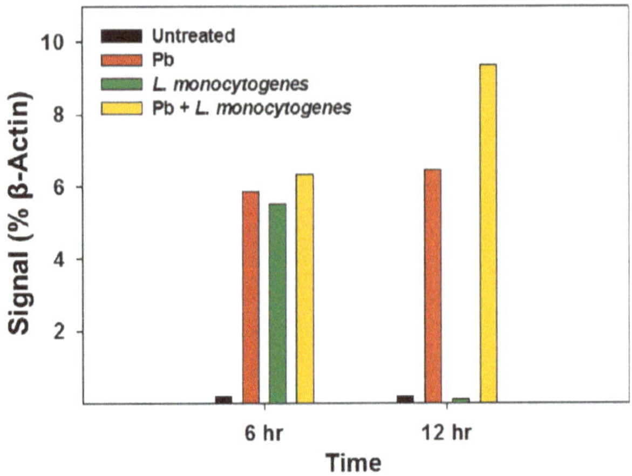

Fig. 3 Interleukin-10 gene expression in the non-perfused brains of BALB/c mice. Mice were exposed to 0.5 mM Pb from pnd0 to pnd21 for the Pb group ± infection with 2×10^4 CFU *Listeria monocytogenes* on pnd22. Control groups were no Pb exposure and *L. monocytogenes* infection alone. Mouse brains were taken and snap frozen at 6 h and 12 h post-infection on pnd22. Brain RNA was isolated by means of Qiagen midi columns according to the manufacturer's instructions. cDNA labeled with ^{32}P was prepared from the purified RNA and hybridized to GEArray mouse common cytokine blots from Super Array, Inc. (Bethesda, MD) according to the manufacturer's specifications. Cytokine transcripts were detected with a Fuji PhosphoImager antified by normalizing the signal to that of β-actin.

subtype by growth in culture medium containing macrophage colony-stimulating factor (M-CSF). Stimulation of these macrophages with LPS will result in secretion of the cytokine IL-10 into the culture medium and addition of a HO-1 inducer to the medium will increase the secretion of IL-10 (Sierra-Filardi et al., 2010). One of the effects of IL-10 in macrophages is to promote autophagy of damaged mitochondria and, in doing so, increase cell viability and health (Ip et al., 2017). Increased autophagy in the presence of IL-10 was coupled with IL-10 signaling via STAT3 inhibition of mTORC1 activation. Further, IL-10 inhibited phosphorylation of Akt targets, PRAS40 and mTOR, and this was found to be due to induction of a negative regulator of mTOR signaling, DDIT4. Although M2 macrophages can be identified by the high-expression of alternative activation markers, arginase-1 (Arg-1) and Chitinase 3-like 3 (YM-1), additional microRNA markers have now been discovered that can assist in characterizing M2 subtypes. These microRNAs are miR-125a-5p and let-7c (Banerjee et al., 2013a, b). Perhaps, these additional markers will be of assistance in delineating the exact impact(s) of Pb on activated macrophage subpopulations.

7. Influence of Pb on the HPA axis and sex hormone effects

The hypothalamus-pituitary-adrenal (HPA) axis reacts to physiological and psychological stress by raising the secretion of glucocorticoids from the adrenal cortex (Fig. 4). Elevated concentrations of glucocorticoids in plasma will signal changes in the function of a number of organs including lungs, bone, brain, skin, liver, and thymus. These glucocorticoids (cortisol in humans or corticosterone in rodents) will then provide a negative feedback to the hippocampus, hypothalamus, and pituitary to restore homeostasis (White et al., 2007). Dysfunction of the HPA axis caused either by physiological stress, such as exposure to a toxicant, such as Pb, or psychological stress, such as low socioeconomic status (SES), will be a contributing factor in a number of disorders and diseases, including cardiovascular disease, osteoporosis, asthma, lupus, Crohn's disease, obesity, diabetes, depression,

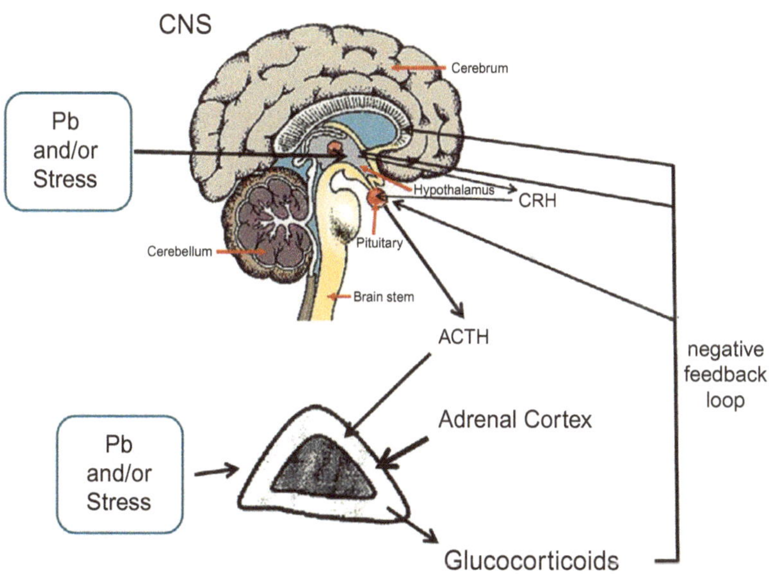

Fig. 4 Diagram of the hypothalamus-pituitary-adrenal (HPA) axis and negative feedback loop. Pb exposure and or environmental stress will affect various areas of the brain and trigger the hypothalamus to secret CRH. The secreted CRH will induce the pituitary to produce and secret ACTH, which in turn will induce the adrenal cortex to produce glucocorticoids in response to the stress. To maintain homeostasis the glucocorticoids will provide a negative feedback to the pituitary and the brain to down-regulate glucocorticoid secretion. Prolonged early life exposure to Pb and/or environmental stress will perturb the function of the HPA axis in a permanent manner.

anxiety, cognitive deficits, and insomnia. Exposure of adult male Albino Wistar rats to 500 ppm Pb for 5 weeks through their drinking water resulted in significantly elevated plasma corticosterone and decreased brain levels of 5-HT (serotonin) compared to absence of Pb (Haider et al., 2013). Behavior testing with the elevated-plus-maze (EPM) test and the forced swim (FS) test designed to measure anxiety and depression, respectively, indicated that the Pb-exposed rats had increased anxiety and depression compared to controls. Recently, a study of 126 human adults, mean age of 65.9 ± 8.1, with a blood Pb range of 0.6–6.1 μg/dL and a mean of 2.1 ± 0.9 μg/dL indicated that subjects with blood levels above the mean had higher saliva cortisol 30 min after awakening and at 14 h and 16 h during the day than the subjects with blood Pb levels below the mean (Souza-Talarico et al., 2017). Prolonged exposure to Pb in humans and rodents will interfere with the function of the HPA axis, especially if the exposure occurs during fetal development (Cory-Slechta et al., 2010, 2013; Gump et al., 2008). Cognitive deficits associated with Pb exposure have been connected to disruption in the function of the HPA axis and the mesocorticolimbic (MESO) system, both of which are critical for stimulation/reward functions and executive/cognitive functions (Sobolewski et al., 2018). In a study with rats exposed to prenatal/lactational Pb, 50 and 150 ppm through the dam's drinking water, it was demonstrated that the Pb exposure significantly increased plasma corticosterone in the females, but not in the males as measured at 2 months of age (Rossi-George et al., 2009, 2011). Measurement of the percent change in monoamine concentrations in the frontal cortex and nucleus accumbens that occurred between 2 and 4 months of age showed a higher degree of changes for dopamine (DA), dihydroxyphenylacetic acid (DOPAC), and 5-hydroxyindoleacetic acid (5-HIAA) in frontal cortex of the females than in the males (Rossi-George et al., 2011). In the males, serotonin displayed the greater degree of changes in the frontal cortex, while in the nucleus accumbens, there were no significant changes observed in the males for any of the monoamines studied. Significant changes in the nucleus accumbens due to Pb exposure and PS were observed for homovanillic acid (HVA) in the females. The negative feedback function of the HPA axis was examined by use of dexamethasone, which acts on the pituitary to suppress secretion of ACTH. While serum corticosterone concentrations indicated a preferential vulnerability of Pb exposure and prenatal stress (PS) in the females, data from the dexamethasone experiments, targeting the negative feedback loop, suggested a greater impact on the males with Pb, PS, and Pb plus PS (Rossi-George et al., 2009). Experiments to measure the effect of stress (barren cage) and Pb exposure (1 and 10 mg/kg Pb acetate administered by gavage from pnd0 to pnd29) indicated that monoamine levels were altered by the Pb exposure or combination of Pb and stress primarily in regions of the female brain (Graham et al., 2011). In the hypothalamus, norepinephrine (NE) was increased by 1 mg/kg Pb exposure only in the females on day pnd29. There were no changes in NE for any of the time points, pnd11, pnd19, and pnd29, or Pb exposure concentrations in the males. Pb in combination with stress raised the level of 5-HIAA in the hypothalamus and hippocampus on pnd29 only for the 1 mg/kg Pb group in the females, while the males did not show any changes in 5-HIAA. Blood Pb concentrations were dose dependent and the 10 mg/kg group had blood Pb concentrations no higher than 14 μg/kg, which indicated that the method of administration employed resulted in environmentally relevant blood Pb concentrations. Interestingly, most significant monoamine changes in the brain were observed in the lower Pb exposure group. Further, it was mentioned by the authors that the mortality rate was higher in the males in the combined Pb and stress groups, especially at pnd19. This might suggest that the effects of Pb and stress in the males were more severe and damaging than in the females; perhaps because in the females Pb up-regulates the expression of genes that code for proteins whose function is to decrease damage to organs and tissues (Kasten-Jolly and Lawrence, 2017) (Fig. 5).

It has been determined that sex hormones modulate glucocorticoid levels with females having higher plasma corticosterone or cortisol concentrations than males (Whitacre, 2001). In an epidemiological study of pregnant women in Mexico City, it was found that blood Pb levels were associated with heightened cortisol awakening response, and it was suggested that Pb could disrupt HPA axis function during pregnancy (Braun et al., 2014). Differences in HPA axis function between males and females could be due to the presence of an estrogen response element within the promoter region of the gene for CRH. CRH produced in the hypothalamus regulates the secretion of ACTH from the pituitary. This regulatory chain controls glucocorticoid release from the adrenal cortex. Raised glucocorticoid concentrations participate in the negative feedback loop to decrease expression of factors associated with an inflammatory response (Homo-Delarche et al., 1991). Gestational Pb exposure via the dam's drinking water indicated that the exposure changed the balance of T_H subsets in a manner that differed by sex of the fetus. Females had a depressed T_{H1} type immune response and a heightened T_{H2} response with elevated production

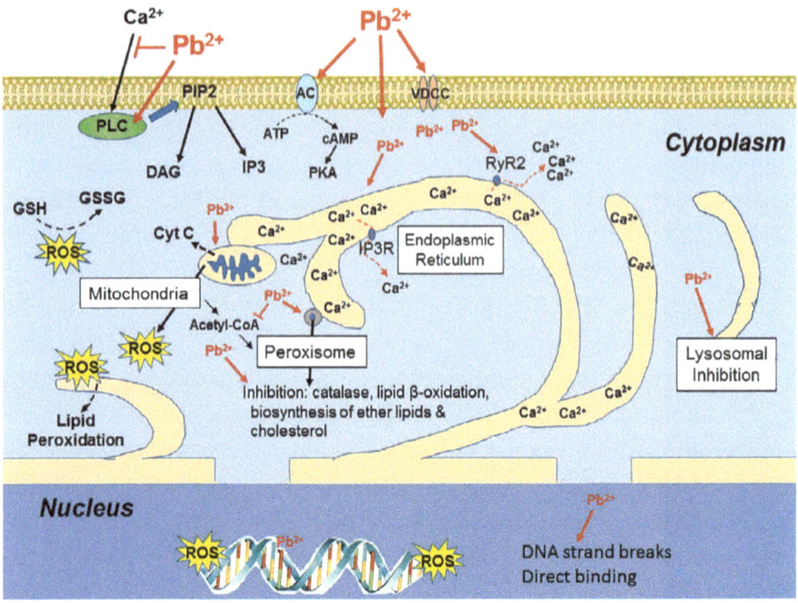

Fig. 5 Pb-induced cell stress via disruption of organelle receptors and enzymes. Pb causes an increase in cytosolic Ca^{2+} by enhancing its release from the endoplasmic reticulum (ER) through the ryanodine receptor (RyR2) and the inositol 1,4,5 trisphosphate receptor (IP3R). Pb enters cells the plasmalemmal voltage gated Ca^{2+} channel (VGCC) and other means. Redistribution of Ca^{2+} within the cell may increase the unfolded protein response (UPR); however, Pb is concomitantly inhibiting lysosomal formation. Pb interferes with mitochondrial and peroxisomal functions. Inhibition of plasmalogen production by peroxisome is especially damaging to CNS which is needed for myelination of nerve cells. Increased Ca^{2+} in mitochondria causes less ATP production and increased release of reactive oxygen species (ROS) and eventual release of cytochrome *c* (cyto *c*) and apoptosis. ROS and Pb directly alter DNA to affect transcription.

of IgE (Chen et al., 1999; Miller et al., 1998; Snyder et al., 2000). To determine if sex hormones had a role in this skewing of the female immune response, hen eggs exposed to Pb on embryonic day 8 (E8) were given testosterone at embryonic day 12 (E12) and the thymuses of the male and female embryos were taken on day 20 (E20) (Hussain et al., 2005). Isolated thymocytes were analyzed by flow cytometry and the results indicated that the testosterone treated embryos had higher percentages of $CD4^{+}CD8^{+}$ double positive cells than the controls or Pb alone treated embryos. Pb exposure alone increased the percentage of $CD4^{+}CD8^{-}$ single positive cells, but testosterone alone or in combination with Pb exposure decreased this cell population. The population of $CD4^{-}CD8^{+}$ cells did not have their percentage changed by Pb or testosterone treatment, but Pb plus testosterone slightly decreased the percentage of the $CD4^{-}CD8^{-}$ cell population. The conclusion from this study was that the sex hormone levels during thymus development influences how Pb exposure affects this early stage of the immune system.

8. Pb effect on endocrine function that may modulate neuroimmune activity

Pb exposure has an impact on blood thyroxine and thyrotropin (thyroid-stimulating hormone or TSH) levels, and these effects are dependent on blood Pb concentration, duration of exposure, and gender (Abdelouahab et al., 2008; Chen et al., 2013; Dunbar et al., 2006; Jurdziak et al., 2018; Meeker et al., 2009; Nie et al., 2017; Singh et al., 2000). Among persons occupationally exposed to Pb no significant changes in TSH levels were observed in individuals with blood Pb concentration of $22.21 \pm 14.1\,\mu g/dL$ (Jurdziak et al., 2018). No association was detected between blood Pb concentrations $<5\,\mu g/dL$ and thyroid hormone and TSH levels in a National Health and Nutrition Examination Survey (Chen et al., 2013). However, with blood Pb levels $>50\,\mu L/dL$ and exposure times of >60 months, TSH levels were elevated and T_3 was decreased (Singh et al., 2000). In a study of male adolescents, mean blood Pb levels of $7.3 \pm 2.92\,\mu g/dL$ and prolonged exposure were associated with an increased incidence of abnormally low free (f)T_4 levels compared to the control group which had a mean blood Pb of $2.08 \pm 1.24\,\mu g/dL$, but there were no differences in the fT3 or TSH (Dunbar et al., 2006). In a study of men and women living at lakeside communities in Quebec (Canada), it was observed that although blood Pb levels were higher in men, 2.42–20.9 μg/dL, than in women, 1.23–12.5 μg/dL, no differences were noted in thyroid hormone or TSH levels in the men, but TSH levels were decreased with increasing blood Pb levels in the women (Abdelouahab et al., 2008). Correlations between thyroid function and blood Pb levels were also observed only in women in a large study of 2278 men and 3350 women in China (Nie et al., 2017). Again, the men had higher blood Pb levels than the women, 4.0 (2.9–6.22) μg/dL for the men and 3.79 (2.5–5.5) μg/dL for the women. It was found that the women had increased thyroid peroxidase antibody (TPOAb) concentrations with increasing blood Pb levels. Relatively higher blood Pb in the women correlated with a higher incidence of hypothyroidism, while the men did not show evidence of blood Pb and compromised thyroid function. A study of 219 men aged between

18 and 55 with blood Pb levels <1.1 to >2.0 μg/dL displayed a non-monotonic decrease in TSH concentration with increases in blood Pb levels (Meeker et al., 2009).

Studies performed in animals on thyroid function associated with Pb exposure indicated that blood Pb levels >30 μg/dL resulted in conditions of hypothyroidism, and this condition affected neuronal development (Miao et al., 2015; Wu et al., 2011). Pb exposure of Wistar rats reduced levels of both T_3 and T_4 with decreased TSH levels (Wu et al., 2011). It was indicated that the hippocampus is sensitive to and dependent on thyroid hormone homeostasis, and this condition was disrupted by the Pb exposure. Treatment of the rats with Levothyroxine was able to overcome the detrimental effects of Pb on impairment of long-term potentiation in the hippocampal CAI region. In a study of Zebrafish, Pb exposure also decreased levels of T_3 and T_4, and this was exacerbated by the addition of titanium dioxide nanoparticles (Miao et al., 2015). The combination of Pb and nanoparticles decreased transcription of genes involved in neuronal development that are connected to thyroid function. Pb is well known to increase oxidative stress in the CNS with loss of protein thiols and increases of malondialdehyde (MDA) from lipid peroxidation (Maiti et al., 2017). These outcomes are linked to Pb effects on mitochondria (Sanders et al., 2009).

Pb was found to have a direct effect on the enzyme that converts T_4 to T_3, iodothyronine 5′-monodeiodinase (5′-D) (Chaurasia et al., 1996). The type-1 5′-D is a membrane bound selenoprotein. The enzyme requires free sulfhydryl (–SH) groups to maintain its three-dimensional active conformation. Pb increased lipid peroxidation and decreased 5′-D enzymatic function as indicated by the increase in MDA and decrease in production of T_3.

Humans and animals exposed to Pb have decreased growth when young and increased weight gain as adults; these aspects of Pb exposure have been linked to the impact of Pb on glucose metabolism. Pb exposure has been observed to reduce linear and ponderal growth in children and studies with weanling rats suggested that this reduction in growth rate is the result of depressed appetite (Hammond and Succop, 1995; Hammond et al., 1989). Treatment of Pb-exposed weanling rats with growth hormone and thyroxine had no effect on rat growth rate, but nutritional supplementation could overcome the Pb depression of growth rate. In another study with rats the activity of gluconeogenesis and glycogenolysis enzymes along with markers for oxidative stress and inflammation was evaluated in the liver after 32 days of Pb exposure (Mostafalou et al., 2015). It was found that the activity of these enzymes increased due to the Pb exposure indicating glucose intolerance. In the pancreas Pb disrupted insulin secretory function by activating glycogen-synthase-kinase-3β (GSK-3β) and increasing endoplasmic reticulum stress. It was concluded that Pb exposure disrupts glucose homeostasis through induction of insulin resistance. In a study of adult mice exposed to 16 or 32 ppm Pb during gestation and lactation it was found that food intake increased both in the females and males over that of controls (Faulk et al., 2014). Both females and males demonstrated increased energy expenditure over controls, but the females showed increased horizontal activity. The males showed increased body weight gain and total body fat differed between Pb-exposed males and females. The study results emphasize that the effects of developmental Pb exposure persist over a lifetime and vary according to sex and age. In a study employing a rodent model of obesity, adult female ZDF rats, Pb exposure was able to induce fasting hyperglycemia after 8 weeks and glucose intolerance after 12 weeks of exposure (Tyrrell et al., 2017). It was suggested that the Pb exposure was promoting diabetes by direct stimulation of hepatic gluconeogenic enzyme gene expression up-regulation. In a study to determine if obesity in Pb-exposed adult rodents is associated with DNA hypermethylation it was found that DNA hypermethylation in the Pb-exposed rats extended along pathways related to glucose and lipid metabolism in the NCD + 0.05% Pb acetate group (Sun et al., 2017). The conclusion from this study was that DNA hypermethylation was associated with a Pb-dose specific weight gain, glucose intolerance, and insulin resistance in the adult rats. In an epidemiological study for the association of heavy metal exposures and metabolic syndrome it was found that causal inferences remained challenging and prospective studies will need to have greater standardization and consistency in the study design and reporting in order to demonstrate potential causal links between heavy metal exposure and metabolic syndrome (Planchart et al., 2018).

9. Summary

An extended network of cross-talk exists between the endocrine, immune and nervous systems; therefore, when Pb perturbs the functioning of one of these systems the other two will be affected as well. For example, Pb exposure decreases host resistance to pathogens and, consequently, has been found to prolong sickness behavior, a response of the hypothalamic-pituitary-adrenal axis. In the immune system, Pb polarizes $CD4^+$

lymphocytes toward T_{H2}, IL-4 secreting cells, which promote production of IgE, an antibody associated with asthma and allergies. Further, the increased presence of pro-inflammatory cytokines in the brain alters neuronal and glial interactions creating increased neuronal damage, via activation of microglia and increased astrocyte proliferation. Pb exposure increases oxidative stress and has been demonstrated to lower the concentration of GSH, a buffer for oxidative stress. By lowering GSH, Pb exposure increases apoptosis of cells in the brain through increased autophagy and down-regulation of the mTOR signaling pathway. In the immune system decreases in GSH, in association with mTOR pathway down-regulation, by Pb could decrease metabolic reprogramming in lymphocytes and polarize macrophage toward M2. Pb exposure during development increases glucocorticoid levels and results in dysfunction of the hypothalamic-pituitary-adrenal axis, which then interferes with hippocampal function. Pb exposure also interferes with thyroid function and in this manner also compromises the function of the hippocampus. A commonality of Pb's influences on different organ systems probably relates to its interference with mitochondrial functions with loss of ATP production and enhanced release of ROS (Sanders et al., 2009), loss of peroxisome's lipid metabolism and catalase activity, and intracellular effects on Ca^{2+}, which mediates endoplasmic reticulum stress and modulation of Ca^{2+}-dependent enzymes (Simons, 1993). In addition to Ca^{2+} dysregulation, Pb may alter Zn and Fe distribution and related enzymes, which would disrupt numerous organ systems. Cross-talk among organs and intracellular organelles is critical for the maintenance of health.

Acknowledgments

The common cytokine gene expression research was supported by NIH grant ES03179. The authors thank Vladimir A. Dyatlov for his technical assistance with respect to Pb exposure and *L. monocytogenes* infection of the mice used for these experiments.

References

Abazyan, B., Dziedzic, J., Hua, K., Abazyan, S., Yang, C., Mori, S., Pletnikov, M.V., Guilarte, T.R., 2014. Chronic exposure of mutant DISC1 mice to lead produces sex-dependent abnormalities consistent with schizophrenia and related mental disorders: a gene-environment interaction study. Schizophr. Bull. 40, 575–584.

Abdelouahab, N., Mergler, D., Takser, L., Vanier, C., St-Jean, M., Baldwin, M., Spear, P.A., Chan, H.M., 2008. Gender differences in the effects of organochlorines, mercury, and lead on thyroid hormone levels in lakeside communities of Quebec (Canada). Environ. Res. 107, 380–392.

Aebisher, I., Stampfli, M.R., Zurcher, A., Miescher, S., Urwyler, A., Frey, B., Luger, T., White, R.R., Stadler, B.M., 1994. Neuropeptides are potent modulators of human in vitro immunoglobulin E synthesis. Eur. J. Immunol. 24, 1908–1913.

Alirezaei, M., Kemball, C.C., Whitton, J.L., 2011. Autophagy, inflammation and neurodegenerative disease. Eur. J. Neurosci. 33, 197–204.

Anisman, H., 2004. Considering cytokine panels. Brain Behav. Immun. 18, 221–222.

Aste-Amezaga, M., Ma, X., Sartori, A., Trinchieri, G., 1998. Molecular mechanisms of the induction of IL-12 and its inhibition by IL-10. J. Immunol. 160, 5936–5944.

Banerjee, S., Cui, H., Xie, N., Tan, Z., Yang, S., Icyuz, M., Thannickal, V.J., Abraham, E., Liu, G., 2013a. miR-125a-5p regulates differential activation of macrophages and inflammation. J. Biol. Chem. 288, 35428–35436.

Banerjee, S., Xie, N., Cui, H., Tan, Z., Yang, S., Icyuz, M., Abraham, E., Liu, G., 2013b. MicroRNA let-7c regulates macrophage polarization. J. Immunol. 190, 6542–6549.

Banks, W.A., 2005. Blood-brain barrier transport of cytokines: a mechanism for neuropathology. Curr. Pharm. Des. 11, 973–984.

Becher, B., Prat, A., Antel, J.P., 2000. Brain-immune connection: immune-regulatory properties of CNS-resident cells. Glia 29, 293–304.

Bellinger, D.C., 2008. Very low lead exposures and children's neurodevelopment. Curr. Opin. Pediatr. 20, 172–177.

Bent, S., Göttsch, C., Braam, U., Schmutzler, W., 1992. The effects of heavy metal ions (Cd^{2+}, Hg^{2+}, Pb^{2+}, Bi^{3+}) on histamine release from human adenoidal and cutaneous mast cells. Agents Actions 36, C321–C324.

Boche, D., Cunningham, C., Docagne, F., Scott, H., Perry, V.H., 2006. TGFβ1 regulates the inflammatory response during chronic neurodegeneration. Neurobiol. Dis. 22, 638–650.

Braun, J.M., Wright, R.J., Just, A.C., Power, M.C., Ortiz, M.T.Y., Schnaas, L., Hu, H., Wright, R.O., Tellez-Rojo, M.M., 2014. Relationships between lead biomarkers and diurnal salivary cortisol indices in pregnant women from Mexico City: a cross-sectional study. Environ. Health 13, 50–60.

Brombacher, T.M., Nono, J.K., deGouveia, K.S., Makena, N., Darby, M., Womersley, J., Tamgue, O., Brombacher, F., 2017. IL-13-mediated regulation of learning and memory. J. Immunol. 198, 2681–2688.

Brunyanszki, A., Olah, G., Coletta, C., Szczesny, B., Szabo, C., 2014. Regulation of mitochondrial poly(ADP-ribose) polymerase activation by the β-adrenoceptor/cAMP/protein kinase A axis during oxidative stress. Mol. Pharmacol. 86, 450–462.

Bunn, T.L., Parsons, P.J., Kao, E., Dietert, R.R., 2001. Exposure to lead during critical windows of embryonic development: differential immunotoxic outcome based on stage of exposure and gender. Toxicol. Sci. 64, 57–66.

Calvo, C.-F., Amigou, E., Desaymard, C., Glowinski, J., 2005. A pro- and an anti-inflammatory cytokine are synthesized in distinct brain macrophage cells during innate activation. J. Neuroimmunol. 170, 21–30.

Calzolari, A., 1898. Recherches experimentals sur un rapport probanle entire la function du thymus et celle des testicules. Arch. Ital. Biol. 30, 71–80.

Campbell, I.L., Abraham, C.R., Masliah, E., Kemper, P., Inglis, J.D., Oldstone, M.B., Mucke, L., 1993. Neurologic disease induced in transgenic mice by cerebral overexpression of interleukin-6. Proc. Natl. Acad. Sci. U. S. A. 90, 10061–10065.

Cao, S., Liu, J., Chesi, M., Bergsagel, P.L., Ho, I.C., Donnelly, R.P., Ma, X., 2002. Differential regulation of IL-12 and IL-10 gene expression in macrophages by the basic leucine zipper transcription factor c-Maf fibrosarcoma. J. Immunol. 169, 5715–5725.

Cao, S., Liu, J., Song, L., Ma, X., 2005. The protooncogene c-Maf is an essential transcription factor for IL-10 gene expression in macrophages. J. Immunol. 174, 3484–3492.

Cappelletti, C., Galbardi, B., Kapetis, D., Vattemi, G., Guglielmi, V., Tonin, P., Salerno, F., Morandi, L., Tomelleri, G., Mantegazza, R., Bernasconi, P., 2014. Autophagy, inflammation and innate immunity in inflammatory myopathies. PLoS One 9 (11), e111490.

Cavaillon, J.M., Annane, D., 2006. Compartmentalization of the inflammatory response in sepsis and SIRS. J. Endotoxin Res. 12, 151–170.
Chaurasia, S.S., Gupta, P., Kar, A., Maiti, P.K., 1996. Free radical mediated membrane perturbation and inhibition of type-1 iodothyronine 5′-monodeiodinase activity by lead and cadmium in rat liver homogenate. Biochem. Mol. Biol. Int. 39, 765–770.
Chen, S., Golemboski, K.A., Sandes, F.S., Dietert, R.R., 1999. Persistent effect of in utero meso-2,3-dimercaptosuccinic acid (DMSA) on immune function and lead induced immunotoxicity. Toxicology 132, 67–79.
Chen, C.-A., Wang, T.-Y., Varadharaj, S., Reyes, L.A., Hemann, C., Talukder, M.A.H., Chen, Y.-R., Druhan, L.J., Zweier, J.L., 2010. S-glutathionylation uncouples eNOS and regulates cellular and vascular function. Nature 468, 1115–1120.
Chen, A., Kim, S.S., Chung, E., Dietrich, K., 2013. Thyroid hormones in relation to lead, mercury, and cadmium exposure in the national health and nutrition examination survey, 2007-2008. Environ. Health Perspect. 121, 181–186.
Conti, P., Caraffa, A.l., Kritas, S.K., Ronconi, G., Lessiani, G., Toniato, E., Theoharides, T.C., 2017. Mast cell, pro-inflammatory and anti-inflammatory: Jekyll and Hyde, the story continues. J. Biol. Regul. Homeost. Agents 31, 263–267.
Corsetti, G., Romano, C., Stacchiotti, A., Pasini, E., Dioguardi, F.S., 2017. Endoplasmic reticulum stress and apoptosis triggered by sub-chronic lead exposure in mice spleen: a histopathological study. Biol. Trace Elem. Res. 178, 86–97.
Cory-Slechta, D.A., Stern, S., Weston, D., Allen, J.L., Liu, S., 2010. Enhanced learning deficits in female rats following lifetime Pb exposure combined with prenatal stress. Toxicol. Sci. 117, 427–438.
Cory-Slechta, D.A., Merchant-Borna, K., Allen, J.L., Liu, S., Weston, D., Conrad, K., 2013. Variations in the nature of behavioral experience can differentially alter the consequences of developmental exposures to lead, prenatal stress, and the combination. Toxicol. Sci. 131, 194–205.
Crabtree, M.J., Brixey, R., Batchelor, H., Hale, A.B., Channon, K.M., 2013. Integrated redox sensor and effector functions for tetrahydrobiopterin- and glutathionylation-dependent endothelial nitric-oxide synthase uncoupling. J. Biol. Chem. 288, 561–569.
Cunningham, C., Konsman, J.-P., Cartmell, T., 2002. Cytokines and the ageing brain. Trends Neurosci. 25, 546–547.
da Silva, E.Z., Jamur, M.C., Oliver, C., 2014. Mast cell function: a new vision of an old cell. J. Histochem. Cytochem. 62, 698–738.
Dantzer, R., Bluthe, R.-M., Castanon, N., Kelley, K.W., Konsman, J.-P., Laye, S., Lestage, J., Parnet, P., 2007. Cytokines, sickness behavior, and depression. In: Psychoneuroimmunology, fourth ed. vol. 1. Elsevier, Inc., London, pp. 281–318
Dantzer, R., O'Conner, J.C., Freund, G.G., Johnson, R.W., Kelley, K.W., 2008. From inflammation to sickness and depression: when the immune system subjugates the brain. Nat. Rev. Neurol. 9, 46–57.
de Souza, I.D., de Andrade, A.S., Dalmolin, R.J.S., 2018. Lead-interacting proteins and their implication in lead poisoning. Crit. Rev. Toxicol. 48, 375–386.
Demarquoy, J., Le Borgne, F., 2015. Crosstalk between mitochondria and peroxisomes. World J. Biol. Chem. 6, 301–309.
Deng, H., Zheng, W., Jankovic, J., 2015. Genetics and molecular biology of brain calcification. Ageing Res. Rev. 22, 20–38.
Derecki, N.C., Cardani, A.N., Yang, C.H., Quinnies, K.M., Crihfield, A., Lynch, K.R., Kipnis, J., 2010. Regulation of learning and memory by meningeal immunity: a key role for IL-4. J. Exp. Med. 207, 1067–1080.
Dietert, R.R., Piepenbrink, M.S., 2006. Lead and immune function. Crit. Rev. Toxicol. 36, 359–385.

Dietert, R.R., Etzel, R.A., Chen, D., Halonen, M., Holladay, S.D., Jarabek, A.M., Landreth, K., Peden, D., Pinkerton, K., Smialowicz, R.J., Zoetis, T., 2000. Workshop to identify critical windows of exposure for children's health: immune and respiratory systems workgroup summary. Environ. Health Perspect. 108 (S3), 483–490.

Dietert, R.R., Lee, J.-E., Hussain, I., Piepenbrink, M., 2004. Developmental immunotoxicology of lead. Toxicol. Appl. Pharmacol. 198, 86–94.

Dunbar, B., Oktem, F., Arslan, M.K., Delibas, N., Baykal, B., Arslan, C., Gultepe, M., Ilhan, I.E., 2006. The effect of long-term low-dose lead exposure on thyroid function in adolescents. Environ. Res. 101, 140–145.

Dustin, M.L., 2012. Signaling at neuro/immune synapses. J. Clin. Invest. 122, 1149–1155.

Dyatlov, V.A., Lawrence, D.A., 2002. Neonatal lead exposure potentiates sickness behavior induced by *Listeria monocytogenes* infection of mice. Brain Behav. Immun. 16, 477–492.

Dyatlov, V., Platoshin, A.V., Lawrence, D.A., Carpenter, D.O., 1998. Lead potentiates cytokine- and glutamate-mediated increases in permeability of the blood-brain barrier. Neurotoxicology 19, 283–292.

Eikelenboom, E., Veerhuis, R., 1996. The role of complement and activated microglia in the pathogenesis of Alzheimer's disease. Neurobiol. Aging 17, 673–680.

Elbaz, Y., Schuldiner, M., 2011. Staying in touch: the molecular era of organelle contact sites. Trends Biochem. Sci. 36, 616–623.

Faith, R.E., Luster, M.I., Kimmel, C.A., 1979. Effect of chronic developmental lead exposure on cell-mediated immune functions. Clin. Exp. Immunol. 35, 413–420.

Faulk, C., Barks, A., Sanchez, B.N., Zhang, Z., Anderson, O.S., Peterson, K.E., Dolinoy, D.C., 2014. Perinatal lead (Pb) exposure results in sex-specific effects on food intake, fat, weight, and insulin response across the murine life-course. PLoS One 9 (8), e104273.

Filippi-Chiela, E.C., Viegas, M.S., Thomé, M.P., Buffon, A., Wink, M.R., Lenz, G., 2016. Modulation of autophagy by calcium signalosome in human disease. Mol. Pharmacol. 90, 371–384.

Finkelman, F.D., Urban Jr., J.F., Beckmann, M.P., Schooley, K.A., Holmes, J.M., Katona, I.M., 1991. Regulation of murine in vivo IgG and IgE responses by a monoclonal anti-IL-4 receptor antibody. Int. Immunol. 3, 599–607.

Fischbein, A., Tsang, P., Luo, J., Robox, J.P., Jiang, J.D., Bekesi, J.G., 1993. Phenotypic aberrations of CD3+ and CD4+ cells and functional impairments of lymphocytes at low-level occupational exposure to lead. Clin. Immunol. Immunopathol. 66 (2), 163–168.

Flora, G., Gupta, D., Tiwari, A., 2012. Toxicity of lead: a review with recent updates. Interdiscip. Toxicol. 5, 47–58.

Flores-Montoya, M.G., Alvarez, J.M., Sobin, C., 2015. Olfactory recognition memory is disrupted in young mice with chronic low-level lead exposure. Toxicol. Lett. 236, 69–74.

Forsythe, P., 2015. The parasympathetic nervous system as a regulator of mast cell function. Methods Mol. Biol. 1220, 141–154.

Fransen, M., Lismont, C., Walton, P., 2017. The peroxisome-mitochondria connection: how and why? Int. J. Mol. Sci. 18, 1126.

Gainer, J.H., 1974. Lead aggravates viral disease and represses the antiviral activity of interferon inducers. Environ. Health Perspect. 7, 113–119.

Garcia-Bueno, B., Caso, J.R., Leza, J.C., 2008. Stress as a neuroinflammatory condition in brain: damaging and protective mechanisms. Neurosci. Behav. Rev. 32, 1136–1151.

Graham, D.L., Grace, C.E., Braun, A.A., Schaefer, T.L., Skelton, M.R., Tang, P.H., Vorhees, C.V., Williams, M.T., 2011. Effects of developmental stress and lead (Pb) on corticosterone after chronic and acute stress, brain monoamines, and blood Pb levels in rats. Int. J. Dev. Neurosci. 29, 45–55.

Grakoui, A., Bromley, S.K., Sumen, C., Davis, M.M., Shaw, A.S., Allen, P.M., Dustin, M.L., 1999. The immunological synapse: a molecular machine controlling T cell activation. Science 285, 221–227.
Grammas, P., Ovase, R., 2002. Cerebrovascular transforming growth factor-β contributes to inflammation in the Alzheimer's disease brain. Am. J. Pathol. 160, 1583–1587.
Gump, B.B., Stewart, P., Reihman, J., Lonky, E., Darvill, T., Parsons, P.J., Granger, D.A., 2008. Low-level prenatal and postnatal blood lead exposure and adrenocortical responses to acute stress in children. Environ. Health Perspect. 116, 249–255.
Gurish, M.F., Austen, K.F., 2001. The diverse roles of mast cells. J. Exp. Med. 194, F1–F5.
Guyon, A., Massa, F., Rovere, C., Nahon, J.-L., 2008. How cytokines can influence the brain: a role for chemokines? J. Neuroimmunol. 198, 46–55.
Haider, S., Saleem, S., Tabassum, S., Khaliq, S., Shamim, S., Batool, Z., Parveen, T., Inam, Q., Haleem, D.J., 2013. Alteration in plasma corticosterone levels following long term oral administration of lead produces depression like symptoms in rats. Metab. Brain Dis. 28, 85–92.
Hammond, P.B., Succop, P.A., 1995. Effects of supplemental nutrition on lead-induced depression of growth and food consumption in weanling rats. Toxicol. Appl. Pharmacol. 131, 80–84.
Hammond, P.B., Chernausek, S.D., Succop, P.A., Shukla, R., Bornschein, R.L., 1989. Mechanisms by which lead depresses linear and ponderal growth in weanling rats. Toxicol. Appl. Pharmacol. 99, 474–486.
Hemphill, F.E., Kaeberle, M.L., Buck, W.B., 1971. Lead suppression of mouse resistance to *Salmonella typhimurium*. Science 172, 1031–1032.
Heo, Y., Parsons, P.J., Lawrence, D.A., 1996. Lead differentially modifies cytokine production *in vitro* and *in vivo*. Toxicol. Appl. Pharmacol. 138, 149–157.
Heo, Y., Lee, W.T., Lawrence, D.A., 1997. *In vivo* the environmental pollutants lead and mercury induce oligoclonal T cell responses skewed toward type-2 reactivities. Cell. Immunol. 179, 185–195.
Heo, Y., Lee, W.T., Lawrence, D.A., 1998. Differential effects of lead and cAMP on development and activities of Th1- and Th2-lymphocytes. Toxicol. Sci. 43, 172–185.
Heo, Y., Mondal, T.K., Gao, D., Kasten-Jolly, J., Kishikawa, H., Lawrence, D.A., 2007. Posttranscriptional inhibition of interferon-gamma production by lead. Toxicol. Sci. 96, 92–100.
Homma, Y., Cao, S., Shi, X., Ma, X., 2007. The Th2 transcription factor c-Maf inhibits IL-12p35 gene expression in activated macrophages by targeting NF-kappaB nuclear translocation. J. Interferon Cytokine Res. 27, 799–808.
Homo-Delarche, F., Fitzpatrick, F., Christeff, N., Nunez, E.A., Bach, J.F., Dardenne, M., 1991. Sex steroids, glucocorticoids, stress and autoimmunity. J. Steroid Biochem. Mol. Biol. 40, 619–637.
Hong, S., Beja-Glasser, V.F., Nfonoyim, B.M., Frouin, A., Li, S., Ramakrishnan, S., Merry, K.M., Shi, Q., Rosenthal, A., Barres, B.A., Lemere, C.A., Selkoe, D.J., Stevens, B., 2016. Complement and microglia mediate early synapse loss in Alzheimer mouse models. Science 352, 712–716.
Huang, Y., Henry, C.J., Dantzer, R., Johnson, R.W., Godbout, J.P., 2008. Exaggerated sickness behavior and brain proinflammatory cytokine expression in aged mice in response to intracerebroventricular lipopolysaccharide. Neurobiol. Aging 29, 1744–1755.
Hull, M., Fiebich, B.L., Lieb, K., Strauss, S., Berger, M., Volk, B., Bauer, J., 1996. Interleukin-6-associated inflammatory processes in Alzheimer's disease: new therapeutic options. Neurobiol. Aging 17, 795–800.
Hunaiti, A.A., Soud, M., 2000. Effect of lead concentration on the level of glutathione, glutathione S-transferase, reductase and peroxidase in human blood. Sci. Total Environ. 248, 45–50.

Hussain, I., Piepenbrink, M.S., Dietert, R.R., 2005. Impact of in ovo-administered lead and testosterone on developing female thymocytes. J. Toxicol. Environ. Health Part A 68, 1309–1319.

Ip, W.K.E., Hoshi, N., Shouvul, D.S., Snapper, S., Medzhitov, R., 2017. Anti-inflammatory effect of IL-10 mediated by metabolic reprogramming of macrophages. Science 256, 513–519.

Jafarian-Tehrani, M., Sternberg, E.M., 1999. Animal models of neuroimmune interactions in inflammatory diseases. J. Neuroimmunol. 100, 13–20.

Jia, Q., Du, G., Li, Y., Wang, Z., Xie, J., Gu, J., Yin, G., Zhang, S., Gao, Y., Zhou, F., Feng, C., Fan, G., 2018. Pb2+ modulates ryanodine receptors from the endoplasmic reticulum in rat brain. Toxicol. Appl. Pharmacol. 338, 103–111.

Johnson, J.D., Cortez, V., Kennedy, S.L., Foley, T.E., Hanson III, H., Fleshner, M., 2008. Role of central β-adrenergic receptors in regulating proinflammatory cytokine responses to a peripheral bacterial challenge. Brain Behav. Immun. 22, 1078–1086.

Jomova, K., Valko, M., 2011. Advances in metal-induced oxidative stress and human disease. Toxicology 283, 65–87.

Jurdziak, M., Gac, P., Poreba, M., Szymanska-Chabowska, A., Mazur, G., Poreba, R., 2018. Concentrations of thyrotropic hormone in persons occupationally exposed to lead, cadmium, and arsenic. Biol. Trace Elem. Res. 182, 196–203.

Karaca, T., Simşek, N., 2007. Effects of spirulina on the number of ovary mast cells in lead-induced toxicity in rats. Phytother. Res. 21, 44–46.

Karmaus, W., Brooks, K.R., Nebe, T., Witten, J., Osi-Osius, N., Kruse, H., 2005. Immune function biomarkers in children exposed to lead and organochlorine compounds: a cross sectional study. Environ. Health 4, 5.

Kasten-Jolly, J., Lawrence, D.A., 2014. Lead modulation of macrophages causes multiorgan detrimental health effects. J. Biochem. Mol. Toxicol. 28, 355–372.

Kasten-Jolly, J., Lawrence, D.A., 2017. Sex-specific effects of developmental lead exposure on the immune-neuroendocrine network. Toxicol. Appl. Pharmacol. 334, 142–157.

Kasten-Jolly, J., Heo, Y., Lawrence, D.A., 2010. Impact of developmental lead exposure on splenic factors. Toxicol. Appl. Pharmacol. 247, 105–115.

Kasten-Jolly, J., Heo, Y., Lawrence, D.A., 2011. Central nervous system cytokine gene expression: modulation by lead. J. Biochem. Mol. Toxicol. 25, 41–54.

Kasten-Jolly, J., Pabello, N., Bolivar, V.J., Lawrence, D.A., 2012. Developmental lead effects on behavior and brain gene expression in male and female BALB/cAnNTac mice. Neurotoxicology 33, 1005–1020.

Kim, D., Lawrence, D.A., 2000. Immunotoxic effects of inorganic lead on host resistance of mice with different circling behavior preferences. Brain Behav. Immun. 14, 305–317.

Kim, B.-H., Shenoy, A.R., Kumar, P., Das, R., Tiwari, S., MacMicking, J.D., 2011. A family of IFN-γ-inducible 65-kD GTPases protects against bacterial infection. Science 332, 717–721.

Kishikawa, H., Song, R., Lawrence, D.A., 1997. Interleukin-12 promotes enhanced resistance to *Listeria monocytogenes* infection of lead-exposed mice. Toxicol. Appl. Pharmacol. 147, 180–189.

Kopf, M., Le Gros, G., Coyle, A.J., Kosco-Vilbois, M., Brombacher, F., 1995. Immune responses of IL-4, IL-5, and IL-6 deficient mice. Immunol. Rev. 148, 45–69.

Kuehn, S., Reinehr, S., Stute, G., Rodust, C., Grotegut, P., Hensel, A.T., Dick, H.B., Joachim, S.C., 2018. Interaction of complement system and microglia activation in retina and optic nerve in a NMDA damage model. Mol. Cell. Neurosci. 89, 95–106.

Kumawat, K.L., Kaushik, D.K., Goswami, P., Basu, A., 2014. Acute exposure to lead acetate activates microglia and induces subsequent bystander neuronal death *via* caspase-3 activation. Neurotoxicology 41, 143–153.

Lalancette-Hebert, M., Gowing, G., Simard, A., Weng, Y.C., Kriz, J., 2007. Selective ablation of proliferating microglial cells exacerbates ischemic injury in the brain. J. Neurosci. 27, 2596–2605.
Lanphear, B.P., Dietrich, K., Auinger, P., Cox, C., 2000. Cognitive deficits associated with blood lead concentrations <10 μg/dL in US children and adolescents. Public Health Rep. 115, 521–529.
Laschi-Loquerie, A., Decotes, J., Tachon, P., Evreux, J.C., 1984. Influence of lead acetate on hypersensitivity. Experimental study. J. Immunopharmacol. 6, 87–93.
Lawrence, D.A., 1985. Immunotoxicity of heavy metals. In: Dean, J.H., Luster, M.J., Munson, R. et al., (Eds.), Immunotoxicology and Immunopharmacology. Raven Press, New York, pp. 341–350.
Lawrence, D.A., Harry, G.J., 2000. Environmental stressors and neuroimmunotoxicological processes. Brain Behav. Immun. 14, 231–238.
Lawrence, D.A., McCabe Jr., M.J., 1995. Immune modulation by toxic metals. In: Goyer, R.A., Klaassen, C.D., Waalkes, M.P. (Eds.), Metal Toxicology. Academic Press, San Diego.
Lawrence, D.A., McCabe Jr., M.J., 2002. Immunomodulation by metals. Int. Immunopharmacol. 2, 293–302.
Lawrence, T., Natoli, G., 2011. Transcriptional regulation of macrophage polarization enabling diversity with identity. Nat. Rev. Immunol. 11, 750–761.
Ledo, J.H., Azevedo, E.P., Beckman, D., Ribeiro, F.C., Santos, L.E., Razolli, D.S., Kincheski, G.C., Melo, H.M., Bellio, M., Teixeira, A.L., Velloso, L.A., Foguel, D., De Felice, F.G., Ferreira, S.T., 2016. Cross talk between brain innate immunity and serotonin signaling underlies depressive-like behavior induced by Alzheimer's amyloid-β oligomers in mice. J. Neurosci. 36, 12106–12116.
Lee, J.-E., Dietert, R.R., 2003. Developmental immunotoxicity of lead: impact on thymic function. Birth Defects Res. Part A 67, 861–867.
Lee, J.-E., Chen, S., Golemboski, K.A., Parsons, P.J., Dietert, R.R., 2001. Developmental windows of differential lead-induced immunotoxicity in chickens. Toxicology 156, 161–170.
Lenz, K.M., Nugent, B.M., Haliyur, R., McCarthy, M.M., 2013. Microglia are essential to masculinization of brain and behavior. J. Neurosci. 33, 2761–2772.
Liu, M.-C., Liu, X.-Q., Wang, W., Shen, X.-F., Che, H.-L., Guo, Y.-Y., Zhao, M.-G., Chen, J.-Y., Luo, W.-J., 2012. Involvement of microglia activation in the lead induced long-term potentiation impairment. PLoS One 7 (8), e43924.
Lodhi, I.J., Semenkovich, C.F., 2014. Peroxisomes: a nexus for lipid metabolism and cellular signaling. Cell Metab. 19, 380–392.
Lu, X., Jin, C., Yang, J., Liu, Q., Wu, S., Li, D., Guan, Y., Cai, Y., 2013. Prenatal and lactational lead exposure enhanced oxidative stress and altered apoptosis status in offspring rats' hippocampus. Biol. Trace Elem. Res. 151, 75–84.
MacKenzie, K.F., Clark, K., Naqvi, S., McGuire, V.A., Nöehren, G., Kristariyanto, Y., van den Bosch, M., Mudaliar, M., McCarthy, P.C., Pattison, M.J., Pedrioli, P.G., Barton, G.J., Toth, R., Prescott, A., Arthur, J.S., 2013. PGE(2) induces macrophage IL-10 production and a regulatory-like phenotype via a protein kinase A-SIK-CRTC3 pathway. J. Immunol. 190, 565–577.
MacMicking, J.D., 2004. IFN-inducible GTPases and immunity to intracellular pathogens. Trends Immunol. 25, 601–609.
Maiti, A.K., Saha, N.C., More, S.S., Panigrahi, A.K., Paul, G., 2017. Neuroprotective efficacy of mitochondrial antioxidant MitoQ in suppressing peroxynitrite-mediated mitochondrial dysfunction inflicted by lead toxicity in the rat brain. Neurotox. Res. 31, 358–372.

Mak, T.W., Grusdat, M., Duncan, G.S., Dostert, C., Nonnenmacher, Y., Cox, M., Binsfeld, C., Hao, Z., Brustle, A., Itsumi, M., Jager, C., Chen, Y., Pinkenburg, O., Camara, B., Ollert, M., Bindslev-Jensen, C., Vasiliou, V., Gorrini, C., Lang, P.A., Lohoff, M., Harris, I.S., Hiller, K., Brenner, D., 2017. Glutathione primes T cell metabolism for inflammation. Immunity 46, 675–689.

Makwana, M., Jones, L.L., Cuthill, D., Heuer, H., Bohatschek, M., Hristova, M., Friedrichsen, S., Ormsby, I., Bueringer, D., Koppius, A., Bauer, K., Doetschman, T., Raivich, G., 2007. Endogenous transforming growth factor β1 suppresses inflammation and promotes survival in adult CNS. J. Neurosci. 27, 11201–11213.

Manley, K., Han, W., Zelin, G., Lawrence, D.A., 2018. Crosstalk between the immune, endocrine, and nervous systems in immunotoxicology. Curr. Opin. Toxicol. 10, 37–45.

Martens, S., Howard, J., 2006. The interferon-inducible GTPases. Annu. Rev. Cell Dev. Biol. 22, 559–589.

Maslinska, D., Laure-Kamionowska, M., Maslinski, K.T., Gujski, M., Maslinski, S., 2007. Distribution of tryptase-containing mast cells and metallothionein reactive astrocytes in human brains with amyloid deposits. Inflamm. Res. 56, S17–S18.

McCabe Jr., M.J., Lawrence, D.A., 1990. The heavy metal lead exhibits B cell stimulatory factor by enhancing B cell Ia expression and differentiation. J. Immunol. 145, 671–677.

McCabe Jr., M.J., Lawrence, D.A., 1991. Lead, a major environmental pollutant, is immunomodulatory by its differential effects on $CD4^+$ T cell subsets. Toxicol. Appl. Pharmacol. 111, 13–23.

McCabe Jr., M.J., Singh, K.P., Reiners Jr., J.J., 1999. Lead intoxication impairs the generation of a delayed type hypersensitivity response. Toxicology 139, 255–264.

McCarthy, M.M., 2016. Sex differences in the developing brain as a source of inherent risk. Dialoques Clin. Neurosci. 18, 361–372.

McCarthy, M.M., Schwarz, J.M., Wright, C.L., Dean, S.L., 2008. Mechanisms mediating oestradiol modulation of the developing brain. J. Neuroendocrinol. 20, 777–783.

Meeker, J.D., Rossano, M.G., Protas, B., Diamond, M.P., Puscheck, E., Daly, D., Paneth, N., Wirth, J.J., 2009. Multiple metals predict prolactin and thyrotropin (TSH) levels in men. Environ. Res. 109, 869–873.

Meng, H., Wang, L., He, J., Wang, Z., 2016. The protective effect of gangliosides on lead (Pb)-induced neurotoxicity is mediated by autophagic pathways. Int. J. Environ. Res. Public Health 13, 365–376.

Miao, W., Zhu, B., Xiao, X., Li, Y., Dirbaba, N.B., Zhou, B., Wu, H., 2015. Effects of titanium dioxide nanoparticles on lead bioconcentration and toxicity on thyroid endocrine system and neural development in zebrafish larvae. Aquat. Toxicol. 161, 117–126.

Michelucci, A., Heurtaux, T., Grandbarbe, L., Morga, E., Heuschling, P., 2009. Characterization of the microglial phenotype under specific pro-inflammatory and anti-inflammatory conditions: effects of oligomeric and fibrillar amyloid-beta. J. Neuroimmunol. 210, 3–12.

Miller, T.E., Golemboski, K.A., Ha, R.S., Bunn, T., Sanders, F.S., Dietert, R.R., 1998. Developmental exposure to lead causes persistent immunotoxicity in Fischer 344 rats. Toxicol. Sci. 42, 129–135.

Miller, A.H., Maletic, V., Raison, C.L., 2009. Inflammation and its discontents: the role of cytokines in the pathophysiology of major depression. Biol. Psychiatry 65, 732–741.

Mondal, T.K., Emeny, R.T., Gao, D., Ault, J.G., Kasten-Jolly, J., Lawrence, D.A., 2015. A physical/psychological and biological stress combine to enhance endoplasmic reticulum stress. Toxicol. Appl. Pharmacol. 289, 313–322.

Mortensen, A., Lykkesfeldt, J., 2014. Does vitamin C enhance nitric oxide bioavailability in a tetrahydrobiopterin-dependent manner? In vitro, in vivo and clinical studies. Nitric Oxide 36, 51–57.

Moskot, M., Montefusco, S., Jakóbkiewicz-Banecka, J., Mozolewski, P., Węgrzyn, A., Di Bernardo, D., Węgrzyn, G., Medina, D.L., Ballabio, A., Gabig-Cimińska, M., 2014. The phytoestrogen genistein modulates lysosomal metabolism and transcription factor EB (TFEB) activation. J. Biol. Chem. 289, 17054–17069.

Mostafalou, S., Baeeri, M., Bahadar, H., Soltany-Rezaee-Rad, M., Gholami, M., Abdollahi, M., 2015. Molecular mechanisms involved in lead-induced disruption of hepatic and pancreatic glucose metabolism. Environ. Toxicol. Pharmacol. 19, 16–26.

Muller, S., Gilbert, K.E., Krause, C., Gross, U., Age-Stehr, J., Diamantstein, T., 1977. Suppression of delayed type hypersensitivity of mice by lead. Experientia 33, 667–668.

Naito, Y., Takagi, T., Higashimura, Y., 2014. Heme oxygenase-1 and anti-inflammatory M2 macrophages. Arch. Biochem. Biophys. 564, 83–88.

Nava-Ruiz, C., Mendez-Armenta, M., Rios, C., 2012. Lead neurotoxicity: effects on brain nitric oxide synthase. J. Mol. Histol. 43, 553–563.

Needleman, H., 2009. Low level lead exposure: history and discovery. Ann. Epidemiol. 19, 235–238.

Nie, X., Chen, Y., Chen, Y., Chen, C., Han, B., Li, Q., Zhu, C., Xia, F., Zhai, H., Wang, N., Lu, Y., 2017. Lead and cadmium exposure, higher thyroid antibodies and thyroid dysfunction in Chinese women. Environ. Pollut. 230, 320–328.

Niederhoffer, N., Levy, R., Sick, E., Andre, P., Coupin, G., Lombard, Y., Gies, J.P., 2009. Amyloid beta peptides trigger CD47-dependent mast cell secretory and phagocytic responses. Int. J. Immunopathol. Pharmacol. 22, 473–483.

Olmo, R., Teijon, C., Muniz, E., Beneit, J.V., Villarino, A.L., Blanco, M.D., 2012. Modulation of lysozyme activity by lead administered by different routes. *In vitro* study and analysis in blood, kidney, and lung. Biol. Trace Elem. Res. 149, 405–411.

Pabello, N.G., Lawrence, D.A., 2006. Neuroimmunotoxicology: modulation of neuroimmune networks by toxicants. Clin. Neurosci. Res. 6, 69–85.

Pinkerton, L.E., Biagini, R.E., Ward, E.M., Hull, R.D., Deddens, J.A., Boeniger, M.F., Schnorr, T.M., MacKenzie, B.A., Luster, M.I., 1998. Immunologic findings among lead-exposed workers. Am. J. Ind. Med. 33, 400–408.

Planchart, A., Green, A., Hoyo, C., Mattingly, C.J., 2018. Heavy metal exposure and metabolic syndrome: evidence from human and model system studies. Curr. Environ. Health Rep. 5, 110–124.

Pollmächer, T., Haack, M., Schuld, A., Reichenberg, A., Yirmiya, R., 2002. Low levels of circulating inflammatory cytokines—do they affect human brain functions? Brain Behav. Immun. 16, 525–532.

Porta, C., Riboldi, E., Ippolito, A., Sica, A., 2015. Molecular and epigenetic basis of macrophage polarized activation. Semin. Immunol. 2, 237–248.

Presumey, J., Bialas, A.R., Carroll, M.C., 2017. Complement system in neural synapse elimination in development and disease. Adv. Immunol. 135, 53–79.

Quan, N., Mhlanga, J.D.M., Whiteside, M.B., McCoy, A.N., Kristensson, K., Herkenham, M., 1999a. Chronic overexpression of proinflammatory cytokines and histopathology in the brains of rats infected with *Trypanosoma brucei*. J. Comp. Neurol. 414, 114–130.

Quan, N., Stern, E.L., Whiteside, M.B., Herkenham, M., 1999b. Induction of proinflammatory cytokine mRNAs in the brain after peripheral injection of subseptic doses of lipopolysaccharide in the rat. J. Neuroimmunol. 93, 72–80.

Raffaello, A., Mammucar, C., Gherardi, G., Rizzuto, R., 2016. Calcium at the center of vell signaling: interplay between endoplasmic reticulum, mitochondria, and lysosomes. Trends Biochem. Sci. 41, 1035–1049.

Rossi-George, A., Virgolini, M.B., Weston, D., Cory-Slechta, D.A., 2009. Alterations in glucocorticoid negative feedback following maternal Pb, prenatal stress and the

combination: a potential biological unifying mechanism for their corresponding disease profiles. Toxicol. Appl. Pharmacol. 234, 117–127.
Rossi-George, A., Virgolini, M.B., Weston, D., Thiruchelvam, M., Cory-Slechta, D.A., 2011. Interactions of lifetime lead exposure and stress: behavioral, neurochemical and HPA axis effects. Neurotoxicology 32, 83–99.
Sanders, T., Liu, Y., Buchner, V., Tchounwou, P.B., 2009. Neurotoxic effects and biomarkers of lead exposure: a review. Rev. Environ. Health 24, 15–45.
Santos, L.E., Beckman, D., Ferreira, S.T., 2016. Microglial dysfunction connects depression and Alzheimer's disease. Brain Behav. Immun. 55, 151–165.
Sarasua, S.M., Vogt, R.F., Henderson, L.O., Jones, P.A., Lybarger, J.A., 2000. Serum immunoglobulins and lymphocyte subset distribution in children and adults living in communities assessed for lead and cadmium exposure. J. Toxicol. Environ. Health A 60, 1–15.
Sata, F., Araki, S., Tanigawa, T., Morita, Y., Sakurai, S., Nakata, A., Katsuno, N., 1998. Changes in T-cells subpopulations in lead workers. Environ. Res. 76, 61–64.
Schrader, M., Costello, J., Godinho, L.F., Islinger, M., 2015. Peroxisome-mitochondria interplay and disease. J. Inherit. Metab. Dis. 38, 681–702.
Selye, H., Tuchweber, B., Gabbiani, G., 1964. Further studies on anacalciphylaxis. J. Amer. Geriatrics Soc. 12 (3), 207–214.
Seo, J., Lee, B.-K., Jin, S.-U., Jang, K.E., Park, J.W., Kim, Y.-T., Park, S.-J., Jeong, K.S., Park, J., Kim, A., Kim, Y., Chang, Y., 2015. Altered executive function in the lead-exposed brain: a functional magnetic resonance imaging study. Neurotoxicity 50, 1–9.
Shinkai, Y., Kaji, T., 2012. Cellular defense mechanisms against lead toxicity in the vascular system. Biol. Pharm. Bull. 35, 1885–1891.
Sierra-Filardi, E., Vega, M.A., Sanchez-Mateos, P., Corbi, A.L., Puig-Kroger, A., 2010. Heme oxygenase-1 expression in M-CSF-polarized M2 macrophages contribute to LPS-induced IL-10 release. Immunobiology 215, 788–795.
Simons, T.J., 1993. Lead-calcium interactions in cellular lead toxicity. Neurotoxicology 14, 77–85.
Singh, B., Chandran, V., Bandhu, H.K., Mittal, B.R., Bhattacharya, A., Jindal, S., Varma, S., 2000. Impact of lead exposure on pituitary-thyroid axis in humans. Biometals 13, 187–192.
Skaper, S.D., Facci, L., Giusti, P., 2014. Mast cells, glia and neuroinflammation: partners in crime? Immunology 141, 314–327.
Smith, S.E.P., Li, J., Garbett, K., Mirnics, K., Patterson, P.H., 2007. Maternal immune activation alters fetal brain development through interleukin-6. J. Neurosci. 27, 10695–10702.
Snyder, J.E., Filipov, N.M., Parsons, P.J., Lawrence, D.A., 2000. The efficiency of maternal transfer of lead and its influence on plasma IgE and splenic cellularity of mice. Toxicol. Sci. 57, 87–94.
Sobin, C., Montoya, M.G.F., Parisi, N., Schaub, T., Cervantes, M., Armijos, R.X., 2013. Microglial disruption in young mice with early chronic lead exposure. Toxicol. Lett. 220, 44–52.
Sobolewski, M., Conrad, K., Marvin, E., Allen, J.L., Cory-Slechta, D.A., 2018. Endocrine active metals, prenatal stress and enhanced neurobehavioral disruption. Horm. Behav. 101, 36–49.
Soltaninejad, K., Kebriaeezadeh, A., Miniaiee, B., Ostad, S.N., Hosseini, R., Azizi, E., Abdollahi, M., 2003. Biochemical and ultrastructural evidences for toxicity of lead through free radicals in rat brain. Hum. Exp. Toxicol. 22, 417–423.
Song, X.-B., Liu, G., Liu, F., Yan, Z.-G., Wang, Z.-Y., Liu, Z.-P., Wang, L., 2017. Autophagy blockade and lysosomal membrane permeabilization contribute to lead-induced nephrotoxicity in primary rat proximal tubular cells. Cell Death Dis. 8, e2863.

Souza-Talarico, J., Suchecki, D., Juster, R.-P., Plusquellec, P., Barbosa Jr., F., Bunscheit, V., Marcourakis, T., de Matos, T.M., Lupien, S.J., 2017. Lead exposure is related to hypercortisolemic profiles and allostatic load in Brazilian older adults. Environ. Res. 154, 261–268.
Spector, N.H., 1996. Neuroimmunomodulation: a brief review. Regul. Toxicol. Pharmacol. 24, S32–S38.
Steffensen, S.C., Campbell, I.L., Henriksen, S.J., 1994. Site-specific hippocampal pathophysiology due to cerebral overexpression of interleukin-6 in transgenic mice. Brain Res. 652, 149–153.
Steinman, L., 2004. Elaborate interactions between the immune and nervous systems. Nat. Immunol. 5 (6), 575–581.
Sternberg, E.M., 1997. Emotions and disease from balance of humors to balance of molecules. Nat. Med. 3, 264–267.
Struzynska, L., 2009. A glutamatergic component of lead toxicity in adult brain: the role of astrocytic glutamate transporters. Neurochem. Int. 55, 151–156.
Su, P., Zhang, J., Wang, S., Aschner, M., Cao, Z., Zhao, F., Wang, D., Chen, J., Luo, W., 2016. Genistein alleviates lead-induced neurotoxicity in vitro and in vivo: involvement of multiple signaling pathways. Neurotoxicology 53, 153–164.
Sun, H., Wang, H., Nie, X., Zhao, L., Li, Q., Cang, Z., Chen, C., Lu, M., Cheng, J., Zhai, H., Xia, F., Ye, L., Lu, Y., 2017. Lead exposure induces weight gain in adult rats, accompanied by DNA hypermethylation. PLoS One 12 (1), e0169958.
Taylor, G.A., Feng, C.G., Sher, A., 2004. P47 GTPases: regulators of immunity to intracellular pathogens. Nat. Rev. Immunol. 4, 100–109.
Tian, L., Lawrence, D.A., 1995. Lead inhibits nitric oxide production *in vitro* by murine splenic macrophages. Toxicol. Appl. Pharmacol. 132, 156–163.
Tian, L., Noelle, R.J., Lawrence, D.A., 1995. Activated T cells enhance nitric oxide production by murine splenic macrophages through gp39 and LFA-1. Eur. J. Immunol. 25, 306–309.
Tokoyoda, K., Tsujikawa, K., Matsushita, H., Ono, Y., Hayashi, T., Harada, Y., Abe, R., Kubo, M., Yamamoto, H., 2004. Up-regulation of IL-4 production by the activated cAMP/cAMP-dependent protein kinase (protein kinase A) pathway in CD3/CD28-stimulated naïve T cells. Int. Immunol. 16, 643–653.
Traina, G., 2017. Mast cells in the brain—old cells new target. J. Integr. Neurosci. 16, S69–S83.
Tsuda, Y., Takahashi, H., Kobayashi, M., Hanafusa, T., David, N., Herndon, D.N., Suzuki, F., 2004. CCL2, a product of mice early after systemic inflammatory response syndrome (SIRS), induces alternatively activated macrophages capable of impairing antibacterial resistance of SIRS mice. J. Leukoc. Biol. 76, 368–373.
Tyrrell, J.B., Hafida, S., Stemmer, P., Adhami, A., Leff, T., 2017. Lead (Pb) exposure promotes diabetes in obese rodents. J. Trace Elem. Med. Biol. 10, 221–226.
Valentino, M., Rapisarda, V., Santarelli, L., Bracci, M., Scorcelletti, M., Di Lorenzo, L., Cassano, F., Soleo, L., 2007. Effect of lead on the levels of some immunoregulatory cytokines in occupationally exposed workers. Hum. Exp. Toxicol. 26, 551–556.
Valle, B.L., Ulmer, D.D., 1972. Biochemical effects of mercury, cadmium, and lead. Annu. Rev. Biochem. 41, 91–128.
VanRyzin, J.W., Pickett, L.A., McCarthy, M.M., 2018. Microglia: driving critical periods and sexual differentiation of the brain. Dev. Neurobiol. 78, 580–592.
Weston, H.I., Weston, D.D., Allen, J.L., Cory-Slechta, D.A., 2014. Sex-dependent impacts of low-level lead exposure and prenatal stress on impulsive choice behavior and associated biochemical and neurochemical manifestations. Neurotoxicology 44, 169–183.
Whitacre, C.C., 2001. Sex differences in autoimmune disease. Nat. Immunol. 2, 777–780.

White, L.D., Cory-Slechta, D.A., Gilbert, M.E., Tiffany-Castiglioni, E., Zawia, N.H., Virgolini, M., Rossi-George, A., Lasley, S.M., Qian, Y.C., Basha, M.R., 2007. New and evolving concepts in the neurotoxicology of lead. Toxicol. Appl. Pharmacol. 225, 1–27.
Wu, C.-Y., Liu, B., Wang, H.-L., Ruan, D.-Y., 2011. Levothyroxine rescues the lead-induced hypothyroidism and impairment of long-term potentiation in hippocampal CAI region of the developmental rats. Toxicol. Appl. Pharmacol. 256, 191–197.
Zhang, J., Cai, t., Zhao, F., Yao, T., Chen, Y., Liu, X., Luo, W., Chen, J., 2012. The role of α-synuclein and tau hyperphosphorylation-mediated autophagy and apoptosis in lead-induced learning and memory injury. Int. J. Biol. Sci. 8, 935–944.
Zhao, Z.-Y., Rong, L., Sun, L., Li, Z.-Y., Yang, R.-L., 2004. Effect of lead exposure on the immune function of lymphocytes and erythrocytes in preschool children. J. Zhejiang Univ. Sci. 5, 1001–1004.
Ziv, Y., Ron, N., Butovsky, O., Landa, G., Sudai, E., Greenberg, N., Cohen, H., Kipnis, J., Schwartz, M., 2006. Immune cells contribute to the maintenance of neurogenesis and spatial learning abilities in adulthood. Nat. Neurosci. 9, 268–275.

Further reading

Rezin, G.T., Amboni, G., Zugno, A.I., Quevedo, J., Streck, E.L., 2009. Mitochondrial dysfunction and psychiatric disorders. Neurochem. Res. 34, 1021–1029.

CHAPTER FOUR

The role of manganese in neuroinflammation

Souvarish Sarkar[†], Emir Malovic, Huajun Jin, Arthi Kanthasamy, Anumantha G. Kanthasamy*

Parkinson Disorders Research Laboratory, Department of Biomedical Sciences, Iowa State University, Iowa Center for Advanced Neurotoxicology, Ames, IA, United States
*Corresponding author: e-mail address: akanthas@iastate.edu

Contents

1. Introduction

Manganese (Mn) is a vital trace element necessary for physiological development and proper functioning of immune, metabolic, and reproductive systems. Mn serves as a cofactor for a multitude of enzyme functions, including superoxide dismutase's antioxidant functions, and glutamine synthetase's conversion of glutamate into glutamine during astrocytic reuptake. Systemic Mn homeostasis can be achieved by consuming diets rich in grains, nuts, legumes, and certain fruits (Horning et al., 2015; Kwakye et al., 2015; Peres et al., 2016), from which less than 5% of the consumed Mn is generally absorbed, suggesting a low dietary risk for Mn toxicity (Karki et al., 2013). However, aerosolized Mn is a toxicological concern stemming from occupational exposure (Horning et al., 2015; Peres et al., 2016; Rodier, 1955). Aerial overexposure to Mn was initially observed in miner populations (Archibald and Tyree, 1987; Rodier, 1955)

[†] Present affiliation: Brigham and Women's Hospital, Harvard Medical School, Department of Pathology, Boston, MA 02115, United States.

Advances in Neurotoxicology, Volume 3
ISSN 2468-7480
https://doi.org/10.1016/bs.ant.2018.10.005

and later supported by observations in other industrial occupations that included welding (Fored et al., 2006; Sadek et al., 2003) and ferroalloy smelting (Bast-Pettersen et al., 2004; Lucchini et al., 1995). Since Mn can exist in 11 different oxidation states, with Mn^{+2} and Mn^{+3} in largest abundance, it can be contained in various aerosolized dusts and particulates (Horning et al., 2015; Kwakye et al., 2015; Ngwa et al., 2017). Such Mn dust particles have been assessed for inhalation risk, with inhalable fractions reaching 301 $\mu g/m^3$, while the respirable fractions that penetrate deeper were 36 $\mu g/m^3$. The respirable fraction is a more conservative estimate of Mn absorption that takes into account the different Mn species whose solubility is affected by their oxidation states and compositions (Ellingsen et al., 2003). Chronic overexposure to Mn, as exhibited by certain welders and Mn-ore miners, can cause a Parkinsonian-like syndrome referred to as manganism, which exhibits clinical features similar to Parkinson's disease patients. Patients suffering from manganism can develop gait impairment (cock-walking), postural instability, dystonia, and neurological signs of hallucinations and psychoses. Animal and human studies reveal that Mn targets the basal ganglia, particularly the globus pallidus (Dobson et al., 2004; Guilarte, 2010, 2013; Huang, 2007; Lee et al., 2018; Lewis et al., 2016; Milatovic et al., 2007; Moberly et al., 2012; Nelson et al., 1993; Racette et al., 2001, 2012). Despite the prevalence and potential risk to human health, the mechanisms by which Mn exerts its neurotoxic effects are not well understood, however, growing evidence suggests oxidative damage, mitochondrial dysfunction, and neuroinflammation contribute to Mn-mediated neuronal cell death.

Although Mn is directly toxic to neurons, impairing their mitochondria and releasing cytochrome *c*, which leads to caspase activation (Latchoumycandane et al., 2005), Mn can also activate astrocytes and microglia (Aschner et al., 2009; Kirkley et al., 2017; Moreno et al., 2009; Sarkar et al., 2017a; Sidoryk-Wegrzynowicz et al., 2009; Yin et al., 2008). Despite the many homeostatic functions attributed to both astrocytes and microglia, under insults they can exhibit immune functions by releasing many inflammatory cytokines and chemokines. When the insult is a chronic condition like occupational exposure to Mn, these glial factors can contribute to a perpetual inflammatory state of glial activation that becomes highly toxic to neurons (Liddelow and Barres, 2017). Inhibiting the activation of these glial cells through various mechanisms can significantly reduce Mn-induced neuronal death (Streifel et al., 2012; Zhao et al., 2009).

In this chapter, we review Mn-mediated mechanisms in both astrocytes and microglia, delineating the important findings, and highlighting potential avenues for therapeutic intervention.

2. Role of manganese in regulating astrocyte function

Mn-mediated neurotoxicity can occur directly through mitochondrial impairment. Astrocytes are the cell type most affected by excessive Mn accumulation since astrocytes preferentially uptake Mn with an affinity 50 times greater than that of neurons (Aschner et al., 1992; Gonzalez et al., 2008). The strong uptake preference astrocytes exhibit at least partly reflects their localization to the vasculature where they act as a blood-brain barrier constituent, allowing them to regulate CNS homeostasis (Pekny and Pekna, 2014). Furthermore, the Mn-targeted basal ganglia structures demonstrate increased surface expression of the divalent metal ion transporter 1 (DMT1), which acts as one of the major cellular Mn transporters (Huang et al., 2004; Sidoryk-Wegrzynowicz and Aschner, 2013). Thus, astrocytes act as buffering cells to divert toxic Mn away from the neurons to slow the degeneration. On the other hand, after prolonged exposure, astrocytes preferentially sequester the Mn in mitochondria (Gunter et al., 2006), leading to dynamic and damaging metabolic changes. Mitochondria sequester Mn with the Ca^{+2} uniporter, but the inherently poor efflux kinetics of Mn is hypothesized to underlie the increase in reactive oxygen species (ROS) observed in Mn-treated astrocytes (Gunter et al., 1975; Gunter and Sheu, 2009; Martinez-Finley et al., 2013). Increasing the dose and/or exposure period of Mn treatments elicits more ROS and coincides with an induction of proinflammatory factors (Filipov et al., 2005; Park and Park, 2010).

The downstream changes in astrocytic homeostasis following Mn treatment have revealed plausible mechanisms of Mn-mediated neuronal toxicity (Karki et al., 2014; Streifel et al., 2012), further supporting that astrocytes indeed are the initial targets of Mn (Henriksson and Tjalve, 2000). In vitro experimentation with Mn-treated astrocytes shows upregulation of various signaling pathways involved in apoptosis, proliferation, and inflammation (Cordova et al., 2012; Tjalkens et al., 2017). Nuclear factor kappa B (NF-κB) activation by Mn is critical for the induction of the *Nos2* gene that codes for inducible nitric oxide synthase (iNOS) (Filipov et al., 2005). Streifel et al. (2012) showed that $Nos2^{-/-}$ mice are resistant to the neurotoxic effects of oral Mn administration. Additionally, co-culturing Mn-treated wild-type primary astrocytes with striatal neurons can induce neuronal apoptosis, whereas Mn-treated *Nos2*$^{-/-}$ astrocytes do not. Such evidence suggests that both inflammatory cascades and oxidative stress lead to Mn-induced neuronal toxicity. Our recent study also shows crosstalk between neuroinflammation and oxidative stress, where inhibiting the Mn-induced production of mitochondrially-generated ROS decreases the expression of proinflammatory genes (Sarkar et al., 2017a). Our findings support previous in vivo results by Cordova et al. (2012), where the antioxidant Trolox™ reduced oxidative stress and concomitantly decreased the phosphorylation of ERK1/2, a MAPK kinase implicated in regulating proinflammatory cytokines such as IL-6 and TNFα (Goral and Kovacs, 2005). Although activation of MAPK signaling is extremely rapid and most likely precedes ROS generation, the continual production of ROS, especially mitochondrially-generated ROS, may be the factor sustaining inflammation. Therefore, utilizing antioxidants to target mitochondrially-generated ROS might therapeutically decrease neuronal apoptosis, astrogliosis, and microgliosis (Ghosh et al., 2016).

3. Role of manganese in regulating microglial biology

In addition to astrocytes, Mn toxicity has been well documented in microglial biology (Dodd and Filipov, 2011; Filipov et al., 2005; Hazell et al., 1999; Milatovic et al., 2007; Sarkar et al., 2017a; Sidoryk-Wegrzynowicz et al., 2009). More specifically, Zhao et al. (2009) revealed that Mn induces neurotoxicity in a rat model of Mn neurotoxicity by modulating microgliosis. They observed activation of microglia in the SN region of the brain as soon as 1 day after Mn administration, while dopaminergic neuronal loss accompanied by significant microgliosis in the SNpc region lasted 7 days post-Mn administration. This neurodegeneration was further accompanied by induction of proinflammatory factors, including but not limited to TNFα, IL-β, and iNOS. Strikingly, treatment with minocycline, an inhibitor of microgliosis, reduced Mn-induced production of proinflammatory factors as well as the loss of TH-positive neurons in this rat model (Zhao et al., 2009). A recent study by Kirkley et al. (2017) further shows that microglia can amplify Mn-induced astrocytic inflammation. They show that microglia conditioned medium (MCM) drastically enhanced Mn-stimulated inflammatory factor production in astrocytes. Mn exposure studies in non-human primates show that Mn induces microgliosis as well as iNOS production (Verina et al., 2011). Moreover, in South African mine workers, the microglial density in the globus pallidus correlated with the number of years of mining, providing further evidence for the effect of Mn on microgliosis in a human population (Gonzalez-Cuyar et al., 2014).

Mn exposure can induce classical activation of microglial cells as shown in Chang and Liu (1999) in the CNS, leading to the production of pro-inflammatory cytokines. Chang and Liu demonstrated that Mn treatment potentiated iNOS production in LPS-activated N9 microglial cells. Interestingly, this effect might be limited to Mn, as other metals, like iron, cobalt, copper, nickel, and zinc, failed to do so. The authors argued that the sustained production of neurotoxic factors by microglial cells can be detrimental to the surrounding neurons. This speculation was elucidated by Zhang et al. (2009), who demonstrated that Mn-induced production of superoxide in microglia drives the neurotoxicity in surrounding neurons. Dodd and Filipov (2011) later revealed that free radical scavengers could reverse Mn's neurotoxicity. Furthermore, the same group reported that Mn-induced ROS generation plays a crucial role in potentiating LPS-induced production of proinflammatory factors in microglial cells. They also identified that the enzyme heme-oxygenase 1 specifically regulates the release of Mn-potentiated proinflammatory factors in microglial cells but not in cells of neuronal origin (Dodd and Filipov, 2011).

Various signaling pathways have been implicated in Mn-induced microglial activation and its potentiation of proinflammatory factors. Crittenden and Filipov (2011) demonstrate that Mn-induced microglial activation is associated with the persistent activation of canonical MAPK cascades in microglial cells, indicating a chronic inflammatory response to Mn. The same group suggests that the potentiation effect of Mn can be attributed to sustained activation of the p38 MAPK pathway (Crittenden and Filipov, 2008). Similarly, Filipov et al. (2005) found that the Mn-induced potentiation of proinflammatory factors was mediated by activation of the NF-κB pathway, which can be MAPK inducible given the considerable crosstalk known to occur between the p38 MAPK and NF-κB pathways (Baeza-Raja and Munoz-Canoves, 2004). Moreno et al. (2011) report that supplementing juvenile male mice with estrogen can reduce Mn-induced NF-κB activation, microgliosis, and the inflammatory response.

Recent studies by our group and others have elucidated the biochemical crosstalk between mitochondrial dysfunction and neuroinflammation (Lawana et al., 2017; Sarkar et al., 2017a, b), including in a mitochondrially defective transgenic mouse model of PD, which we used to show that Mn potentiates microglia activation (Langley et al., 2017). LRRK2, one of the major genes associated with familial PD, also plays a role in Mn-induced neuroinflammation in microglial cells and autophagy as demonstrated by Chen et al.'s (2018) knockdown of LRRK2 in microglia, which decreased Mn-induced production of proinflammatory factors and modulated autophagy. These studies together demonstrate a gene-environment crosstalk regulating microglial inflammation in disease pathology. Mn-treated microglial cells can also exhibit changes in autophagy, lysosomal dysfunction, and NLRP3 inflammasome activation mediated by cathepsin B, a lysosomal cysteine protease (Wang et al., 2017). Collectively, the current evidence strongly implicates Mn in microglial activation and the production of proinflammatory factors. The mechanism of this activation is not well understood. Though a lot of studies have been performed on Mn transporters in neuronal and astrocyte cell lineages (Hutchens et al., 2017; Zogzas and Mukhopadhyay, 2017), microglial biology with respect to metal transporters is not well studied. Further studies on metal transporters may provide key insights into the mechanism of Mn-induced microglial activation.

Our group has previously shown that Mn can interact with α-synuclein and protects against Mn-induced neurotoxicity at early time points. But chronic exposure of Mn leads to aggregation of α-synuclein, resulting in the exacerbation of Mn neurotoxicity (Harischandra et al., 2015). Furthermore, aggregated α-synuclein has been previously shown to induce astroglial and microglial activation (Gordon et al., 2016; Sarkar et al., 2017a). Recently, we further demonstrated that Mn alters the exosomal release of miRNA from neuronal cells (Harischandra et al., 2017). Previously, miRNA has been shown to affect microglial morphology and inflammation (Saika et al., 2017). Furthermore, exosomes have been shown to regulate cell-to-cell transmission of aggregated α-synuclein (Rokad et al., 2016). Together, these studies may indicate the probable role of Mn in regulating exosomal content, which can lead to microglial and astroglial activation. Further studies using a co-culture system are required to understand the neuron-glia interaction regulated by exosomes in response to Mn.

4. Concluding remarks

Though Mn has been shown to induce astroglial and microglial inflammation through ROS generation and the activation of various proinflammatory cascades, it is not quite clear what intervention strategy or combination of therapeutics might be the best option for treating neurodegenerative and neuroinflammatory conditions. This complexity still stands, especially in light of recent research demonstrating that microglia activate astrocytes through specific signaling factors (Liddelow et al., 2017). Cell-to-cell signaling is another layer of complexity that must be addressed. Commonalties between the behaviors of these two cell types need additional elucidation. Mechanistic research as well as the use of omics will yield a better understanding of the key molecular pathways affected by Mn in regulating these inflammatory cascades.

Acknowledgments

This chapter was supported by National Institutes of Health (NIH) grants ES026892, ES027245, NS100090, and NS088206. The W. Eugene and Linda Lloyd Endowed Chair and Eminent Scholar and Armbrust Endowment to A.G.K. and the Salisbury chair to A.K. are also acknowledged. The authors acknowledge Gary Zenitsky for his assistance in editing this manuscript.

References

Archibald, F.S., Tyree, C., 1987. Manganese poisoning and the attack of trivalent manganese upon catecholamines. Arch. Biochem. Biophys. 256, 638–650.

Aschner, M., Gannon, M., Kimelberg, H.K., 1992. Manganese uptake and efflux in cultured rat astrocytes. J. Neurochem. 58, 730–735.

Aschner, M., Erikson, K.M., Hernández, E.H., Tjalkens, R., 2009. Manganese and its role in Parkinson's disease: from transport to neuropathology. NeuroMolecular Med. 11, 252–266. Humana Press.

Baeza-Raja, B., Munoz-Canoves, P., 2004. p38 MAPK-induced nuclear factor-kappaB activity is required for skeletal muscle differentiation: role of interleukin-6. Mol. Biol. Cell 15, 2013–2026.

Bast-Pettersen, R., Ellingsen, D.G., Hetland, S.M., Thomassen, Y., 2004. Neuropsychological function in manganese alloy plant workers. Int. Arch. Occup. Environ. Health 77, 277–287.

Chang, J.Y., Liu, L.Z., 1999. Manganese potentiates nitric oxide production by microglia. Brain Res. Mol. Brain Res. 68, 22–28.

Chen, J., Su, P., Luo, W., Chen, J., 2018. Role of LRRK2 in manganese-induced neuroinflammation and microglial autophagy. Biochem. Biophys. Res. Commun. 498, 171–177.

Cordova, F.M., Aguiar Jr., A.S., Peres, T.V., Lopes, M.W., Goncalves, F.M., Remor, A.P., Lopes, S.C., Pilati, C., Latini, A.S., Prediger, R.D., Erikson, K.M., Aschner, M., Leal, R.B., 2012. In vivo manganese exposure modulates Erk, Akt and Darpp-32 in the striatum of developing rats, and impairs their motor function. PLoS One 7, e33057.

Crittenden, P.L., Filipov, N.M., 2008. Manganese-induced potentiation of in vitro proinflammatory cytokine production by activated microglial cells is associated with persistent activation of p38 MAPK. Toxicol. In Vitro 22, 18–27.

Crittenden, P.L., Filipov, N.M., 2011. Manganese modulation of MAPK pathways: effects on upstream mitogen activated protein kinase kinases and mitogen activated kinase phosphatase-1 in microglial cells. J. Appl. Toxicol. 31, 1–10.

Dobson, A.W., Erikson, K.M., Aschner, M., 2004. Manganese neurotoxicity. Ann. N.Y. Acad. Sci. 1012, 115–128.

Dodd, C.A., Filipov, N.M., 2011. Manganese potentiates LPS-induced heme-oxygenase 1 in microglia but not dopaminergic cells: role in controlling microglial hydrogen peroxide and inflammatory cytokine output. Neurotoxicology 32, 683–692.

Ellingsen, D.G., Hetland, S.M., Thomassen, Y., 2003. Manganese air exposure assessment and biological monitoring in the manganese alloy production industry. J. Environ. Monit. 5, 84–90.

Filipov, N.M., Seegal, R.F., Lawrence, D.A., 2005. Manganese potentiates in vitro production of proinflammatory cytokines and nitric oxide by microglia through a nuclear factor kappa B-dependent mechanism. Toxicol. Sci. 84, 139–148.

Fored, C.M., Fryzek, J.P., Brandt, L., Nise, G., Sjogren, B., McLaughlin, J.K., Blot, W.J., Ekbom, A., 2006. Parkinson's disease and other basal ganglia or movement disorders in a large nationwide cohort of Swedish welders. Occup. Environ. Med. 63, 135–140.

Ghosh, A., Langley, M.R., Harischandra, D.S., Neal, M.L., Jin, H., Anantharam, V., Joseph, J., Brenza, T., Narasimhan, B., Kanthasamy, A., Kalyanaraman, B., Kanthasamy, A.G., 2016. Mitoapocynin treatment protects against neuroinflammation and dopaminergic neurodegeneration in a preclinical animal model of Parkinson's disease. J. NeuroImmune Pharmacol. 11, 259–278.

Gonzalez, L.E., Juknat, A.A., Venosa, A.J., Verrengia, N., Kotler, M.L., 2008. Manganese activates the mitochondrial apoptotic pathway in rat astrocytes by modulating the expression of proteins of the Bcl-2 family. Neurochem. Int. 53, 408–415.

Gonzalez-Cuyar, L.F., Nelson, G., Criswell, S.R., Ho, P., Lonzanida, J.A., Checkoway, H., Seixas, N., Gelman, B.B., Evanoff, B.A., Murray, J., Zhang, J., Racette, B.A., 2014. Quantitative neuropathology associated with chronic manganese exposure in South African mine workers. Neurotoxicology 45, 260–266.

Goral, J., Kovacs, E.J., 2005. In vivo ethanol exposure down-regulates TLR2-, TLR4-, and TLR9-mediated macrophage inflammatory response by limiting p38 and ERK1/2 activation. J. Immunol. 174, 456–463.

Gordon, R., Singh, N., Lawana, V., Ghosh, A., Harischandra, D.S., Jin, H., Hogan, C., Sarkar, S., Rokad, D., Panicker, N., Anantharam, V., Kanthasamy, A.G., Kanthasamy, A., 2016. Protein kinase Cdelta upregulation in microglia drives neuroinflammatory responses and dopaminergic neurodegeneration in experimental models of Parkinson's disease. Neurobiol. Dis. 93, 96–114.

Guilarte, T.R., 2010. Manganese and Parkinson's disease: a critical review and new findings. Environ. Health Perspect. 118, 1071–1080.

Guilarte, T.R., 2013. Manganese neurotoxicity: new perspectives from behavioral, neuroimaging, and neuropathological studies in humans and non-human primates. Front. Aging Neurosci. 5, 23.

Gunter, T.E., Sheu, S.S., 2009. Characteristics and possible functions of mitochondrial Ca(2+) transport mechanisms. Biochim. Biophys. Acta 1787, 1291–1308.

Gunter, R.E., Puskin, J.S., Russell, P.R., 1975. Quantitative magnetic resonance studies of manganese uptake by mitochondria. Biophys. J. 15, 319–333.

Gunter, T.E., Gavin, C.E., Aschner, M., Gunter, K.K., 2006. Speciation of manganese in cells and mitochondria: a search for the proximal cause of manganese neurotoxicity. Neurotoxicology 27, 765–776.

Harischandra, D.S., Jin, H., Anantharam, V., Kanthasamy, A., Kanthasamy, A.G., 2015. Alpha-synuclein protects against manganese neurotoxic insult during the early stages of exposure in a dopaminergic cell model of Parkinson's disease. Toxicol. Sci. 143, 454–468.

Harischandra, D.S., Ghaisas, S., Rokad, D., Zamanian, M., Jin, H., Anantharam, V., Kimber, M., Kanthasamy, A., Kanthasamy, A.G., 2017. Environmental neurotoxicant manganese regulates exosome-mediated extracellular miRNAs in cell culture model of Parkinson's disease: relevance to alpha-synuclein misfolding in metal neurotoxicity. Neurotoxicology 64, 267–277.

Hazell, A.S., Desjardins, P., Butterworth, R.F., 1999. Chronic exposure of rat primary astrocyte cultures to manganese results in increased binding sites for the 'peripheral-type' benzodiazepine receptor ligand 3H-PK 11195. Neurosci. Lett. 271, 5–8.

Henriksson, J., Tjalve, H., 2000. Manganese taken up into the CNS via the olfactory pathway in rats affects astrocytes. Toxicol. Sci. 55, 392–398.

Horning, K.J., Caito, S.W., Tipps, K.G., Bowman, A.B., Aschner, M., 2015. Manganese is essential for neuronal health. Annu. Rev. Nutr. 35, 71–108.

Huang, C.-C., 2007. Parkinsonism induced by chronic manganese intoxication—an experience in Taiwan. Chang Gung Med. J. 30, 385–395.

Huang, E., Ong, W.Y., Connor, J.R., 2004. Distribution of divalent metal transporter-1 in the monkey basal ganglia. Neuroscience 128, 487–496.

Hutchens, S., Liu, C., Jursa, T., Shawlot, W., Chaffee, B.K., Yin, W., Gore, A.C., Aschner, M., Smith, D.R., Mukhopadhyay, S., 2017. Deficiency in the manganese efflux transporter SLC30A10 induces severe hypothyroidism in mice. J. Biol. Chem. 292, 9760–9773.

Karki, P., Lee, E., Aschner, M., 2013. Manganese neurotoxicity: a focus on glutamate transporters. Ann. Occup. Environ. Med. 25, 4.

Karki, P., Webb, A., Smith, K., Johnson Jr., J., Lee, K., Son, D.S., Aschner, M., Lee, E., 2014. Yin Yang 1 is a repressor of glutamate transporter EAAT2, and it mediates manganese-induced decrease of EAAT2 expression in astrocytes. Mol. Cell. Biol. 34, 1280–1289.

Kirkley, K.S., Popichak, K.A., Afzali, M.F., Legare, M.E., Tjalkens, R.B., 2017. Microglia amplify inflammatory activation of astrocytes in manganese neurotoxicity. J. Neuroinflammation 14, 99.

Kwakye, G.F., Paoliello, M.M., Mukhopadhyay, S., Bowman, A.B., Aschner, M., 2015. Manganese-induced parkinsonism and Parkinson's disease: shared and distinguishable features. Int. J. Environ. Res. Public Health 12, 7519–7540.

Langley, M.R., Ghaisas, S., Ay, M., Luo, J., Palanisamy, B.N., Jin, H., Anantharam, V., Kanthasamy, A., Kanthasamy, A.G., 2017. Manganese exposure exacerbates progressive motor deficits and neurodegeneration in the MitoPark mouse model of Parkinson's disease: relevance to gene and environment interactions in metal neurotoxicity. Neurotoxicology 64, 240–255.

Latchoumycandane, C., Anantharam, V., Kitazawa, M., Yang, Y., Kanthasamy, A., Kanthasamy, A.G., 2005. Protein kinase Cdelta is a key downstream mediator of manganese-induced apoptosis in dopaminergic neuronal cells. J. Pharmacol. Exp. Ther. 313, 46–55.

Lawana, V., Singh, N., Sarkar, S., Charli, A., Jin, H., Anantharam, V., Kanthasamy, A.G., Kanthasamy, A., 2017. Involvement of c-Abl kinase in microglial activation of NLRP3 inflammasome and impairment in autolysosomal system. J. NeuroImmune Pharmacol. 12, 624–660.

Lee, E.Y., Flynn, M.R., Lewis, M.M., Mailman, R.B., Huang, X., 2018. Welding-related brain and functional changes in welders with chronic and low-level exposure. Neurotoxicology 64, 50–59.

Lewis, M.M., Flynn, M.R., Lee, E.Y., Van Buren, S., Van Buren, E., Du, G., Fry, R.C., Herring, A.H., Kong, L., Mailman, R.B., Huang, X., 2016. Longitudinal T1 relaxation rate (R1) captures changes in short-term Mn exposure in welders. Neurotoxicology 57, 39–44.

Liddelow, S.A., Barres, B.A., 2017. Reactive astrocytes: production, function, and therapeutic potential. Immunity 46, 957–967.

Liddelow, S.A., Guttenplan, K.A., Clarke, L.E., Bennett, F.C., Bohlen, C.J., Schirmer, L., Bennett, M.L., Munch, A.E., Chung, W.S., Peterson, T.C., Wilton, D.K., Frouin, A., Napier, B.A., Panicker, N., Kumar, M., Buckwalter, M.S., Rowitch, D.H., Dawson, V.L., Dawson, T.M., Stevens, B., Barres, B.A., 2017. Neurotoxic reactive astrocytes are induced by activated microglia. Nature 541, 481–487.

Lucchini, R., Selis, L., Folli, D., Apostoli, P., Mutti, A., Vanoni, O., Iregren, A., Alessio, L., 1995. Neurobehavioral effects of manganese in workers from a ferroalloy plant after temporary cessation of exposure. Scand. J. Work Environ. Health 21, 143–149.

Martinez-Finley, E.J., Gavin, C.E., Aschner, M., Gunter, T.E., 2013. Manganese neurotoxicity and the role of reactive oxygen species. Free Radic. Biol. Med. 62, 65–75.

Milatovic, D., Yin, Z., Gupta, R.C., Sidoryk, M., Albrecht, J., Aschner, J.L., Aschner, M., 2007. Manganese induces oxidative impairment in cultured rat astrocytes. Toxicol. Sci. 98, 198–205.

Moberly, A.H., Czarnecki, L.A., Pottackal, J., Rubinstein, T., Turkel, D.J., Kass, M.D., Mcgann, J.P., 2012. Intranasal exposure to manganese disrupts neurotransmitter release from glutamatergic synapses in the central nervous system in vivo. Neurotoxicology 33, 996–1004.

Moreno, J.A., Yeomans, E.C., Streifel, K.M., Brattin, B.L., Taylor, R.J., Tjalkens, R.B., 2009. Age-dependent susceptibility to manganese-induced neurological dysfunction. Toxicol. Sci. 112, 394–404.

Moreno, J.A., Streifel, K.M., Sullivan, K.A., Hanneman, W.H., Tjalkens, R.B., 2011. Manganese-induced NF-kappaB activation and nitrosative stress is decreased by estrogen in juvenile mice. Toxicol. Sci. 122, 121–133.

Nelson, K., Golnick, J., Korn, T., Angle, C., 1993. Manganese encephalopathy: utility of early magnetic resonance imaging. Br. J. Ind. Med. 50, 510–513.

Ngwa, H.A., Ay, M., Jin, H., Anantharam, V., Kanthasamy, A., Kanthasamy, A.G., 2017. Neurotoxicity of vanadium. Adv Neurobiol. 18, 287–301.

Park, E.J., Park, K., 2010. Induction of oxidative stress and inflammatory cytokines by manganese chloride in cultured T98G cells, human brain glioblastoma cell line. Toxicol. In Vitro 24, 472–479.

Pekny, M., Pekna, M., 2014. Astrocyte reactivity and reactive astrogliosis: costs and benefits. Physiol. Rev. 94, 1077–1098.

Peres, T.V., Schettinger, M.R., Chen, P., Carvalho, F., Avila, D.S., Bowman, A.B., Aschner, M., 2016. Manganese-induced neurotoxicity: a review of its behavioral consequences and neuroprotective strategies. BMC Pharmacol. Toxicol. 17, 57.

Racette, B.A., McGee-Minnich, L., Moerlein, S.M., Mink, J.W., Videen, T.O., Perlmutter, J.S., 2001. Welding-related parkinsonism: clinical features, treatment, and pathophysiology. Neurology 56, 8–13.

Racette, B.A., Aschner, M., Guilarte, T.R., Dydak, U., Criswell, S.R., Zheng, W., 2012. Pathophysiology of manganese-associated neurotoxicity. Neurotoxicology 33, 881–886.

Rodier, J., 1955. Manganese poisoning in Moroccan miners. Br. J. Ind. Med. 12, 21–35.

Rokad, D., Ghaisas, S., Harischandra, D.S., Jin, H., Anantharam, V., Kanthasamy, A., Kanthasamy, A.G., 2016. Role of neurotoxicants and traumatic brain injury in alpha-synuclein protein misfolding and aggregation. Brain Res. Bull. 133, 60–70.

Sadek, A.H., Rauch, R., Schulz, P.E., 2003. Parkinsonism due to manganism in a welder. Int. J. Toxicol. 22, 393–401.

Saika, R., Sakuma, H., Noto, D., Yamaguchi, S., Yamamura, T., Miyake, S., 2017. MicroRNA-101a regulates microglial morphology and inflammation. J. Neuroinflammation 14, 109.

Sarkar, S., Malovic, E., Harischandra, D.S., Ngwa, H.A., Ghosh, A., Hogan, C., Rokad, D., Zenitsky, G., Jin, H., Anantharam, V., Kanthasamy, A.G., Kanthasamy, A., 2017a. Manganese exposure induces neuroinflammation by impairing mitochondrial dynamics in astrocytes. Neurotoxicology 64, 204–218.

Sarkar, S., Malovic, E., Harishchandra, D.S., Ghaisas, S., Panicker, N., Charli, A., Palanisamy, B.N., Rokad, D., Jin, H., Anantharam, V., Kanthasamy, A., Kanthasamy, A.G., 2017b. Mitochondrial impairment in microglia amplifies NLRP3 inflammasome proinflammatory signaling in cell culture and animal models of Parkinson's disease. NPJ Parkinsons Dis. 3, 30.

Sidoryk-Wegrzynowicz, M., Aschner, M., 2013. Role of astrocytes in manganese mediated neurotoxicity. BMC Pharmacol. Toxicol. 14, 23.
Sidoryk-Wegrzynowicz, M., Lee, E., Albrecht, J., Aschner, M., 2009. Manganese disrupts astrocyte glutamine transporter expression and function. J. Neurochem. 110, 822–830.
Streifel, K.M., Moreno, J.A., Hanneman, W.H., Legare, M.E., Tjalkens, R.B., 2012. Gene deletion of nos2 protects against manganese-induced neurological dysfunction in juvenile mice. Toxicol. Sci. 126, 183–192.
Tjalkens, R.B., Popichak, K.A., Kirkley, K.A., 2017. Inflammatory activation of microglia and astrocytes in manganese neurotoxicity. Adv. Neurobiol. 18, 159–181.
Verina, T., Kiihl, S.F., Schneider, J.S., Guilarte, T.R., 2011. Manganese exposure induces microglia activation and dystrophy in the substantia nigra of non-human primates. Neurotoxicology 32, 215–226.
Wang, D., Zhang, J., Jiang, W., Cao, Z., Zhao, F., Cai, T., Aschner, M., Luo, W., 2017. The role of NLRP3-CASP1 in inflammasome-mediated neuroinflammation and autophagy dysfunction in manganese-induced, hippocampal-dependent impairment of learning and memory ability. Autophagy 13, 914–927.
Yin, Z., Aschner, J.L., Dos Santos, A.P., Aschner, M., 2008. Mitochondrial-dependent manganese neurotoxicity in rat primary astrocyte cultures. Brain Res. 1203, 1–11.
Zhang, P., Wong, T.A., Lokuta, K.M., Turner, D.E., Vujisic, K., Liu, B., 2009. Microglia enhance manganese chloride-induced dopaminergic neurodegeneration: role of free radical generation. Exp. Neurol. 217, 219–230.
Zhao, F., Cai, T., Liu, M., Zheng, G., Luo, W., Chen, J., 2009. Manganese induces dopaminergic neurodegeneration via microglial activation in a rat model of manganism. Toxicol. Sci. 107, 156–164.
Zogzas, C.E., Mukhopadhyay, S., 2017. Inherited disorders of manganese metabolism. Adv. Neurobiol. 18, 35–49.

CHAPTER FIVE

Inflammatory mechanisms of abused drugs

Carmen M. Mitchell, Omar El Jordi, Bryan K. Yamamoto*
Department of Pharmacology and Toxicology, Indiana University School of Medicine, Indianapolis, IN, United States
*Corresponding author: e-mail address: brkyama@iu.edu

Contents

1. Introduction

A total of 8.4% of the population has a substance use disorder defined in the *DSM-5* as mild, moderate, or severe with multiple criteria that determine pharmacological and psychological dependence (American Psychiatric Association, 2013; Lipari and Van Horn, 2017). The neuronal mechanisms underlying substance dependence involve the reward pathway comprised of dopaminergic projections from the ventral tegmental area to the nucleus accumbens (NAc) and frontal cortex; however, the factors that contribute to dependence have not been fully identified and targeted for effective treatment.

Advances in Neurotoxicology, Volume 3
ISSN 2468-7480
https://doi.org/10.1016/bs.ant.2018.10.006

Recent evidence implicates inflammation in the action of drugs of abuse. Users dependent on nicotine or alcohol have high blood concentrations of inflammatory marker C-reactive protein (CRP) (Costello et al., 2014). Likewise, cocaine addicts were found to have high levels of the pro-inflammatory cytokine TNFα and low levels of the anti-inflammatory cytokine IL-10 (Fox et al., 2012). Importantly, administration of an anti-inflammatory drug, ibudilast, reduces drug reward and reduces drug-seeking in methamphetamine and opioid dependent subjects (Beardsley et al., 2010; Metz et al., 2017; Worley et al., 2016). Collectively, these data are suggestive of a common inflammatory mechanism involved in drug abuse. The goal of this chapter will be to highlight common inflammatory mechanisms shared by different drugs of abuse that contribute to their abuse and neurotoxicity profiles.

2. Drugs of abuse and inflammation

2.1 Psychostimulants

2.1.1 Mechanism of action

Psychostimulants are defined by their ability to produce stimulatory effects, causing increased movement, hyperarousal, and euphoria (O'Brien, 2011). They function as indirect sympathomimetics by increasing synaptic concentrations of epinephrine, norepinephrine, dopamine, and serotonin. Specifically, cocaine, amphetamine (Amph), and methamphetamine (Meth) interact at neurotransmitter reuptake transporters. Cocaine binds and blocks dopamine, serotonin and norepinephrine transporters to increase the synaptic availability of the monoamine transmitters (Shoemaker et al., 1985). In contrast, Meth and Amph cause the efflux of monoamine transmitters through dopamine transporters (DAT) and serotonin transporters (SERT), thereby causing an increase in dopamine (DA) and serotonin (5-HT) into the synaptic cleft (Johnson et al., 1998; Jones et al., 1999; Liang and Rutledge, 1981). In addition, Meth releases DA from vesicular stores within the cytoplasm of the presynaptic terminal and increases the availability of dopamine for efflux via the dopamine transporter (Volz et al., 2006). Psychostimulants can also increase extracellular concentrations of glutamate (GLU) in a brain region dependent manner such that Meth acutely increases the release of GLU within the striatum (Mark et al., 2004) whereas cocaine increases GLU release within the frontal cortex and nucleus accumbens (Reid et al., 1997) but not in the striatum.

The neurotoxicity profiles also differ between the psychostimulants. The details of the neurotoxicity related to Meth have been reviewed elsewhere (Halpin et al., 2014a; Yamamoto et al., 2010) but generally, Meth produces long-term depletions of DA and 5-HT in striatum, hippocampus and frontal cortex in rodent models (Ricaurte et al., 1982; Seiden et al., 1988; Wagner et al., 1980) and in humans (McCann et al., 1998). The depletions observed in rodents and humans occur with axonal degeneration in the striatum (Ricaurte et al., 1982), and with cell body loss in the substantia nigra (Ares-Santos et al., 2014). The mechanisms contributing to the depletions and axonal degeneration are a result of oxidative stress, GLU-mediated excitotoxicity, and inflammation (Halpin et al., 2014a). The striatum appears to be vulnerable to Meth-induced toxicity compared to other DA-rich regions due to a dense glutamatergic innervation from the cortex and the release of GLU in the striatum (Mark et al., 2004) resulting in excitotoxicity (Staszewski and Yamamoto, 2006). In contrast, Amph and cocaine do not increase dorsal striatal GLU release and do not result in long-term depletions in the striatum (Goodman and Solviter, 1993; Kleven et al., 1988; Reid et al., 1997).

2.1.2 Psychostimulants and neuroinflammation

Psychostimulants have parallel neuroinflammatory profiles in addition to their similar actions on neurotransmitter systems. Meth increases pro-inflammatory cytokines (IL-6, IL-1β, TNFα) in the systemic circulation and within DA- and 5-HT-rich brain regions (Kelly et al., 2012; Yamaguchi et al., 1991). Amph and cocaine also increase pro-inflammatory cytokines (IL-6, IL-1β, TNFα) in the blood and within microglia and astrocytes in DA-rich brain regions (Northcutt et al., 2015; Periyasamy et al., 2016; Valvassori et al., 2015). The mechanisms underlying the ability of psychostimulants to precipitate inflammation in the periphery and central nervous system (CNS) are not completely understood but may be linked to their neurotoxic and behavioral consequences.

2.1.2.1 Peripheral sources of inflammatory mediators

Macrophages and lymphocytes: Cells in the peripheral blood such as macrophages, lymphocytes as well as T cells and B cells are capable of releasing cytokines. Meth, Amph, and cocaine acutely increase the pro-inflammatory cytokines IL-6, TNFα, and IL-1β in the blood but their effects on the anti-inflammatory cytokine IL-10 are more variable such that Amph and cocaine decrease IL-10, whereas Meth is without effect (Table 1).

Table 1 Summary of cytokine and immune cell changes produced by drugs of abuse.

Drug of abuse	Model	Time point after drug exposure	Cytokine changes	Immune cell population changes	References
Cocaine	Human addicts	Unknown	↑ IL-6, ↓ IL-10 (serum)	N/A	Moreira et al. (2016)
Cocaine	Human addicts	Cocaine use within 72 h	↑ IL-6, no change TNFα, IL-1β (saliva)	N/A	Ersche et al. (2014)
Methamphetamine	Human macrophages	2, 6 h	↑ TNFα, IL-1β	N/A	Burns and Ciborowski (2016)
Methamphetamine	Rat	2 h	↑ TNFα, IL-1β, IL-6	N/A	Gonçalves et al. (2008)
Methamphetamine	Rat	24 h	Stimulation ↑ IFNγ, ↓ TNFα, no change IL-6 (serum)	↑ CD8 cells, ↓ CD4 cells	Mata et al. (2015)
Methamphetamine	Rat	18 days	↓ TNF, IL-6, IL-10, ↓ IL-1β (in frontal cortex, no change in amygdala or striatum)	N/A	Stolyarova et al. (2014)
D-Amphetamine	Mouse	1 h	↑ IL-4, no change IFNγ, IL-10	N/A	Kubera et al. (2002)
D-Amphetamine	Rat	2 h	↑ IL-4, IL-6,IL-10, TNFα (frontal cortex, striatum, and serum)	N/A	Valvassori et al. (2015)
D-Amphetamine	Rat	24 h	↑ IL-4, ↓ IL-10	↑ Macrophages and granulocytes	Ligeiro de Oliveira et al. (2012)

Morphine	Murine splenocyte cultures	3, 20, 40 min	↑ IL-1β, IL-2, TNFα, IFNγ	N/A	Pacifici et al. (2000)
Heroin	Human	Unknown	↑ IL-6, IL-1β (plasma)	N/A	Chan et al. (2015)
Heroin	Murine splenocyte cultures	3, 20, 40 min	↑ IL-1β, IL-2, TNFα, IFNγ	N/A	Pacifici et al. (2000)
Opium	Lymphocyte cells	6, 24 h	↑ IL-6 (6 and 24 h), ↑ IFNγ (24 h), ↓ TGFβ (6, 24 h)	N/A	Asadikaram et al. (2015)
Opium	Rat	10 days	↑ IFNγ, ↓ IL-4, no change IL-10, TNFα (plasma)	N/A	Lashkarizadeh et al. (2016)
CBD	Rat	Unknown	↓ IFNγ, TNFα, ↑ IL-4, ↑ IL-10 (plasma)	↓ Th1 T cells, ↑ Th2 cells (within CD4 cells)	Weiss et al. (2006)
CBD	Mice	10 day treatment	↓ TNFα, IL-1β (spinal cord)	N/A	Mecha et al. (2013)
Δ9-THC	Microglia cell line	6 h pre-treatment	↓ IL-1β, IL-6, IFNγ	N/A	Kozela et al. (2010)

These differences could be due to variations in the tissue source of the cytokines and the time when they are measured after drug exposure. For example, cytokine changes are not evident immediately during withdrawal from Meth, Amph, or cocaine but only begin to increase at 3 days after drug exposure and remain elevated at 18 days of withdrawal. In circumstances where no changes in basal circulating cytokines are observed after drug exposure, challenges with either LPS, stress, or an allergen after drug exposure increase cytokines (Ligeiro de Oliveira et al., 2012; Liu et al., 2012) and suggest a sensitization of the inflammatory response by psychostimulants.

Psychostimulants can produce changes in circulating cell populations such that they enable T cells to contribute to either a pro- or anti-inflammatory state. For example, T cell populations are distinguished based on their expression of glycoprotein CD4 or CD8 that produce either pro- or anti-inflammatory responses. CD4-expressing cells are capable of releasing pro-inflammatory cytokines, such as IL-6 and the anti-inflammatory cytokine IL-4. Increases in peripheral IL-6 and IFNγ can further induce a pro-inflammatory CD4 subpopulation that preferentially releases IL-6, TNFα and IFNγ (Korn et al., 2008) to augment a pro-inflammatory response. In fact, CD4 cells are increased in cocaine addicts (Ersche and Döffinger, 2017) that can differentiate to be pro-inflammatory and release IFNγ and TNFα (Babu and Nutman, 2003; Wakil et al., 1998). Similarly, CD8 T cells can also differentiate in response to cytokines increased by psychostimulants such that IFNγ can drive the differentiation of CD8 to a pro-inflammatory subpopulation that releases TNFα and IFNγ (Slifka and Whitton, 2000; Stoycheva et al., 2014). Along these lines, Meth increases TNFα and IFNγ and can shift the T cell population to a more pro-inflammatory CD8 subpopulation to produce IFNγ (Mata et al., 2015). Moreover, cytokines elevated by this pro-inflammatory shift produced by psychostimulants can interact with their respective receptors to induce NF-κB or CREB-dependent transcription resulting in increases in other inflammatory mediators such as chemokines. The chemokines can then invoke additional cytokine release, influence migration of white blood cells to an inflammatory site, and produce further inflammation (Sokol and Luster, 2015; Wong et al., 2005). Given that the initiation and consequences of this persistent feed-forward inflammatory cycle are dependent on the source of cytokine and chemokine release and the distribution of receptors, organs such as the liver and brain that contain these receptors are particularly vulnerable.

Liver: The liver detoxifies chemicals and drugs within the blood but also is vulnerable to toxicity produced by drugs of abuse as evidenced by increases in serum ALT, AST, ammonia, and changes in hepatocellular morphology (Halpin et al., 2013; Labib et al., 2003; Roberts et al., 1995). The hepatotoxic effects of drugs of abuse appear to be mediated by cytokine release from liver Kupffer cells as Kupffer cell destruction produced by gadolinium chloride protects against Meth-induced liver damage (Roberts et al., 1995). During acute injury, Kupffer cells release TNFα, resulting in the autocrine stimulation of TNFR on Kupffer cells and the synthesis and release of IL-6 (Trautwein et al., 1996). Consequently, IL-6 release stimulates hepatocyte proliferation, release of acute phase proteins that regulate inflammation, immune cell recruitment, cell lysis, and eventual Kupffer cell death with prolonged stimulation (Schmidt-Arras and Rose-John, 2016). In contrast, hepatotoxicity from cocaine is not prevented with gadolinium chloride (Labib et al., 2003; Roberts et al., 1995) and suggests another mechanism that produces liver inflammation and damage. Regardless, Meth and cocaine shift the T cell population to a more pro-inflammatory T cell population (Ersche and Döffinger, 2017; Mata et al., 2015) and can promote inflammation through the release of IL-6 from circulating monocytes and macrophages (Table 1). Alternatively, there could be a decrease in anti-inflammatory signaling from liver endothelial cells as these cells normally maintain an anti-inflammatory protective barrier between hepatocytes and incoming nutrient-rich blood (Carambia et al., 2013).

The liver damage produced by drugs of abuse can also exacerbate inflammation in other organs such as the brain. The existence of a liver-brain inflammatory axis reveals a contribution of liver damage to neuroinflammation. Liver inflammation can result in sickness behavior that includes behavioral disturbances such as lethargy and anhedonia (D'Mello et al., 2017) that also result in microglial activation, all of which are suggestive of a periphery-to-brain inflammatory relationship (Qin et al., 2008). Conversely, brain injury can damage the liver as brain trauma can increase serum amyloid A1 protein (SAA1) within the liver (Villapol et al., 2015). SAA1 is only present in the periphery and is not normally found in the brain, except under inflammatory conditions (Meek and Benditt, 1986).

Although the above evidence implicates liver inflammation in neuroinflammation, the mediators of psychostimulant-induced liver damage and subsequent neuroinflammation and neurotoxicity are not well understood. However, recent evidence suggests that high levels of ammonia due to liver failure may play a role. For example, ammonia is known to rise in the blood of individuals with acute liver failure due to the inability of the liver to convert circulating ammonia to urea. Consequently, hyperammonemia develops and has been demonstrated to mediate Meth-induced astrogliosis, GLU excitotoxicity, and long-term neurotransmitter depletions (Halpin et al., 2013, 2014b). It is unknown if the hepatotoxicity due to Meth exposure leads to increases in other inflammatory mediators that could also produce changes in the CNS.

2.1.2.2 Central sources of inflammatory mediators

Microglia: Induction of peripheral inflammation with LPS results in activation of inflammatory microglia in the brain (Hoogland et al., 2015) and further supports a major role for peripheral inflammation contributing to central inflammation. Although this has not been addressed directly within the context of drugs of abuse, inflammation in the brain occurs in parallel with peripheral inflammation induced by drugs of abuse (Table 1). For instance, Meth and Amph produce morphological changes indicative of microglial activation in the striatum 2–6 days after drug, with more persistent increases localized to the striatum where degeneration occurs (Lavoie et al., 2004; Thomas et al., 2004). Microglial activation is also present in abstinent human Meth addicts in striatal, as well as cortical and thalamic brain regions (Sekine et al., 2008). Similarly, cocaine produces morphological changes in microglia and leads to an activated microglial phenotype and increased cytokine expression (Lewitus et al., 2016; Liao et al., 2016). Microglia activated by Meth can increase IL-1β and TNFα (Yue et al., 2012) that in turn, increase prostaglandin synthesis through cyclooxygenase-2 (COX-2), the latter demonstrated to be increased in the brain by Meth, Amph, and cocaine (Northrop and Yamamoto, 2012; Samikkannu et al., 2014; Thomas and Kuhn, 2004). IL-1β and TNFα can also signal in an autocrine fashion onto microglia, increase COX-2, and produce prostaglandins. Prostaglandins in turn can activate EP1 (Carlson et al., 2009; Lan et al., 2017) or EP2 receptors to further increase COX-2, IL-1β, and IL-6 from microglia and astrocytes (Quan et al., 2013). Moreover, brain extracellular concentrations of PGE2 are increased after the Meth derivative, MDMA (Collins et al., 2016) indicative of enhanced release from tissue stores and were higher in dendritic cells of cocaine users

(Samikkannu et al., 2014) that in turn, can stimulate EP2 receptors on astrocytes to induce pro-inflammatory astrogliosis (Zhang et al., 2009). In contrast, other studies that have measured brain tissue stores of PGE2 showed no change or a decrease in PGE2 after psychostimulants (Avila et al., 2018; Thomas and Kuhn, 2004). Taken together, these studies suggest redistributed stores of PGE2 after psychostimulants.

An additional signaling mechanism by which COX-2 can produce inflammation is through phosphodiesterase (PDE) signaling. More specifically, PDE4 modulates the propagation of microglial inflammation by regulating IL-1β production (You et al., 2017) such that inhibition of microglial PDE4 decreases IL-1β, COX-2, and TNFα through an increase in cAMP (Jin et al., 2005). Thus, PDE4 inhibitors can be used as anti-inflammatory drugs to inhibit microglial inflammation from psychostimulants (You et al., 2017).

Microglia activation is most prominent during periods of withdrawal from Meth, Amph, and cocaine (LaVoie et al., 2004; Lewitus et al., 2016; Sekine et al., 2008). For instance, microglial activation occurs within 2–6 days after drug exposure and is preceded by initial acute changes in cytokines within 24–48 h after withdrawal from drug. Increases in COX-2 align with this time course wherein Meth increases striatal COX-2 as early as 24 h after drug exposure and persists for 7 days (Northrop and Yamamoto, 2012; Thomas and Kuhn, 2004). The increases in COX-2 after Amph also occur relatively early at 1 h whereas increases in COX-2 in human cocaine users occur within 72 h (Samikkannu et al., 2014; Toledano et al., 2008). Consequently, COX-2 and other inflammatory signaling mediators can also stimulate other glial cells in the brain such as astrocytes.

Astrocytes: Peripheral inflammation and microglia derived inflammation induced by drugs of abuse appear to act on astrocytes. In fact, microglia can directly activate astrocytes to cause inflammation and neurotoxicity (Liddelow et al., 2017). Psychostimulants activate astrocytes as evidenced by increases in glial fibrillary acidic protein (GFAP) at time points subsequent to microglia activation (Bowers and Kalivas, 2003) and remain elevated 2–30 days after Meth exposure (Friend and Keefe, 2013; Gonçalves et al., 2008). This increase in GFAP produced by Meth can be exacerbated by prior exposure to stress (Northrop and Yamamoto, 2012) and upregulates nitric oxide and IL-1β (Brahmachari et al., 2006). Chronic, but not acute Amph, can also increase GFAP in hippocampus but does not change GFAP in frontal cortex (Frey et al., 2006). Cocaine also increases GFAP, but only in the frontal cortex and NAc and no sooner than 3 weeks after cocaine (Bowers and Kalivas, 2003). These brain region differences in response to different psychostimulants are not understood but could be due to the subtype of astrocyte stimulated that cannot be differentiated by measures of GFAP. Specifically, stimulation of the A1 compared to the A2 subtype can be toxic and result in upregulation of pro-inflammatory genes. In fact, A1 astrocytes are activated by microglia-derived TNF and IL-1, and from LPS (Liddelow et al., 2017), the former which are known to be associated with increases in GFAP and increased by psychostimulants (Table 1). Moreover, astrocytes in the A1 state have been shown to reduce synapse formation and increase neuronal cell death, consequences which also have been observed with psychostimulants (Ares-Santos et al., 2014; Lorez, 1981; Rasakham et al., 2014). Therefore, it is plausible that psychostimulants can produce neurotoxicity through microglia and activation of astrocytes to an A1 state.

An important feature that allows astrocytes to control inflammatory responses is the presence of multiple types of Toll-like receptors (TLRs). These receptors bind cytokines and chemokines and activate NF-κB and cAMP-CREB-signaling pathways that increase pro-inflammatory cytokine synthesis. Along these lines, inhibition of phosphodiesterases (PDEs) and the increase in accumulation of cAMP inhibit inflammation due to the stimulated release of the anti-inflammatory cytokine IL-10, thus reducing TNFα and IL-1β release (Guo et al., 2016; Platzer et al., 1995). More specifically, PDE4 inhibition reduces increases in GFAP induced by Meth (Charntikov et al., 2015) and can also block subsequent IL-1β and TNFα release (Guo et al., 2016; Schwenkgrub et al., 2017).

2.1.3 *Neurotoxic consequences of psychostimulants*

Based on the evidence that psychostimulants produce neuroinflammation through pro-inflammatory cytokine signaling, COX-2 and PDE in microglia, and PDE in astrocytes, their pro-inflammatory effects may underlie the neurodegeneration, GLU excitotoxicity, and cerebrovascular damage produced by these drugs.

2.1.3.1 Neurodegeneration

Neurodegeneration produced by Meth has been linked directly to COX-mediated inflammation. For instance, COX-2 knockout mice but not COX-1 knockout mice exhibit abrogated striatal dopamine depletions after

Meth (Thomas and Kuhn, 2004). An upstream inhibitor of COX signaling, salicylic acid, also prevents Meth-induced striatal dopamine depletions (Thrash-Williams et al., 2016). Although COX has been linked to Meth-induced neurodegeneration, some caveats remain. While COX-2 knockout prevents Meth-induced dopamine depletions, COX inhibitors do not prevent striatal dopamine depletions (Northrop and Yamamoto, 2012; Thomas and Kuhn, 2004). It remains to be determined how COX-2 per se contributes to neurodegeneration from Meth. In contrast and despite the finding that Amph treatment increases COX-2 in cortical layers of the rat brain and associated increases in lipid peroxidation (Toledano et al., 2008), Amph is not neurotoxic at doses comparable to those that produce neurotoxicity to Meth. Thus, activation of COX-2 alone may not be sufficient to cause neurotoxicity to the amphetamines.

2.1.3.2 Glutamate excitotoxicity

Excitotoxicity is associated with high concentrations of GLU. In this regard, Meth acutely increases striatal GLU release through disinhibition of thalamocortical GLU, resulting in increased corticostriatal GLU release (Halpin et al., 2014b; Mark et al., 2004; Stephans and Yamamoto, 1994). Corticostriatal GLU can act on NMDA and AMPA receptors to increase calcium influx and result in excitotoxicity through the cleavage of the cytoskeletal protein, spectrin (Halpin et al., 2013; Staszewski and Yamamoto, 2006). Astrocytes may also be a source of extracellular GLU as they can uptake and release GLU (Bezzi et al., 2004) and contribute to Meth-induced increases in GLU (Halpin et al., 2014b). Moreover, GLU release can also be stimulated by TNFα activation of chemokine CXCR4 receptors on astrocytes (Bezzi et al., 2001). In addition, downregulation of GLU transporters can be produced by LPS-induced inflammation, likely mediated through microglia activation (Bezzi et al., 2001; Tilleux and Hermans, 2008), thus increasing extracellular GLU through the decreased removal of GLU (Tilleux and Hermans, 2008). Overall, peripherally and centrally derived TNFα can contribute to excitotoxicity and subsequent neuron loss from psychostimulants.

2.1.3.3 Cerebrovascular damage

Psychostimulants can also induce damage to the cerebrovasculature evidenced by cerebral blood vessel damage and breakdown of the blood-brain barrier (BBB). Hemorrhagic stroke that ruptures blood vessels in the brain poses a major risk factor for Meth, Amph, and cocaine abusers. In fact, Amph abuse appears to result in a high risk for stroke and stroke fatalities whereas cocaine accounts for the majority of non-traumatic intracranial hemorrhage fatalities in humans (Huang et al., 2016; Nolte et al., 1996; Westover et al., 2007). Hemorrhagic stroke is also highly prevalent among Meth abusers and is also associated with poor recovery and survival outcomes (Lappin et al., 2017). Causative factors for stroke caused by psychostimulants include platelet aggregation, cardioembolism, cerebral vasculitis and hypertensive surges (Pozzi et al., 2008). These cerebrovascular injuries have been associated with and are initiated by cytokines and COX signaling. For example, psychostimulants enhance the risk for stroke by cytokine-induced increases in platelet aggregation (Heesch et al., 2000; Lumadue et al., 1996). Cardioembolism is also associated with psychostimulant-induced stroke (Fehnel et al., 2014) and high TNFα and IL-6 are correlated with deterioration in patients with cardioembolic stroke (Vila et al., 2000) that could account for poor stroke recovery and survival. Therefore, both the increased stroke risk and decreased stroke survival and recovery could be a result of psychostimulant-induced inflammation.

BBB breakdown is another cerebrovascular insult consequent to inflammation induced by psychostimulants. The BBB is responsible for selectively excluding toxins from the brain by restricting molecules according to molecular weight and water solubility. Small molecules can pass through endothelial cell gaps that line cerebral blood vessels, or be actively transported. However, larger molecules are restricted primarily due to tight junction proteins between the endothelial cells. Meth alone or in combination with stress can cause a loss of cerebral blood vessel tight junction proteins occludin and claudin and enhance BBB permeability (Eugenin et al., 2013; Northrop and Yamamoto, 2012). The tight junction protein loss and enhanced BBB permeability are prevented with COX-2 inhibition, suggesting that the mechanism by which psychostimulants produce BBB breakdown is mediated by prostaglandins (Northrop and Yamamoto, 2012). The source of COX-2 has yet to be determined, although both microglia from the brain parenchyma and endothelial cells that line the cerebral blood vessels can produce and release COX-2 (Matsumura et al., 1998; Xia et al., 2015). COX-2 could also be derived from peripheral macrophages that can traverse the BBB (Giroux and Descoteaux, 2000). Overall, evidence supports a role for COX-mediated inflammation in BBB breakdown regardless of the source of COX-2 release in response to psychostimulants.

2.2 Opioids

2.2.1 Mechanism of action

Opioids are utilized therapeutically to relieve pain. Similar to psychostimulants, they also produce euphoria and reward, resulting in opioid abuse (Al-Hasani and Bruchas, 2011; Gavériaux-Ruff and Kieffer, 2002; Mcnally and Akil, 2002; Pradhan et al., 2012). Opioids bind to receptors belonging to the G-protein coupled receptor (GPCR) family, which include the μ-opioid receptor (MOR), κ-opioid receptor (KOR) and δ-opioid receptor (DOR). These receptors interact with a number of naturally occurring and synthetic abused opioids including morphine, codeine, heroin, oxycodone, hydrocodone, fentanyl, and methadone. The direct consequence of opioid-receptor interaction depends on the specific GPCR that is bound, however the abused opioids have preferential selectivity for MOR (Chahl, 1996) and generally result in decreased neuronal excitability through decreasing cAMP, activating potassium channels, or decreasing calcium channel conductance (Tahamtan et al., 2016). These downstream signaling changes result in suppressed neurotransmitter release (Chahl, 1996). Furthermore, opioid receptor signaling is diversified by the oligomerization of these receptors with other receptors such as the cannabinoid receptor (CB1), which can synergistically interact to suppress neurotransmitter release (Cichewicz, 2004).

Extended opioid abuse has been associated with long-term neurotoxicity in human addicts. For instance, heroin abuse can increase proteins such as α-synuclein (Horvath et al., 2007), P62 (Kovacs et al., 2015), and eotaxin-1 (Kuo et al., 2018) at early ages, and is suggested to accelerate aging and predispose opioid addicts to neurodegenerative disorders like Alzheimer's. However, opioids do not result in an acute neurotoxic profile, such as the dopaminergic cell loss and GLU-mediated excitotoxicity observed with some psychostimulants but both opioids and psychostimulants have a high abuse potential (Epstein et al., 2006; Romach et al., 2014) and appear to share similar mechanisms related to inflammation that mediate their abuse liability (Table 1).

2.2.2 Opioids and neuroinflammation

Emerging evidence suggests that chronic opioid use increases pro-inflammatory cytokines TNFα, IL-1β, and IFNγ in the periphery in a manner mediated by MOR, but not KOR or DOR (Asadikaram et al., 2015; Lashkarizadeh et al., 2016; Pacifici et al., 2000; Wang et al., 2002). It is also well accepted that opioids produce pro-inflammatory effects in the CNS, wherein opioid-induced glial activation has been observed (Amaral et al., 2016; Doyle et al., 2017; Song and Zhao, 2001; Taylor et al., 2016). Specific sources for peripheral and central inflammation are similar to those of the psychostimulants including macrophages, lymphocytes, microglia and astrocytes.

2.2.2.1 Peripheral sources of inflammatory mediators

Peripheral immune cells: There is a dichotomy between the peripheral pro- and anti-inflammatory effects of opioid abuse; however, chronic opioid use in humans and rat models produces a pro-inflammatory state in the periphery. For instance, opium increases pro-inflammatory IL-6 and IFNγ release, and decreases anti-inflammatory TGFβ in both human addicts and opioid-dependent rats (Asadikaram et al., 2015; Lashkarizadeh et al., 2016). Acute and chronic administration of heroin and morphine also increases pro-inflammatory TNFα, IL-1β, IL-6, and IFNγ, with complementary decreases in anti-inflammatory IL-10 and TGFβ in serum and lymphocytes from human addicts (Chan et al., 2015; Pacifici et al., 2000). Morphine has also been demonstrated to potentiate IL-6 increases in the presence of IL-1β (Bertolucci et al., 1996), suggesting additive or synergistic effects with morphine in combination with other pro-inflammatory cytokines. Although the specific immune cells and/or organs which produce pro-inflammatory cytokines in response to opioids have not completely been identified, T lymphocyte cells have been linked to a pro-inflammatory state. For instance, chronic morphine can shift the T cell population to a differentiated, pro-inflammatory T cell population that secretes pro-inflammatory IL-17 and IFNγ (Cornwell et al., 2013). In contrast, the inflammatory response from other peripheral immune cells in response to opioids, such as natural killer (NK) cells and B lymphocytes, has either not been investigated or have no role in opioid-induced inflammation. For instance, B lymphocytes do not express opioid receptors MOR, KOR, and DOR, and the NF-κB IκB subunit phosphorylation is not changed in response to morphine, suggesting that B lymphocytes are not involved in the pro-inflammatory effects of opioids (Vassou et al., 2008).

Opioids can be paradoxically immunosuppressive through anti-inflammatory cytokines. For instance, methadone, fentanyl, morphine, and buprenorphine can upregulate and release the anti-inflammatory cytokine, IL-4 (Liang et al., 2016a) but IL-4 can induce T cells to differentiate into a pro-inflammatory subtype that is capable of releasing pro-inflammatory IL-6 (Romagnani, 2000). In contrast, heroin and morphine can increase the anti-inflammatory cytokines IL-10 and TGFβ in serum (Pacifici et al., 2000). However, the increased peripheral inflammation and lymphocyte infiltration into the CNS observed after morphine (Grace et al., 2015; Olin et al., 2011) are consistent with its pro-inflammatory effects that, in turn, can cause neuroinflammation.

Liver: Hepatotoxicity is a noted consequence of opioid abuse (Marks and Chapple, 1967; Weller et al., 1984) and contributes to peripheral inflammation through the release of cytokines (Ju and Tacke, 2016). Opioid-exposed animals and human heroin users show evidence of hepatotoxicity through increases in ALT and AST (Marks and Chapple, 1967; Weller et al., 1984). This hepatotoxicity is attributed to oxidative stress damage such that antioxidants block opioid-induced liver damage (Zhang et al., 2004). Additionally, the opioid receptor antagonist, naltrexone, also blocks the opioid-induced liver damage and suggests that the hepatotoxicity is opioid-receptor dependent (Payabvash et al., 2006). While it is unknown what specific pro-inflammatory mediators in the liver are affected by opioids, hepatotoxicity can cause release of TNFα (Blazka et al., 1995; Trautwein et al., 1996) but further study is warranted regarding the contribution of opioid-induced liver dysfunction on peripheral inflammation.

2.2.2.2 Central sources of inflammatory mediators

Unlike the periphery, opioid use has primarily pro-inflammatory effects in the brain resulting in microglia and astrocyte activation after acute and chronic exposure. It is not entirely clear if glial cell activation precedes increases in pro-inflammatory cytokines within the brain after opioid use. However, it is known that activation of glial cells contributes to a pro-inflammatory state that can persist even after cessation of opioid use and contribute to withdrawal symptoms and relapse.

Microglia and astrocytes: Opioids activate both microglia and astrocytes as evidenced by an amoeboid microglial morphology (Doyle et al., 2017; Taylor et al., 2016) and increased GFAP, respectively (Amaral et al., 2016; Song and Zhao, 2001). Once activated by opioids, microglia and astrocytes release pro-inflammatory cytokines and increase neuroinflammation (Sama et al., 2008). For instance, acute and chronic morphine and methadone increase microglia-derived pro-inflammatory cytokines IL-1β, IL-6, and TNFα in the CSF (Hutchinson et al., 2008). These increases can be blocked by COX-2 inhibitors and opioid receptor antagonists, suggesting that opioid-induced neuroinflammation is mediated by microglia-derived COX-2 and opioid receptor stimulation (Hayashi et al., 2016). Alternatively, morphine can also act independent of opioid receptors through direct interaction with TLR4 on CNS endothelial cells to increase IL-1β (Wang et al., 2012). Although opioids increase GFAP in select brain regions (Amaral et al., 2016; Hao et al., 2011; Song and Zhao, 2001), astrocytes and their resident pro-inflammatory mediators have not been studied specifically in response to opioids but studies suggest a correlation between opioid-induced GFAP and increases in TNFα (Hao et al., 2011).

2.2.3 *Neuroinflammatory consequences of opioids*

2.2.3.1 Opioid analgesia, withdrawal, and tolerance

Accumulating evidence suggests that inflammation and glial activation during and after opioid use mediate the analgesic effects, tolerance and withdrawal symptoms to opioids. For example, opioid-induced analgesia is increased by neutralizing pro-inflammatory IL-6 (Hutchinson et al., 2008; Johnston et al., 2004). Similarly, opioid withdrawal symptoms in rats (teeth chattering, wet-dogshakes, jumping, etc.) and activation of astrocytes and microglia are reduced by neutralizing TNFα (Hao et al., 2011) and inhibition of COX-2 (Trang et al., 2002), respectively. Morphine tolerance is also prevented by microglial inhibition (Matsushita et al., 2013). Other anti-inflammatory agents, such as the PDE inhibitor ibudilast, or glial inhibitor fluorocitrate, can also attenuate tolerance and withdrawal symptoms from chronic opioid administration (Bland et al., 2009; Romero-Sandoval et al., 2008; Song and Zhao, 2001). Furthermore, ablation of peripheral immune cells after chronic morphine also attenuates naloxone-precipitated withdrawal symptoms (Dafny and Pellis, 1986), supporting a major influence of peripheral inflammation in opioid withdrawal symptoms. It remains to be determined however, if withdrawal symptoms are due solely to cytokines released from immune cells (Hutchinson et al., 2008), chemokines, or inflammation derived from glial cells (Johnston et al., 2004). Regardless, opioids cause neuroinflammation mediated directly by their action on glial cells and through peripheral inflammation, suggesting cytokines and chemokines in both the periphery and brain can contribute to withdrawal symptoms. By extension, mitigation of inflammation during the withdrawal period may be an effective treatment strategy for relapse to opioid addiction through normalization of the analgesia as well as the suppression of withdrawal and tolerance associated with opioid dependency.

2.2.3.2 Opioid modulation of neurotransmission

Inflammation-induced microglia and astrocyte activation can also affect neurotransmission (Hewett et al., 2012) associated with opioid abuse (Ferrini et al., 2013). For example, chronic morphine treatment upregulates NMDA receptor (NR2A) subunit expression in the periaqueductal gray (PAG) that is blocked by microglia depletion, suggesting a microglia-dependent effect on NMDA receptor expression (Matsushita et al., 2013; Takayama and Ueda, 2005). Morphine-activated microglia can also decrease the K^+-Cl^- cotransporter, causing accumulation of Cl^- in GABA neurons and reduced postsynaptic activation of GABA neurons (Ferrini et al., 2013). Morphine-activated astrocytes similarly decrease GABA neuron excitability within the NAc (Wu et al., 2016) in a NMDA-receptor mediated manner (Martin et al., 1999) via the internalization of AMPAR in astrocytes and decreases in NMDAR-mediated GABAergic neuron excitability in the NAc (Wu et al., 2016). Morphine also decreases extracellular levels of GLU in the hippocampus (Guo et al., 2005) as well as the expression of the GLU transporters GLAST and GLT-1 in the spinal cord (Tai et al., 2007), the latter a possible result of decreased extracellular concentrations of synaptic GLU. Taken together, opioid-activated microglia and astrocytes contribute to suppression of GABA and glutamatergic neurotransmission. However, it remains unclear if pro-inflammatory mediators derived from microglia and astrocytes contribute solely and directly to GABAergic and glutamatergic transmission but both microglia-derived COX-2 and GABA/glutamatergic suppression are necessary for the adverse hyperalgesic effects during morphine withdrawal (Liang et al., 2016b).

2.2.3.3 Opioid-induced cerebrovascular changes

The permeability of the BBB is particularly susceptible during opioid withdrawal due to the enhanced pain and inflammation. For example, increases in inflammatory pain compromise the BBB and decrease tight junction proteins, specifically occludin (Brooks et al., 2005). Inflammatory pain can also induce transient increases in the permeability of the blood-spinal cord barrier (BSCB) (Beggs et al., 2010). It also has been demonstrated that morphine results in BBB breakdown in rats with withdrawal symptoms, demonstrating the important role of withdrawal in mediating BBB breakdown (Sharma and Ali, 2006). The exact mechanisms of BBB breakdown during opioid withdrawal have not been identified, but the roles of inflammation and pain during opioid withdrawal suggest inflammation as a possible mechanism of BBB permeability in response to opioids

2.3 Cannabinoids

2.3.1 Mechanism of action

Cannabinoids are isolated from the cannabis plant and include a number of different derivatives that interact with cannabinoid 1 (CB1) and cannabinoid 2 (CB2) GPCR receptors located peripherally and centrally. These derivatives include partial agonists for CB1/CB2 receptors Δ9-tetrahydrocannabinol (Δ9-THC), Δ8-tetrahydrocannabinol (Δ8-THC), and cannabinol (CBN), and CB1/CB2 antagonist cannabidiol (CBD). In a manner similar to the opioid receptors, CB1/CB2 receptor activation by Δ9-THC, Δ8-THC, or CBN inhibits adenylyl cyclase, activates potassium channels, or reduces calcium channel conductance (Reggio, 2010), resulting in suppression of neurotransmission. In addition, cannabinoid use can result in the internalization of CB1/CB2 receptors, which is thought to underlie tolerance to cannabinoids (Rinaldi-Carmona et al., 1998; Svizenska et al., 2008). Cannabinoids can also more broadly interact with non-CB1/CB2 receptors including TRPV1, GPR55, PPARγ, and GPR18, (Morales and Reggio, 2017; O'Sullivan, 2016; Parker, 2017) that can also mediate inflammation.

2.3.2 Cannabinoids and inflammation

Cannabinoids, unlike the abuse potential and neuroinflammatory effects of the psychostimulants and opioids, are anti-inflammatory and indirectly support the overall hypothesis that inflammation is associated with abuse potential. In general, the cannabinoid system is anti-inflammatory and counteracts pro-inflammatory cytokines through upregulation of anti-inflammatory cytokines (Nagarkatti et al., 2009). For instance, CBD treatment decreases serum pro-inflammatory cytokines IL-4, IL-6, and TNFα in rat models of inflammation (Vuolo et al., 2015; Weiss et al., 2006) and increases the anti-inflammatory cytokine, IL-10. Because of their anti-inflammatory effects, cannabinoids are an attractive drug class for inflammatory diseases including psychostimulant and opioid abuse.

2.3.2.1 Peripheral sources of inflammatory mediators

Peripheral immune cells: The suppression of the immune system by cannabinoids is reflected by changes in cytokines/chemokines across multiple subpopulations of peripheral immune cells. For instance, Δ9-THC decreases pro-inflammatory chemokines (IL-8, MIP-1α, MIP-1β) and cytokines (TNFα, IFNγ, and GM-CSF) in NK cells (Srivastava et al., 1998). Δ9-THC also decreases the pro-inflammatory cytokines TNFα, IL-1β, and IL-6 in human macrophages (Fischer-Stenger et al., 1993; Zhu et al., 1994). Lastly, multiple T cell lines treated with CBD and Δ9-THC exhibit decreases in pro-inflammatory IL-2 and IFNγ (Eisenstein, 2015; Kaplan et al., 2008). In addition to reducing the T cell inflammatory response, cannabinoids polarize the T cell population toward a more anti-inflammatory population. For instance, Δ9-THC treatment decreases IL-2, favoring the differentiation of T cells into an anti-inflammatory Th2 subpopulation of CD4 that, in turn, increases the release of anti-inflammatory TGFβ (Eisenstein, 2015). Δ9-THC also suppresses macrophage chemotaxis (Raborn et al., 2008), suggesting that cannabinoids could block infiltration of peripheral immune cells into the brain. In fact, CBD treatment blocks peripheral leukocyte infiltration in a viral model of multiple sclerosis, further suggestive of the conclusion that the peripheral anti-inflammatory effects of cannabinoids can impact the brain (Mecha et al., 2013). Overall these data are supportive of the role for cannabinoids in counteracting the peripheral pro-inflammatory environment, resulting in a shift of T cells to anti-inflammatory cytokine-releasing cells and protection against neuroinflammation.

Liver: Cannabinoids have been demonstrated to have protective effects in the liver. For instance, CBD treatment attenuates cocaine-induced liver damage and inflammation (Vilela et al., 2015). Δ9-THC administration during Concanavalin A-induced hepatitis also reduces markers of hepatotoxicity, including reductions in serum ALT, AST and histological evidence of liver damage (Hegde et al., 2008). As CB1 and CB2 receptors are upregulated during liver disease compared to a healthy liver, CB antagonists block the therapeutic effects of Δ9-THC (Parfieniuk and Flisiak, 2008). In addition, Δ9-THC was found to reduce hepatotoxicity through suppression of pro-inflammatory cytokines and increases in anti-inflammatory cytokines (Hegde et al., 2008). While conflicting studies demonstrate that chronic marijuana use or synthetic analogs of marijuana, such as Spice/K2, can actually induce hepatotoxicity (Borini et al., 2004; Sheikh et al., 2014), the majority of evidence favors a protective effect of cannabinoid receptor signaling on liver disease, demonstrating a therapeutic effect of cannabinoids through their anti-inflammatory action.

2.3.2.2 Central sources of neuroinflammatory and anti-neuroinflammatory mediators

Cannabinoids are also implicated as anti-inflammatory and protective in several neurological diseases that have an underlying inflammatory component (Katzman et al., 2016). Both Δ9-THC and CBD acutely decrease inflammatory cytokines in the brain and mitigate neuroinflammation by reducing microglial and astrocyte activation (Kozela et al., 2010). However, both dose and duration of exposure of Δ9-THC and CBD are important considerations for their protective effects.

Microglia and astrocytes: Microglia and astrocytes are activated in various inflammatory disease models and are counteracted by cannabinoid treatment (Martin-Moreno et al., 2011; Mecha et al., 2013). For example, CBD administration reduces microglia activation as evidenced by reduced microglial calcium influx in an Alzheimer model wherein mice receive glial-activating amyloid-beta injections into the brain (Martin-Moreno et al., 2011). CBD also blocks microglial activation and decreases TNFα and IL-1β in a viral model of multiple sclerosis (Mecha et al., 2013). Similar effects are observed with Δ9-THC which attenuates the release of IL-1β, IL-6, and IFNγ from LPS-stimulated microglia (Kozela et al., 2010) in a manner that is likely mediated by the anti-inflammatory cytokine, IL-10 (Weiss et al., 2006). In addition, cannabinoids exhibit similar anti-inflammatory effects on astrocytes. For instance, Δ9-THC can block increases in striatal GFAP induced by Meth (Castelli et al., 2014), and CBD can block cortical, hippocampal, and striatal GFAP in models of artery occlusion (Kozela et al., 2010).

Despite numerous reports of an anti-inflammatory effect of CBD and Δ9-THC, chronic and high doses of Δ9-THC can upregulate CB2 receptors, activate microglia, increase TNFα, and reduce IL-10 leading to a pro-inflammatory state (Zamberletti et al., 2015). This suggests that the pro- and anti-inflammatory actions of cannabinoids are dependent on dose and duration of exposure.

3. Implications

3.1 Addiction and neurotoxicity

Inflammation produced by drugs of abuse has been implicated in mediating drug-dependency and neurodegeneration. For example, inflammation induced with LPS injections augments the locomotor effects of psychostimulants and withdrawal from alcohol (Breese et al., 2007; Tien et al., 2011), effects associated with increases in dopamine transmission. LPS also has been shown to reinstate cocaine seeking in cocaine-dependent rats, an effect blocked by injection of a TLR4 or IL-1β receptor inhibitor directly into the ventral tegmental area (Brown et al., 2018). Thus, it can be speculated that therapeutics that target and effectively reduce drug-induced inflammation may be a viable treatment for drug-dependency and may also prevent the neurotoxicity associated with psychostimulants and opioids.

3.2 Cannabinoids

There have been numerous reviews supporting the anti-inflammatory role of CBD in the brain (Nagarkatti et al., 2009). For example, CBD treatment can reduce increases in neuronal GFAP, NO, and IL-1β induced in various animal models of disease (Zuardi, 2008). Δ9-THC decreases METH-induced increases in striatal GFAP (Castelli et al., 2014) and decreases psychostimulant-induced GFAP and DA depletions (Touriño et al., 2010). In addition, the anti-inflammatory effects of CBD have been attributed to its ability to reduce drug-dependency (Prud'homme et al., 2015). Along these lines, CBD reduces relapse to opioids (Ren et al., 2009) and both CBD and Δ9-THC reduce conditioned place preference (CPP) to cocaine and Amph (Parker et al., 2004), but in a manner independent of CB1/CB2 receptor activation. Cannabinoids can also reduce neurotoxicity associated with drugs of abuse through their anti-oxidant action. For instance, excitotoxicity in neurons is prevented by CBD independent of its action on CB1/CB2 receptors and is attributed to the potent antioxidant activity of CBD (Hampson et al., 1998). Thus, there is emerging evidence that cannabinoids can decrease drug-induced inflammation, reduce drug-dependency, and prevent neurotoxicity associated with psychostimulants and opioids.

3.3 PDE, COX and drug-induced inflammation

PDE inhibitors have been shown to prevent drug-induced inflammation, as demonstrated by the ability of the PDE4 inhibitor ibudilast to reduce Meth-induced increases in GFAP (Charntikov et al., 2015). Ibudilast has also been reported to enhance extinction behavior to Meth (Charntikov et al., 2015). Similarly, the PDE9 inhibitor BAY-73-6691 enhances extinction of conditioned place preference (CPP) to psychostimulants (Liddie et al., 2012), and PDE10A inhibitors prevent morphine-induced CPP (Mu et al., 2014). PDE inhibitors also enhance morphine analgesia in models of neuropathic pain thereby permitting the use of lower doses to circumvent the adverse effects of opioids (Romero-Sandoval et al., 2008).

Similar effects have been observed in studies with drug addicted humans where intervention with anti-inflammatory drugs has been studied. Humans addicted to opioids were assessed for craving and treated with PDE4 inhibitor ibudilast during opioid detoxification and withdrawal (Metz et al., 2017). Individuals treated with ibudilast showed reduced "drug-liking" after oxycodone administration and showed lower drug-breakpoint. They also reported less pain and less "bothersome" feelings, similar to other reports of reduced opioid withdrawal symptoms produced by ibudilast (Cooper et al., 2015). Moreover, ibudilast also reduced feelings of reward associated with Meth infusion in humans addicted to Meth (Worley et al., 2016). Collectively, the use of PDE inhibitors to combat drug-dependency appears to hold promise as a therapy for drug addiction but it remains unclear whether their effects can be attributed directly to their anti-inflammatory effects.

Analogous to PDE inhibitors, COX inhibitors have also been demonstrated to reduce drug-induced inflammation and neurotoxicity associated with drugs of abuse including Meth and alcohol (Blaker and Yamamoto, 2017; Northrop and Yamamoto, 2012) but the effects of COX inhibition on drug-dependency are lacking. While COX-2 inhibitors can prevent withdrawal effects from ethanol abuse (Dhir et al., 2005), no studies to date have examined the efficacy of COX inhibition on psychostimulant and opioid drug-dependency.

As noted above, damage to the liver can cause the release of inflammatory mediators; however, surprisingly few studies have examined the ability of drugs of abuse to cause hepatotoxicity. Along these lines, binge dosing of Meth causes hepatotoxicity (Halpin et al., 2013) and subsequent increases in ammonia within the brain and periphery to produce pro-inflammatory and neurotoxic effects. Although the prevention of increases in ammonia by lactulose, which binds ammonia and enhances its elimination, blocks the excitotoxic effects of Meth (Halpin et al., 2013, 2014b), no studies have examined if lactulose and its prevention of the liver damage-induced increases in ammonia can alter drug dependency.

4. Summary and conclusions

Psychostimulant and opioid abuse can result in drug dependency and neurotoxicity in humans but no current therapeutics exist. Given that both drug classes have common underlying mechanisms that contribute to

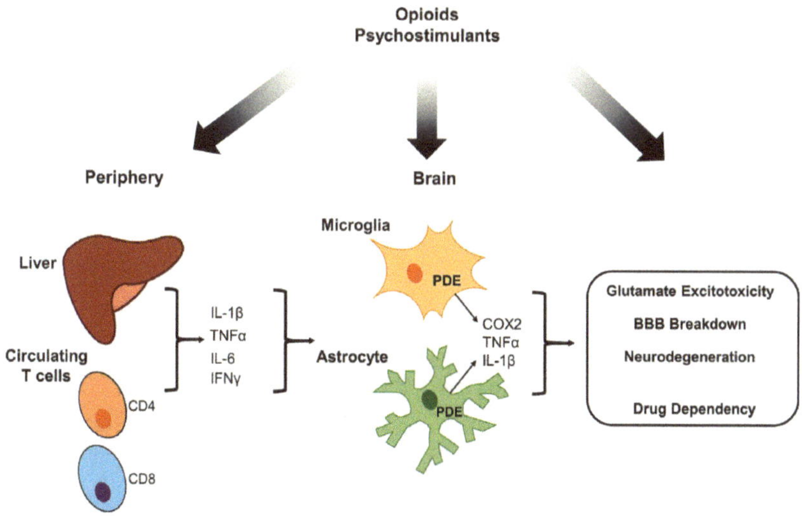

Fig. 1 Schematic of the peripheral and central inflammatory effects produced by opioids and psychostimulants that contribute to neurotoxicity and drug-dependency. Opioids and psychostimulants produce inflammation in the periphery by acting on the liver and circulating T cells, in addition to other circulating immune cells. The liver and T cells increase production and release of inflammatory cytokines in the blood, including IL-1β, TNFα, IL-6, and IFNγ. Increases in peripheral cytokines and the drugs of abuse themselves result in microglia and astrocyte activation. Consequently, COX-2, TNFα, and IL-1β increase in the brain in a PDE-specific manner. Ultimately, the peripheral and central inflammation induced by drugs of abuse result in the neurotoxicity and drug dependency associated with abuse opioids and psychostimulants.

inflammation (Fig. 1) and that inflammation mediates multiple aspects of drug abuse, these mechanisms may be potential targets for treatment (Fig. 2). Along these lines, both psychostimulants and opioids have overlapping inflammatory effects through increases in cytokines and shifts in T cell populations capable of releasing pro-inflammatory cytokines. Subsequent to the immediate increases in cytokines, microglial and astrocyte activation propagate the inflammatory cascade through interaction at cytokine receptors on microglia and astrocytes, and via the synthesis of additional inflammatory mediators through PDE and COX-dependent mechanisms. The consequences of the inflammatory cascade in the brain can result in withdrawal and addiction as well as neurotoxicity and BBB permeability. Based on the above described mechanisms, points of intervention may include inhibition of PDE in the periphery and CNS, prevention of COX-2 activation in microglia within the CNS, and the anti-inflammatory effects of CBD that can serve as targets for treatment of the behavioral and neurological consequences associated with the abuse of psychostimulants and opioids.

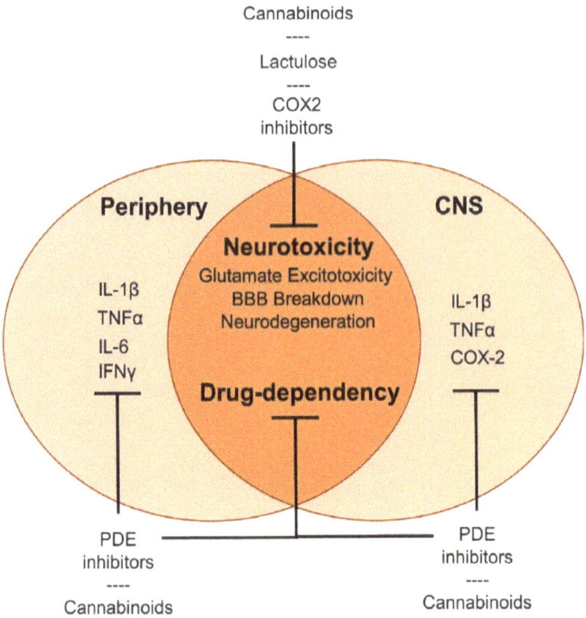

Fig. 2 Inhibition of peripheral and central inflammation and their overlapping effects to alleviate toxicity and drug-dependency. Opioids and psychostimulants produce peripheral and central inflammation, demonstrated by increases in peripheral IL-1β, TNFα, IL-6, and IFNγ, and central COX-2, TNFα, and IL-1β. The combined effects of these inflammatory increases produce neurotoxicity and drug-dependency. PDE inhibitors can block peripheral and central increases in inflammatory mediators and have also been shown to alleviate drug-dependency. Cannabinoids can similarly reduce peripheral and central inflammatory mediators and drug-dependency, and could also play a role in reducing neurotoxicity associated with drugs of abuse. While COX-2 inhibitors and lactulose have not yet been demonstrated to reduce drug-dependency, they have been shown to reduce neurotoxicity to drugs of abuse.

Acknowledgments

This work was supported by NIH grants DA035499 and DA042737.

References

Al-Hasani, R., Bruchas, M.R., 2011. Molecular mechanisms of opioid receptor-dependent signaling and behavior. Anesthesiology 115, 1363–1381.

Amaral, G.F., Dossa, P.D., Viebig, L.B., Konno, F.T.C., Consoli, A., Martins, M.F.M., Viani, F.C., Bondan, E.F., 2016. Astrocytic expression of GFAP and serum levels of IL-1β and TNF-α in rats treated with different pain relievers. Braz. J. Pharm. Sci. 52 (4), 623–633.

American Psychiatric Association, 2013. Diagnostic and Statistical Manual of Mental Disorders, fifth ed. Arlington, Virginia.

Ares-Santos, S., Granado, N., Espadas, I., Martinez-Murillo, R., Moratalla, R., 2014. Methamphetamine causes degeneration of dopamine cell bodies and terminals of the nigrostriatal pathway evidenced by silver staining. Neuropsychopharmacology 39 (5), 1066–1080.

Asadikaram, G., Igder, S., Jamali, Z., Shahrokhi, N., Najafipour, H., Shokoohi, M., Jafarzadeh, A., Kazemi-Arababadi, M., 2015. Effects of different concentrations of opium on the secretion of interleukin-6, interferon-γ and transforming growth factor beta cytokines from jurkat cells. Addict. Health. 7 (1–2), 47–53.

Avila, J.A., Zanca, R.M., Shor, D., Paleologos, N., Alliger, A.A., Figueiredo-Pereira, M.E., Serrano, P.A., 2018. Chronic voluntary oral methamphetamine induces deficits in spatial learning and hippocampal protein kinase Mzeta with enhanced astrogliosis and cyclooxygenase-2 levels. Heliyon 4 (2), e00509.

Babu, S., Nutman, T.B., 2003. Proinflammatory cytokines dominate the early immune response to filarial parasites. J. Immunol. 171 (12), 6723–6732.

Beardsley, P.M., Shelton, K.L., Hendrick, E., Johnson, K.W., 2010. The glial cell modulator and phosphodiesterase inhibitor, AV411 (ibudilast), attenuates prime- and stress-induced methamphetamine relapse. Eur. J. Pharmacol. 637, 102–108.

Beggs, S., Liu, X.J., Kwan, C., Salter, M.W., 2010. Peripheral nerve injury and TRPV1-expressing primary afferent C-fibers cause opening of the blood–brain barrier. Mol. Pain 6, 74.

Bertolucci, M., Perego, C., De Simoni, M.G., 1996. Central opiate modulation of peripheral IL-6 in rats. Neuroreport 7 (6), 1181–1184.

Bezzi, P., Domercq, M., Brambilla, L., Galli, R., Schols, D., De Clercq, E., Vescovi, A., Bagetta, G., Kollias, G., Meldolsi, J., Volterra, A., 2001. CXCR4-activated astrocyte glutamate release via TNFα: amplification by microglia triggers neurotoxicity. Nat. Neurosci. 4 (7), 702–710.

Bezzi, P., Gundersen, V., Galbete, J.L., Seifert, G., Steinhäuser, C., Pilati, E., Volterra, A., 2004. Astrocytes contain a vesicular compartment that is competent for regulated exocytosis of glutamate. Nat. Neurosci. 7 (6), 613–620.

Blaker, A.L., Yamamoto, B.K., 2017. Methamphetamine-induced brain injury and alcohol drinking. J. Neuroimmune Pharmacol. 13 (1), 53–63.

Bland, S.T., Hutchinson, M.R., Maier, S.F., Watkins, L.R., Johnson, K.W., 2009. The glial activation inhibitor AV411 reduces morphine-induced nucleus accumbens dopamine release. Brain Behav. Immun. 23 (4), 492–497.

Blazka, M.E., Wilmer, J.L., Holladay, S.D., Wilson, R.E., Luster, M.I., 1995. Role of proinflammatory cytokines in acetaminophen hepatotoxicity. Toxicol. Appl. Pharmacol. 133 (1), 43–52.

Borini, P., Guimarães, R.C., Borini, S.B., 2004. Possible hepatotoxicity of chronic marijuana usage. Sao Paulo Med. J. 122 (3), 110–116.

Bowers, M.S., Kalivas, P.W., 2003. Forebrain astroglial plasticity is induced following withdrawal from repeated cocaine administration. Eur. J. Neurosci. 17 (6), 1273–1278.

Brahmachari, S., Fung, Y.K., Pahan, K., 2006. Induction of glial fibrillary acidic protein expression in astrocytes by nitric oxide. J. Neurosci. 26 (18), 4930–4939.

Breese, G.R., Knapp, D.J., Overstreet, D.H., Navarro, M., Wills, T.A., Angel, R.A., 2007. Repeated lipopolysaccharide (LPS) or cytokine treatments sensitize ethanol withdrawal-induced anxiety-like behavior. Neuropsychopharmacology 33 (4), 867–876.

Brooks, T.A., Hawkins, B.T., Huber, J.D., Egleton, R.D., Davis, T.P., 2005. Chronic inflammatory pain leads to increased blood–brain barrier permeability and tight junction protein alterations. Am. J. Physiol. Heart Circ. Physiol. 289 (2), H738–H743.

Brown, K.T., Levis, S.C., O'Neill, C.E., Northcutt, A.L., Fabisiak, T.J., Watkins, L.R., Bachtell, R.K., 2018. Innate immune signaling in the ventral tegmental area contributes to drug-primed reinstatement of cocaine seeking. Brain Behav. Immun. 67, 130–138.

Burns, A., Ciborowski, P., 2016. Acute exposure to methamphetamine alters TLR9-mediated cytokine expression in human macrophage. Immunobiology 221 (2), 199–207.

Carambia, A., Frenzel, C., Bruns, O.T., Schwinge, D., Reimer, R., Hohenberg, H., Huber, S., Tiegs, G., Schramm, C., Lohse, A.W., Herkel, J., 2013. Inhibition of inflammatory CD4 T cell activity by murine liver sinusoidal endothelial cells. J. Hepatol. 58 (1), 112–118.

Carlson, N.G., Rojas, M.A., Black, J.D., Redd, J.W., Hille, J., Hill, K.E., Rose, J.W., 2009. Microglial inhibition of neuroprotection by antagonists of the EP1 prostaglandin E2 receptor. J. Neuroinflammation 6 (1), 5.

Castelli, M.P., Madeddu, C., Casti, A., Casu, A., Casti, P., Scherma, M., Fattore, L., Fadda, P., Ennas, M.G., 2014. Δ9-Tetrahydrocannabinol prevents methamphetamine-induced neurotoxicity. PloS One 9 (5), 1–11.

Chahl, L., 1996. Experimental and clinical pharmacology: opioids—mechanisms of action. Aust. Prescr. 19, 63–65.

Chan, Y.Y., Yang, S.N., Lin, J.C., Chang, J.L., Lin, J.G., Lo, W.Y., 2015. Inflammatory response in heroin addicts undergoing methadone maintenance treatment. Psychiatry Res. 226 (1), 230–234.

Charntikov, S., Pittenger, S.T., Thapa, I., Bastola, D.R., Bevins, R.A., Pendyala, G., 2015. Ibudilast reverses the decrease in the synaptic signaling protein phosphatidylethanolamine-binding protein 1 (PEBP1) produced by chronic methamphetamine intake in rats. Drug Alcohol Depend. 152, 15–23.

Cichewicz, D.L., 2004. Synergistic interactions between cannabinoid and opioid analgesics. Life Sci. 74 (11), 1317–1324.

Collins, S.A., Huff, C., Chiaia, N., Gudelsky, G.A., Yamamoto, B.K., 2016. 3,4-Methylenedioxymethamphetamine increases excitability in the dentate gyrus: role of 5HT2A receptor-induced PGE2 signaling. J. Neurochem. 136, 1074–1084.

Cooper, Z.D., Johnson, K.W., Pavlicova, M., Glass, A., Vosburg, S.K., Sullivan, M.A., Manubay, J.M., Martinez, D.M., Jones, J.D., Saccone, P.A., Comer, S.D., 2015. The effects of ibudilast, a glial activation inhibitor, on opioid withdrawal symptoms in opioid-dependent volunteers. Addict. Biol. 21, 895–903.

Cornwell, W.D., Lewis, M.G., Fan, X., Rappaport, J., Rogers, T.J., 2013. Effect of chronic morphine administration on circulating T cell population dynamics in rhesus macaques. J. Neuroimmunol. 265 (1), 43–50.

Costello, E.J., Copeland, W.E., Shanahan, L., Worthman, C.M., Angold, A., 2014. C-reactive protein and substance use disorders in adolescence and early adulthood: a prospective analysis. Drug Alcohol Depend. 133 (2), 712–717.

Dafny, N., Pellis, N., 1986. Evidence that opiate addiction is in part an immune response. Neuropharmacology 25 (8), 815–818.

Dhir, A., Naidu, P., Kulkarni, S., 2005. Protective effect of cyclooxygenase-2 (COX-2) inhibitors but not non-selective cyclooxygenase (COX)-inhibitors on ethanol withdrawal-induced behavioural changes. Addict. Biol. 10, 329–335.

D'Mello, C., Almishri, W., Liu, H., Swain, M.G., 2017. Interactions between platelets and inflammatory monocytes affect sickness behavior in mice with liver inflammation. Gastroenterology 153 (5), 1416–1428.e2.

Doyle, H.H., Eidson, L.N., Sinkiewicz, D.M., Murphy, A.Z., 2017. Sex differences in microglia activity within the periaqueductal gray of the rat: a potential mechanism driving the dimorphic effects of morphine. J. Neurosci. 37 (12), 3202–3214.

Eisenstein, T.K., 2015. Effects of cannabinoids on T-cell function and resistance to infection. J. Neuroimmune Pharmacol. 10 (2), 204–216.

Epstein, D.H., Preston, K.L., Jasinski, D.R., 2006. Abuse liability, behavioral pharmacology, and physical-dependence potential of opioids in humans and laboratory animals: lessons from tramadol. Biol. Psychol. 73 (1), 90–99.

Ersche, K.D., Döffinger, R., 2017. Inflammation and infection in human cocaine addiction. Curr. Opin. Behav. Sci. 13, 203–209.

Ersche, K.D., Hagan, C.C., Smith, D.G., Abbott, S., Jones, P.S., Apergis-Schoute, A.M., Döffinger, R., 2014. Aberrant disgust responses and immune reactivity in cocaine-dependent men. Biol. Psychiatry 75 (2), 140–147.

Eugenin, E.A., Greco, J.M., Frases, S., Nosanchuk, J.D., Martinez, L.R., 2013. Methamphetamine alters blood brain barrier protein expression in mice, facilitating central nervous system infection by neurotropic cryptococcus neoformans. J. Infect. Dis. 208 (4), 699–704.

Fehnel, C.R., Ayres, A.M., Rost, N.S., 2014. Socioeconomic status does not predict cocaine use among ischemic stroke patients: a nested case-control study. JRSM Cardiovasc. Dis. 3, 1–5.

Ferrini, F., Trang, T., Mattioli, T.-A.M., Laffray, S., Del'Guidice, T., Lorenzo, L.-E., Castonguay, A., Doyon, N., Zhang, W., Godin, A.G., Mohr, D., Beggs, S., Vandal, K., Beaulieu, J.-M., Cahill, C.M., Salter, M.W., De Koninck, Y., 2013. Morphine hyperalgesia gated through microglia-mediated disruption of neuronal Cl^- homeostasis. Nat. Neurosci. 16, 183–192.

Fischer-Stenger, K., Dove Pettit, D.A., Cabral, G.A., 1993. Δ9-Tetrahydrocannabinol inhibition of tumor necrosis factor-α: suppression of post-translational events. J. Pharmacol. Exp. Ther. 267 (3), 1558–1565.

Fox, H.C., D'Sa, C., Kimmerling, A., Siedlarz, K.M., Tuit, K.L., Stowe, R., Sinha, R., 2012. Immune system inflammation in cocaine dependent individuals: implications for medications development. Hum. Psychopharmacol. 27 (2), 156–166.

Frey, B.N., Andreazza, A.C., Cereser, K.M., Martins, M.R., Petronilho, F.C., de Souza, D.F., Tramontina, F., Gonçalves, C.A., Quevedo, J., Kapczinski, F., 2006. Evidence of astrogliosis in rat hippocampus after D-amphetamine exposure. Prog. Neuro-Psychopharmacol. Biol. Psychiatry 30 (7), 1231–1234.

Friend, D.M., Keefe, K.A., 2013. Glial reactivity in resistance to methamphetamine-induced neurotoxicity. J. Neurochem. 125, 566–574.

Gavériaux-Ruff, C., Kieffer, B., 2002. Opioid receptor genes inactivated in mice: the highlights. Neuropeptides 36 (2–3), 62–71.

Giroux, M., Descoteaux, A., 2000. Cyclooxygenase-2 expression in macrophages: modulation by protein kinase C. J. Immunol. 165 (7), 3985–3991.

Gonçalves, J., Martins, T., Ferreira, R., Milhazes, N., Borges, F., Ribeiro, C.F., Malva, J.O., Macedo, T.R., Silva, A.P., 2008. Methamphetamine-induced early increase of IL-6 and TNF-α mRNA expression in the mouse brain. Ann. N. Y. Acad. Sci. 1139, 103–111.

Goodman, J.H., Solviter, R.S., 1993. Cocaine neurotoxicity and altered neuropeptide Y immunoreactivity in the rat hippocampus; a silver degeneration and immunocytochemical study. Brain Res. 616 (1–2), 263–272.

Grace, P.M., Maier, S.F., Watkins, L.R., 2015. Opioid-induced central immune signaling: implications for opioid analgesia. Headache 55, 475–489.

Guo, M., Xu, N., Li, Y., Yang, J., Wu, C., Pei, G., 2005. Morphine modulates glutamate release in the hippocampal CA1 area in mice. Neurosci. Lett. 381 (1–2), 12–15.

Guo, C.H., Bai, L., Wu, H.H., Yang, J., Cai, G.H., Wang, X., Wu, S.X., Ma, W., 2016. The analgesic effect of rolipram is associated with the inhibition of the activation of the spinal astrocytic JNK/CCL2 pathway in bone cancer pain. Int. J. Mol. Med. 38, 1433–1442.

Halpin, L.E., Gunning III, W.T., Yamamoto, B.K., 2013. Methamphetamine causes acute hyperthermia-dependent liver damage. Pharmacol. Res. Perspect. 1 (1), e00008.

Halpin, L.E., Northrop, N.A., Yamamoto, B.K., 2014a. Ammonia mediates methamphetamine-induced increases in glutamate and excitotoxicity. Neuropsychopharmacology 39, 1031–1038.

Halpin, L.E., Collins, S.A., Yamamoto, B.K., 2014b. Neurotoxicity of methamphetamine and 3,4-methylenedioxymethamphetamine. Life Sci. 97 (1), 37–44.

Hampson, A.J., Grimaldi, M., Axelrod, J., Wink, D., 1998. Cannabidiol and (−)Δ9-tetrahydrocannabinol are neuroprotective antioxidants. Proc. Natl. Acad. Sci. U.S.A. 95 (14), 8268–8273.
Hao, S., Liu, S., Zheng, X., Zheng, W., Ouyang, H., Mata, M., Fink, D.J., 2011. The role of TNFα in the periaqueductal gray during naloxone-precipitated morphine withdrawal in rats. Neuropsychopharmacology 36, 664–676.
Hayashi, Y., Morinaga, S., Zhang, J., Satoh, Y., Meredith, A., Nakata, T., Wu, Z., Kohsaka, S., Inoue, K., Nakanishi, H., 2016. BK channels in microglia are required for morphine-induced hyperalgesia. Nat. Commun. 7, 11697.
Heesch, C.M., Wilhelm, C.R., Ristich, J., Adnane, J., Bontempo, F.A., Wagner, W.R., 2000. Cocaine activates platelets and increases the formation of circulating platelet containing microaggregates in humans. Heart 83, 688–695.
Hegde, V.L., Hegde, S., Cravatt, B.F., Hofseth, L.J., Nagarkatti, M., Nagarkatti, P.S., 2008. Attenuation of experimental autoimmune hepatitis by exogenous and endogenous cannabinoids: involvement of regulatory T cells. Mol. Pharmacol. 74 (1), 20–33.
Hewett, S.J., Jackman, N.A., Claycomb, R.J., 2012. Interleukin-1β in central nervous system injury and repair. Eur. J. Neurodegener. Dis. 1 (2), 195–211.
Hoogland, I.C.M., Houbolt, C., van Westerloo, D.J., van Gool, W.A., van de Beek, D., 2015. Systemic inflammation and microglial activation: systematic review of animal experiments. J. Neuroinflammation 12, 114.
Horvath, M., Kovacs, G., Kovari, V., Majtenyi, K., Hurd, Y., Keller, E., 2007. Heroin abuse is characterized by discrete mesolimbic dopamine and opioid abnormalities and exaggerated nuclear receptor-related 1 transcriptional decline with age. J. Neurosci. 27 (49), 13371–13375.
Huang, M.C., Yang, S.Y., Lin, S.K., Chen, K.Y., Chen, Y.Y., Kuo, C.J., Hung, Y.N., 2016. Risk of cardiovascular diseases and stroke events in methamphetamine users. J. Clin. Psychiatry 77, 1396–1403.
Hutchinson, M., Coats, B., Lewis, S.S., Zhang, Y., Sprunger, D.B., Rezvani, N., Baker, E.M., Jekich, B.M., Wieseler, J.L., Somogyi, A.A., Martine, D., Poole, S., Judd, C.M., Maier, S.F., Watkins, L.R., 2008. Proinflammatory cytokines oppose opioid-induced acute and chronic analgesia. Brain Behav. Immun. 22 (8), 1178–1189.
Jin, S.L., Lan, L., Zoudilova, M., Conti, M., 2005. Specific role of phosphodiesterase 4B in lipopolysaccharide-induced signaling in mouse macrophages. J. Immunol. 175 (3), 1523–1531.
Johnson, R.A., Eshelman, A.J., Meyers, T., Neve, K.A., Janowsky, A., 1998. [3H]substrate- and cell-specific effects of uptake inhibitors on human dopamine and serotonin transporter-mediated efflux. Synapse 30 (1), 97–106.
Johnston, I.N., Milligan, E.D., Wieseler-Frank, J., Frank, M.G., Zapata, V., Campisi, J., Langer, S., Martin, D., Green, P., Fleshner, M., Leinwand, L., Maier, S.F., Watkins, L.R., 2004. A role for proinflammatory cytokines and fractalkine in analgesia, tolerance, and subsequent pain facilitation induced by chronic intrathecal morphine. J. Neurosci. 24 (33), 7353–7365.
Jones, S.R., Joseph, J.D., Barak, L.S., Caron, M.G., Wightman, R.M., 1999. Dopamine neuronal transport kinetics and effects of amphetamine. J. Neurochem. 73 (6), 2406–2414.
Ju, C., Tacke, F., 2016. Hepatic macrophages in homeostasis and liver diseases: from pathogenesis to novel therapeutic strategies. Cell. Mol. Immunol. 13, 316–327.
Kaplan, B.L.F., Springs, A.E.B., Kaminski, N.E., 2008. The profile of immune modulation by cannabidiol (CBD) involves deregulation of nuclear factor of activated t cells (NFAT). Biochem. Pharmacol. 76 (6), 726–737.
Katzman, M., Furtado, M., Anand, L., 2016. Targeting the endocannabinoid system in psychiatric illness. J. Clin. Psychopharmacol. 36 (6), 691–703.

Kelly, K.A., Miller, D.B., Bowyer, J.F., O'Callaghan, J.P., 2012. Chronic exposure to corticosterone enhances the neuroinflammatory and neurotoxic responses to methamphetamine. J. Neurochem. 122 (5), 995–1009.

Kleven, M.S., Wolverton, W.L., Seiden, L.S., 1988. Lack of long-term monoamine depletions following repeated or continuous exposure to cocaine. Brain Res. Bull. 21 (2), 233–237.

Korn, T., Mitsdoerffer, M., Croxford, A.L., Awasthi, A., Dardalhon, V.A., Galileos, G., Vollmar, P., Stritesky, G.L., Kaplan, M.H., Waisman, A., Kuchroo, V.K., Oukka, M., 2008. IL-6 controls Th17 immunity in vivo by inhibiting the conversion of conventional T cells into Foxp3+ regulatory T cells. Proc. Natl. Acad. Sci. U.S.A. 105 (47), 18460–18465.

Kovacs, G.G., Horvath, M.C., Majtenyi, K., Lutz, M.I., Hurd, Y.L., Keller, E., 2015. Heroin abuse exaggerates age-related deposition of hyperphosphorylated tau and p62-positive inclusions. Neurobiol. Aging 36 (11), 3100–3107.

Kozela, E., Pietr, M., Juknat, A., Rimmerman, N., Levy, R., Vogel, Z., 2010. Cannabinoids Δ9-tetrahydrocannabinol and cannabidiol differentially inhibit the lipopolysaccharide-activated NF-κB and interferon-β/STAT proinflammatory pathways in BV-2 microglial cells. J. Biol. Chem. 285 (3), 1616–1626.

Kubera, M., Filip, M., Basta-Kaim, A., Nowak, E., Budziszewska, B., Tetich, M., … Przegalinski, E., 2002. The effect of amphetamine sensitization on mouse immunoreactivity. J. Physiol. Pharmacol. 53 (2), 233–242.

Kuo, H.-W., Liu, T.-H., Tsou, H.-H., Hsu, Y.-T., Wang, S.-C., Fang, C.-P., Liu, C.-C., Chen, A.C.H., Liu, Y.-L., 2018. Inflammatory chemokine eotaxin-1 is correlated with age in heroin dependent patients under methadone maintenance therapy. Drug Alcohol Depend. 183, 19–24.

Labib, R., Turkall, R., Abdel-Rahman, M.S., 2003. Endotoxin potentiates cocaine-mediated hepatotoxicity by nitric oxide and reactive oxygen species. Int. J. Toxicol. 22 (4), 305–316.

Lan, X., Han, X., Li, Q., Yang, Q.-W., Wang, J., 2017. Modulators of microglial activation and polarization after intracerebral hemorrhage. Nat. Rev. Neurol. 13, 420–433.

Lappin, J.M., Darke, S., Farrell, M., 2017. Stroke and methamphetamine use in young adults: a review. J. Neurol. Neurosurg. Psychiatry 88, 1079–1091.

Lashkarizadeh, M.R., Garshasbi, M., Shabani, M., Dabiri, S., Hadavi, H., Manafi-Anari, H., 2016. Impact of opium addiction on levels of pro- and anti-inflammatory cytokines after surgery. Addict. Health. 8 (1), 9–15.

LaVoie, M.J., Card, J.P., Hastings, T.G., 2004. Microglial activation precedes dopamine terminal pathology in methamphetamine-induced neurotoxicity. Exp. Neurol. 187 (1), 47–57.

Lewitus, G.M., Konefal, S.C., Greenhalgh, A.D., Pribiag, H., Augereau, K., Stellwagen, D., 2016. Microglial TNF-α suppresses cocaine-induced plasticity and behavioral sensitization. Neuron 90 (3), 483–491.

Liang, N.Y., Rutledge, C.O., 1981. Comparison of the release of [3H]dopamine from isolated corpus striatum by amphetamine, fenfluramine and unlabeled dopamine. Biochem. Pharmacol. 31 (6), 983–992.

Liang, Y., Chu, H., Jiang, Y., Yuan, L., 2016a. Morphine enhances IL-1β release through toll-like receptor 4-mediated endocytic pathway in microglia. Purinergic Signal. 12 (4), 637–645.

Liang, X., Liu, R., Chen, C., Ji, F., Li, T., 2016b. Opioid system modulates the immune function: a review. Transl. Perioper. Pain Med. 1 (1), 5–13.

Liao, K., Guo, M., Niu, F., Yang, L., Callen, S.E., Buch, S., 2016. Cocaine-mediated induction of microglial activation involves the ER stress-TLR2 axis. J. Neuroinflammation 13 (1), 33.

Liddelow, S.A., Guttenplan, K.A., Clarke, L.E., Bennett, F.C., Bohlen, C.J., Schirmer, L., Bennett, M.L., Munch, A.E., Chung, W.S., Peterson, T.C., Wilton, D.K., Frouin, A., Napier, B.A., Panicker, N., Kumar, M., Buckwalter, M.S., Rowitch, D.H., Dawson, V.L., Dawson, T.M., Stevens, B., Barres, B.A., 2017. Neurotoxic reactive astrocytes are induced by activated microglia. Nature 541, 481–487.

Liddie, S., Anderson, K.L., Paz, A., Itzhak, Y., 2012. The effect of phosphodiesterase inhibitors on the extinction of cocaine-induced conditioned place preference in mice. J. Psychopharmacol. 26, 1375–1382.

Ligeiro de Oliveira, A.P., Lino-dos-Santos-Franco, A., Acceturi, B.G., Hamasato, E.K., Machado, I.D., Gimenes Júnior, J.A., Vieira, R., de, P., Damazo, A.S., Farsky, S.H.P., Tavares-de-Lima, W., Palermo-Neto, J., 2012. Long-term amphetamine treatment exacerbates inflammatory lung reaction while decreases airway hyper-responsiveness after allergic stimulus in rats. Int. Immunopharmacol. 14, 523–529.

Lipari, R.N., Van Horn, S.L., 2017. Trends in Substance Use Disorders Among Adults Aged 18 or Older. The CBHSQ ReportCenter for Behavioral Health Statistics and Quality, Substance Abuse and Mental Health Services Administration, Rockville, Maryland.

Liu, X., Silverstein, P.S., Singh, V., Shah, A., Qureshi, N., Kumar, A., 2012. Methamphetamine increases LPS-mediated expression of IL-8, TNF-α and IL-1β in human macrophages through common signaling pathways. PLoS One 7, e33822.

Lorez, H., 1981. Fluorescence histochemistry indicates damage of striatal dopamine nerve terminals in rats after multiple doses of methamphetamine. Life Sci. 28 (8), 911–916.

Lumadue, J.A., Lanzkron, S.M., Kennedy, S.D., Kuhl, D.T., Kickler, T.S., 1996. Cytokine induction of platelet activation. Am. J. Clin. Pathol. 106 (6), 795–798.

Mark, K.A., Soghomonian, J.J., Yamamoto, B.K., 2004. High-dose methamphetamine acutely activates the striatonigral pathway to increase striatal glutamate and mediate long-term dopamine toxicity. J. Neurosci. 24 (50), 11449–11456.

Marks, V., Chapple, P., 1967. Hepatic dysfunction in heroin and cocaine users. Addiction 62, 189–195.

Martin, G., Ahmed, S., Blank, T., Spiess, J., Koob, G., Siggins, G., 1999. Chronic morphine treatment alters NMDA receptor-mediated synaptic transmission in the nucleus accumbens. J. Neurosci. 19 (20), 9081–9089.

Martin-Moreno, A.M., Reigada, D., Ramirez, B.G., Mechoulam, R., Innamorato, N., Cuadrado, A., de Ceballos, M.L., 2011. Cannabidiol and other cannabinoids reduce microglial activation in vitro and in vivo: relevance to Alzheimer's disease. Mol. Pharmacol. 79 (6), 964–973.

Mata, M.M., Napier, T.C., Graves, S.M., Mahmood, F., Raeisi, S., Baum, L.L., 2015. Methamphetamine decreases CD4 T cell frequency and alters pro-inflammatory cytokine production in a model of drug abuse. Eur. J. Pharmacol. 752, 26–33.

Matsumura, K., Cao, C., Ozaki, M., Morii, H., Nakadate, K., Watanabe, Y., 1998. Brain endothelial cells express cyclooxygenase-2 during lipopolysaccharide-induced fever: light and electron microscopic immunocytochemical studies. J. Neurosci. 18 (16), 6279–6289.

Matsushita, Y., Omotuyi, I., Mukae, T., Ueda, H., 2013. Microglia activation precedes the anti-opioid BDNF and NMDA receptor mechanisms underlying morphine analgesic tolerance. Curr. Pharm. Des. 19 (42), 7355–7361.

McCann, U.D., Wong, D.F., Yokoi, F., Villemagne, V., Dannals, R.F., Ricaurte, G.A., 1998. Reduced striatal dopamine transporter density in abstinent methamphetamine and methcathinone users: evidence from positron emission tomography studies with [11C]WIN-35,428. J. Neurosci. 18 (20), 8417–8422.

Mcnally, G., Akil, H., 2002. Opioid peptides and their receptors: overview and function in pain modulation. In: Davis, K., Charney, D., Coyle, J., Nemeroff, C. (Eds.),

Neuropsychopharmacology: The Fifth Generation of Progress. American College of Neuropsychopharmacology, pp. 35–46.
Mecha, M., Feliú, A., Iñigo, P.M., Mestre, L., Carrillo-Salinas, F.J., Guaza, C., 2013. Cannabidiol provides long-lasting protection against the deleterious effects of inflammation in a viral model of multiple sclerosis: a role for A2A receptors. Neurobiol. Dis. 59, 141–150.
Meek, R.L., Benditt, E.P., 1986. Amyloid A gene family expression in different mouse tissues. J. Exp. Med. 164 (6), 2006–2017.
Metz, V.E., Jones, J.D., Manubay, J., Sullivan, M.A., Mogali, S., Segoshi, A., Madera, G., Johnson, K.W., Comer, S.D., 2017. Effects of ibudilast on the subjective, reinforcing, and analgesic effects of oxycodone in recently detoxified adults with opioid dependence. Neuropsychopharmacology 42 (9), 1825–1832.
Morales, P., Reggio, P., 2017. An update on non-CB1, non-CB2 cannabinoid related g-protein-coupled receptors. Cannabis Cannabinoid Res. 2 (1), 265–273.
Moreira, F.P., Medeiros, J.R.C., Lhullier, A.C., de Mattos Souza, L.D., Jansen, K., Portela, L.V., Lara, D.R., da Silva, R.A., Wiener, C.D., Oses, J.P., 2016. Cocaine abuse and effects in the serum levels of cytokines IL-6 and IL-10. Drug Alcohol Depend. 158, 181–185.
Mu, Y., Ren, Z., Jia, J., Gao, B., Zheng, L., Wang, G., Friedman, E., Zhen, X., 2014. Inhibition of phosphodiesterase10A attenuates morphine-induced conditioned place preference. Mol. Brain 7, 70.
Nagarkatti, P., Pandey, R., Rieder, S.A., Hegde, V.L., Nagarkatti, M., 2009. Cannabinoids as novel anti-inflammatory drugs. Future Med. Chem. 1 (7), 1333–1349.
Nolte, K.B., Brass, L.M., Fletterick, C.F., 1996. Intracranial hemorrhage associated with cocaine abuse: a prospective autopsy study. Neurology 46 (5), 1291–1296.
Northcutt, A.L., Hutchinson, M.R., Wang, X., Baratta, M.V., Hiranita, T., Cochran, T.A., Pomrenze, M.B., Galer, E.L., Kopajtic, T.A., Li, C.M., Amat, J., Larson, G., Cooper, D.C., Huang, Y., O'Neill, C.E., Yin, H., Zahniser, N.R., Katz, J.L., Rice, K.C., Maier, S.F., Bachtell, R.K., Watkins, L.R., 2015. DAT isn't all that: cocaine reward and reinforcement requires toll like receptor 4 signaling. Mol. Psychiatry 20 (12), 1525–1537.
Northrop, N.A., Yamamoto, B.K., 2012. Persistent neuroinflammatory effects of serial exposure to stress and methamphetamine on the blood-brain barrier. J. Neuroimmune Pharmacol. 7 (4), 951–968.
O'Brien, C.P., 2011. Drug addiction. In: Brunton, L.L., Chabner, B.A., Knollmann, B.C. (Eds.), Goodman & Gilman's: The Pharmacological Basis of Therapeutics, 12th ed. New York, NY.
Olin, M., Oh, S., Roy, S., Peterson, P., Molitor, T., 2011. Morphine induces splenocyte trafficking into the CNS. J. Neuroimmune Pharmacol. 7 (2), 436–443.
O'Sullivan, S., 2016. An update on PPAR activation by cannabinoids. Br. J. Pharmacol. 173 (12), 1899–1910.
Pacifici, R., Carlo, S., Bacosi, A., Pichini, S., Zuccaro, P., 2000. Pharmacokinetics and cytokine production in heroin and morphine-treated mice. Int. J. Immunopharmacol. 22 (8), 603–614.
Parfieniuk, A., Flisiak, R., 2008. Role of cannabinoids in chronic liver diseases. World J. Gastroenterol. 14 (40), 6109–6114.
Parker, L., 2017. Cannabinoids and the Brain. MIT Press, Cambridge, MA.
Parker, L.A., Burton, P., Sorge, R.E., Yakiwchuk, C., Mechoulam, R., 2004. Effect of low doses of delta9-tetrahydrocannabinol and cannabidiol on the extinction of cocaine-induced and amphetamine-induced conditioned place preference learning in rats. Psychopharmacology (Berl) 175 (3), 360–366.

Payabvash, S., Beheshtian, A., Salmasi, A., Kiumehr, S., Ghahremani, M., Tavangar, S., Sabzevari, O., Dehpour, A.R., 2006. Chronic morphine treatment induces oxidant and apoptotic damage in the mice liver. Life Sci. 79 (10), 972–980.

Periyasamy, P., Guo, M.L., Buch, S., 2016. Cocaine induces astrocytosis through ER stress-mediated activation of autophagy. Autophagy 12 (8), 1310–1329.

Platzer, C., Meisel, C., Vogt, K., Platzer, M., Volk, H.D., 1995. Up-regulation of monocytic IL-10 by tumor necrosis α and cAMP elevating drugs. Int. Immunol. 7 (4), 517–523.

Pozzi, M., Roccatagliata, D., Sterzi, R., 2008. Drug abuse and intracranial hemorrhage. Neurol. Sci. 29 (S2), 269–270.

Pradhan, A., Smith, M., Kieffer, B., Evans, C., 2012. Ligand-directed signaling within the opioid receptor family. Br. J. Pharmacol. 167 (5), 960–969.

Prud'homme, M., Cata, R., Jutras-Aswad, D., 2015. Cannabidiol as an intervention for addictive behaviors: a systematic review of the evidence. Subst. Abuse 9, 33–38.

Qin, L., He, J., Hanes, R.N., Pluzarev, O., Hong, J.S., Crews, F.T., 2008. Increased systemic and brain cytokine production and neuroinflammation by endotoxin following ethanol treatment. J. Neuroinflammation 5, 10.

Quan, Y., Jiang, J., Dingledine, R., 2013. EP2 receptor signaling pathways regulate classical activation of microglia. J. Biol. Chem. 288 (13), 9293–9302.

Raborn, E.S., Marciano-Cabral, F., Buckley, N.E., Martin, B.R., Cabral, G.A., 2008. The cannabinoid delta-9-tetrahydrocannabinol mediates inhibition of macrophage chemotaxis to RANTES/CCL5: linkage to the CB2 receptor. J. Neuroimmune Pharmacol. 3 (2), 117–129.

Rasakham, K., Schmidt, H.D., Kay, K., Huizenga, M.N., Calcagno, N., Pierce, R.C., Spires-Jones, T.L., Sadri-Vakili, G., 2014. Synapse density and dendritic complexity are reduced in the prefrontal cortex following seven days of forced abstinence from cocaine self-administration. PLoS One 9 (7), e102524.

Reggio, P.H., 2010. Endocannabinoid binding to the cannabinoid receptors: what is known and what remains unknown. Curr. Med. Chem. 17 (14), 1468–1486.

Reid, M.S., Hsu, K.J., Berger, S.P., 1997. Cocaine and amphetamine preferentially stimulate glutamate release in the limbic system: studies on the involvement of dopamine. Synapse 27 (2), 95–105.

Ren, Y., Whittard, J., Higuera-Matas, A., Morris, C.V., Hurd, Y.L., 2009. Cannabidiol, a nonpsychotropic component of cannabis, inhibits cue-induced heroin seeking and normalizes discrete mesolimbic neuronal disturbances. J. Neurosci. 29 (47), 14764–14769.

Ricaurte, G.A., Guillery, R.W., Seiden, L.S., Schuster, C.R., Moore, R.Y., 1982. Dopamine nerve terminal degeneration produced by high doses of methylamphetamine in the rat brain. Brain Res. 2 (35), 93–103.

Rinaldi-Carmona, M., Le Duigou, A., Oustric, D., Barth, F., Bouaboula, M., Carayon, P., Casellas, P., Le Fur, G., 1998. Modulation of CB1 cannabinoid receptor functions after a long-term exposure to agonist or inverse agonist in the Chinese hamster ovary cell expression system. J. Pharmacol. Exp. Ther. 287 (3), 1038–1047.

Roberts, S.M., Harbison, R.D., James, R.C., 1995. Mechanistic studies on the potentiation of carbon tetrachloride hepatotoxicity by methamphetamine. Toxicology 97 (1–3), 49–57.

Romach, M.K., Schoedel, K.A., Sellers, E.M., 2014. Human abuse liability evaluation of CNS stimulant drugs. Neuropharmacology 87, 81–90.

Romagnani, S., 2000. T-cell subsets (Th1 versus Th2). Ann. Allergy Asthma Immunol. 85 (1), 9–18.

Romero-Sandoval, E.A., Horvath, R.J., DeLeo, J.A., 2008. Neuroimmune interactions and pain: focus on glial-modulating targets. Curr. Opin. Investig. Drugs 9 (7), 726–734.

Sama, M.A., Mathis, D.M., Furman, J.L., Abdul, H.M., Artiushin, I.A., Kraner, S.D., Norris, C.M., 2008. Interleukin-1β-dependent signaling between astrocytes and neurons depends critically on astrocytic calcineurin/NFAT activity. J. Biol. Chem. 283 (32), 21953–21964.

Samikkannu, T., Rao, K.V.K., Ding, H., Agudelo, M., Raymond, A.D., Yoo, C., Nair, M.P.N., 2014. Immunopathogenesis of HIV infection in cocaine users: role of arachidonic acid. PLoS One 9 (8), e106348.

Schmidt-Arras, D., Rose-John, S., 2016. IL-6 pathway in the liver: from physiopathology to therapy. J. Hepatol. 64 (6), 1403–1415.

Schwenkgrub, J., Zaremba, M., Joniec-Maciejak, I., Cudna, A., Mirowska-Guzel, D., Kurkowska-Jastrzębska, I., 2017. The phosphodiesterase inhibitor, ibudilast, attenuates neuroinflammation in the MPTP model of Parkinson's disease. PLoS One 12 (7), e0182019.

Seiden, L.S., Commins, D., Vosmer, G., Axt, K., Marek, G., 1988. Neurotoxicity in dopamine and serotonin terminal fields: a regional analysis in nigrostriatal and mesolimbic projections. Ann. N. Y. Acad. Sci. 537, 161–172.

Sekine, Y., Ouchi, Y., Sugihara, G., Takei, N., Yoshikawa, E., Nakamura, K., Iwata, Y., Tsuchiya, K.J., Suda, S., Suzuki, K., Kawai, M., Takebayashi, K., Yamamoto, S., Matsuzaki, H., Ueki, T., Mori, N., Gold, M.S., Cadet, J.L., 2008. Methamphetamine causes microglial activation in the brains of human abusers. J. Neurosci. 28 (22), 5756–5761.

Sharma, H.S., Ali, S.F., 2006. Alterations in blood-brain barrier function by morphine and methamphetamine. Ann. N. Y. Acad. Sci. 1074, 198–224.

Sheikh, I.A., Lukšič, M., Ferstenberg, R., Culpepper-Morgan, J.A., 2014. SPICE/K2 synthetic marijuana-induced toxic hepatitis treated with n-acetylcysteine. Am. J. Case Rep. 15, 584–588.

Shoemaker, H., Pimoule, C., Arbilla, S., Scatton, B., Jovoy-Agid, F., Langer, S.Z., 1985. Sodium dependent [3H]cocaine binding associated with dopamine uptake sites in the rat striatum and human putamen decrease after dopaminergic denervation and in Parkinsons disease. Naunyn Schmiedebergs Arch. Pharmacol. 329 (3), 227–235.

Slifka, M.K., Whitton, J.L., 2000. Antigen-specific regulation of T cell-mediated cytokine production. Immunity 12 (5), 451–457.

Sokol, C.L., Luster, A.D., 2015. The chemokine system in innate immunity. Cold Spring Harb. Perspect. Biol. 7, a016303.

Song, P., Zhao, Z., 2001. The involvement of glial cells in the development of morphine tolerance. Neurosci. Res. 39 (3), 281–286.

Srivastava, M., Srivastava, B., Brouhard, B., 1998. Δ9 Tetrahydrocannabinol and cannabidiol alter cytokine production by human immune cells. Immunopharmacology 40 (3), 179–185.

Staszewski, R.D., Yamamoto, B.K., 2006. Methamphetamine-induced spectrin proteolysis in the rat striatum. J. Neurochem. 96 (5), 1267–1276.

Stephans, S.E., Yamamoto, B.K., 1994. Methamphetamine-induced neurotoxicity: roles for glutamate and dopamine efflux. Synapse 17 (3), 203–209.

Stolyarova, A., Thompson, A.B., Barrientos, R.M., Izquierdo, A., 2014. Reductions in frontocortical cytokine levels are associated with long-lasting alterations in reward valuation after methamphetamine. Neuropsychopharmacology 40 (5), 1234–1242.

Stoycheva, D., Deiser, K., Stärck, L., Nishanth, G., Schlüter, D., Uckert, W., Schüler, T., 2014. IFN-γ regulates CD8+memory T cell differentiation and survival in response to weak, but not strong, TCR signals. J. Immunol. 194 (2), 553–559.

Svizenska, I., Dubovy, P., Sulcova, A., 2008. Cannabinoid receptors 1 and 2 (CB1 and CB2), their distribution, ligands and functional involvement in nervous system structures—a short review. Pharmacol. Biochem. Behav. 90 (4), 501–511.

Tahamtan, A., Tavakoli-Yaraki, M., Mokhtari-Azad, T., Teymoori-Rad, M., Bont, L., Shokri, F., Salimi, V., 2016. Opioids and viral infections: a double-edged sword. Front. Microbiol. 7, 970.

Tai, Y.H., Wang, Y.H., Tsai, R.Y., Wang, J.J., Tao, P.L., Liu, T.M., Wang, Y.C., Wong, C.S., 2007. Amitriptyline preserves morphine's antinociceptive effect by regulating the glutamate transporter GLAST and GLT-1 trafficking and excitatory amino acids concentration in morphine-tolerant rats. Pain 129 (3), 343–354.

Takayama, N., Ueda, H., 2005. Morphine-induced chemotaxis and brain-derived neurotrophic factor expression in microglia. J. Neurosci. 25 (2), 430–435.

Taylor, A.M.W., Castonguay, A., Ghogha, A., Vayssiere, P., Pradhan, A.A.A., Xue, L., Mehrabani, S., Wu, J., Levitt, P., Olmstead, M.C., Koninck, Y.D., Evans, C.J., Cahill, C.M., 2016. Neuroimmune regulation of GABAergic neurons within the ventral tegmental area during withdrawal from chronic morphine. Neuropsychopharmacology 41, 949–959.

Thomas, D.M., Kuhn, D.M., 2004. Cyclooxygenase-2 is an obligatory factor in methamphetamine-induced neurotoxicity. J. Pharmacol. Exp. Ther. 313 (2), 870–876.

Thomas, D.M., Dowgiert, J., Geddes, T.J., Francescutti-Verbeem, D., Liu, X., Kuhn, D.M., 2004. Microglial activation is a pharmacologically specific marker for the neurotoxic amphetamines. Neurosci. Lett. 367 (3), 349–354.

Thrash-Williams, B., Karuppagounder, S.S., Bhattacharya, D., Ahuja, M., Suppiramaniam, V., Dhanasekaran, M., 2016. Methamphetamine-induced dopaminergic toxicity prevented owing to the neuroprotective effects of salicylic acid. Life Sci. 154, 24–29.

Tien, L.T., Cai, Z., Rhodes, P.G., Fan, L.W., 2011. Neonatal exposure to lipopolysaccharide enhances methamphetamine-induced reinstated behavioral sensitization in adult rats. Behav. Brain Res. 224 (1), 166–173.

Tilleux, S., Hermans, E., 2008. Down-regulation of astrocytic GLAST by microglia-related inflammation is abrogated in dibutyryl cAMP-differentiated cultures. J. Neurochem. 105 (6), 2224–2236.

Toledano, A., Álvarez, M.I., Caballero, I., Carmona, P., De Miguel, E., 2008. Immunohistochemical increase in cyclooxygenase-2 without apoptosis in different brain areas of subchronic nicotine- and d-amphetamine-treated rats. J. Neural Transm. 115 (8), 1093–1108.

Touriño, C., Zimmer, A., Valverde, O., 2010. THC prevents MDMA neurotoxicity in mice. PLoS One 5 (2), e9143.

Trang, T., Sutak, M., Quirion, R., Jhamandas, K., 2002. The role of spinal neuropeptides and prostaglandins in opioid physical dependence. Br. J. Pharmacol. 136 (1), 37–48.

Trautwein, C., Rakemann, T., Niehof, M., Rose-John, S., Manns, M.P., 1996. Acute-phase response factor, increased binding, and target gene transcription during liver regeneration. Gastroenterology 110 (6), 1854–1862.

Valvassori, S.S., Tonin, P.T., Varela, R.B., Carvalho, A.F., Mariot, E., Amboni, R.T., Bianchini, G., Andersen, M.L., Quevedo, J., 2015. Lithium modulates the production of peripheral and cerebral cytokines in an animal model of mania induced by dextroamphetamine. Bipolar Disord. 17, 507–517.

Vassou, D., Bakogeorgou, E., Kampa, M., Dimitriou, H., Hatzoglou, A., Castanas, E., 2008. Opioids modulate constitutive B-lymphocyte secretion. Int. Immunopharmacol. 8, 634–644.

Vila, N., Castillo, J., Davalos, A., Chamorro, A., 2000. Proinflammatory cytokines and early neurological worsening in ischemic stroke. Stroke 31 (10), 2325–2329.

Vilela, L., Gomides, L., David, B., Antunes, M., Diniz, A., Moreira, F., Menezes, G., 2015. Cannabidiol rescues acute hepatic toxicity and seizure induced by cocaine. Mediators Inflamm. 2015, 1–12, Article ID 534418.

Villapol, S., Kryndushkin, D., Balarezo, M.G., Campbell, A.M., Saavedra, J.M., Shewmaker, F.P., Symes, A.J., 2015. Hepatic expression of serum amyloid a1 is induced by traumatic brain injury and modulated by telmisartan. Am. J. Pathol. 185 (10), 2641–2652.

Volz, T.J., Hanson, G.R., Fleckenstein, A.E., 2006. Measurement of kinetically resolved vesicular dopamine uptake and efflux using rotating disk electrode voltammetry. J. Neurosci. Methods 155 (1), 109–115.

Vuolo, F., Petronilho, F., Sonai, B., Ritter, C., Hallak, J.E.C., Zuardi, A.W., Crippa, J.A., Dal-Pizzol, F., 2015. Evaluation of serum cytokines levels and the role of cannabidiol treatment in animal model of asthma. Mediators Inflamm. 2015, 1–5, Article ID 538670.

Wagner, G.C., Ricaurte, G.A., Seiden, L.S., Schuster, C.R., Miller, R.J., Westley, J., 1980. Long-lasting depletions of striatal dopamine and loss of dopamine uptake sites following repeated administration of methamphetamine. Brain Res. 181, 151–160.

Wakil, A.E., Wang, Z.E., Ryan, J.C., Fowell, D.J., Locksley, R.M., 1998. Interferon gamma derived from CD4(+) T cells is sufficient to mediate T helper cell type 1 development. J. Exp. Med. 188 (9), 1651–1656.

Wang, J., Charboneau, R., Balasubramanian, S., Barke, R.A., Loh, H.H., Roy, S., 2002. The immunosuppressive effects of chronic morphine treatment are partially dependent on corticosterone and mediates by the μ-opioid receptor. J. Leukoc. Biol. 71 (5), 782–790.

Wang, X., Loram, L., Ramos, K., de Jesus, A., Thomas, J., Cheng, K., Reddy, A., Somogyi, A.A., Hutchinson, M.R., Watkins, L.R., Yin, H., 2012. Morphine activates neuroinflammation in a manner parallel to endotoxin. Proc. Natl. Acad. Sci. U.S.A. 109 (16), 6325–6330.

Weiss, L., Zeira, M., Reich, S., Har-Noy, M., Mechoulam, R., Slavin, S., Gallily, R., 2006. Cannabidiol lowers incidence of diabetes in non-obese diabetic mice. Autoimmunity 39 (2), 143–151.

Weller, I., Cohn, D., Sierralta, A., Mitcheson, M., Ross, M., Montano, L., Scheuer, P., Thomas, H.C., 1984. Clinical, biochemical, serological, histological and ultrastructural features of liver disease in drug abusers. Gut 25, 417–423.

Westover, A.N., McBride, S., Haley, R.W., 2007. Stroke in young adults who abuse amphetamines or cocaine: a population-based study of hospitalized patients. Arch. Gen. Psychiatry 64 (4), 495–502.

Wong, C.K., Wang, C.B., Ip, W.K., Tian, Y.P., Lam, C.W.K., 2005. Role of p38 MAPK and NF-kB for chemokine release in coculture of human eosinophils and bronchial epithelial cells. Clin. Exp. Immunol. 139, 90–100.

Worley, M.J., Heinzerling, K.G., Roche, D.J.O., Shoptawa, S., 2016. Ibudilast attenuates subjective effects of methamphetamine in a placebo-controlled inpatient study. Drug Alcohol Depend. 162, 245–250.

Wu, J., Zhao, R., Guo, L., Zhen, X., 2016. Morphine-induced inhibition of Ca2+ -dependent d-serine release from astrocytes suppresses excitability of GABAergic neurons in the nucleus accumbens. Addict. Biol. 22, 1289–1303.

Xia, Q., Hu, Q., Wang, H., Yang, H., Gao, F., Ren, H., Chen, D., Fu, C., Zheng, L., Zhen, X., Ying, Z., Wang, G., 2015. Induction of COX-2-PGE2 synthesis by activation of the MAPK/ERK pathway contributes to neuronal death triggered by TDP-43-depleted microglia. Cell Death Dis. 6, e1702.

Yamaguchi, T., Kuraishi, Y., Yabuuchi, K., Minami, M., Satoh, M., 1991. In situ hybridization analysis of the induction of interleukin-1beta mRNA by methamphetamine in the rat hypothalamus. Mol. Cell. Neurosci. 2 (3), 259–265.

Yamamoto, B.K., Moszczynska, A., Gudelsky, G.A., 2010. Amphetamine toxicities. Ann. N. Y. Acad. Sci. 1187 (1), 101–121.

You, T., Cheng, Y., Zhong, J., Bi, B., Zeng, B., Zheng, W., Wang, H., Xu, J., 2017. Roflupram, a phosphodiesterase 4 inhibitior, suppresses inflammasome activation through autophagy in microglial cells. ACS Chem. Nerosci. 8 (11), 2381–2392.
Yue, X., Qiao, D., Wang, A., Tan, X., Li, Y., Liu, C., Wang, H., 2012. CD200 attenuates methamphetamine-induced microglial activation and dopamine depletion. J. Huazhong Univ. Sci. Technolog. Med. Sci. 32 (3), 415–421.
Zamberletti, E., Gabaglio, M., Prini, P., Rubino, T., Parolaro, D., 2015. Cortical neuroinflammation contributes to long-term cognitive dysfunctions following adolescent delta-9-tetrahydrocannabinol treatment in female rats. Eur. Neuropsychopharmacol. 25 (12), 2404–2415.
Zhang, Y., Zheng, Q., Pan, J., Zheng, R., 2004. Oxidative damage of biomolecules in mouse liver induced by morphine and protected by antioxidants. Basic Clin. Paharmacol. Toxicol. 95 (2), 53–58.
Zhang, D., Hu, X., Qian, L., Wilson, B., Lee, C., Flood, P., Langenbach, R., Hong, J.S., 2009. Prostaglandin E2 released from activated microglia enhances astrocyte proliferation in vitro. Toxicol. Appl. Pharmacol. 238 (1), 64–70.
Zhu, W., Newton, C., Dka, Y., Friedman, H., Klein, T.W., 1994. Δ9-Tetrahydrocannabinol enhances the secretion of interleukin 1 from endotoxin-stimulated macrophages. J. Pharmacol. Exp. Ther. 270 (3), 1334–1339.
Zuardi, A., 2008. Cannabidiol: from an inactive cannabinoid to a drug with wide spectrum of action. Rev. Bras. Psiquiatr. 30 (3), 271–280.

Further reading

U.S. Department of Justice, 2017. Drug Enforcement Administration. Drugs of Abuse. Drug Enforcement Administration, U.S. Department of Justice. (Edition: a DEA Resource Guide).

CHAPTER SIX

Overview of peripheral and central inflammatory responses and their contribution to neurotoxicity

Nikolay M. Filipov*
Department of Physiology and Pharmacology, College of Veterinary Medicine, University of Georgia, Athens, GA, United States
*Corresponding author: e-mail address: filipov@uga.edu

Contents

1. Overview of the inflammatory process

Historically, the definition of inflammation as a process initiated by a causative agent in order to heal wounded/damaged tissues has withstood the test of time. Inflammation's cardinal signs of heat (*calor*), pain (*dolor*), redness

Advances in Neurotoxicology, Volume 3
ISSN 2468-7480
https://doi.org/10.1016/bs.ant.2018.10.001

(*rubor*), and swelling (*tumor*) were defined in the first century AD by the Roman scholar Celsus (Serhan et al., 2010). Later on, a fifth sign, loss of function (*functio laesa*), was added; these signs are still used at present time to describe the acute form of the process (Gilroy and De Maeyer, 2015). In short, the acute inflammatory process can be divided into vascular and immune phases. During the vascular phase, the permeability of the blood vessels is increased, the vessels become dilated, and the blood flow to the affected site is increased. During the immune phase, circulating immune cells (monocytes, neutrophils, T-cells; stimulus-dependent) are recruited to the site. During both phases, a host of molecules, including cytokines, chemokines, lipids, and enzymes, in a coordinate fashion, work to deal with the insult and, eventually, to proceed to the resolution phase (Serhan et al., 2010). Once resolved, acute inflammation will have no major lasting negative consequences on the body. However, in instances where the reaction to an inflammatory stimulus is excessive and/or unresolved, the consequences to such inflammatory response could be detrimental (Gilroy and De Maeyer, 2015; Kumar and Robbins, 2007; Serhan et al., 2010). Uncontrolled inflammation has emerged as an underlying pathophysiology for a plethora of diseases, ranging from gastrointestinal disturbances, liver disease, metabolic diseases and arthritis, to cardiovascular and respiratory diseases and cancer (Serhan et al., 2010). Variety of factors are considered as contributing in the transition to uncontrolled or chronic inflammation and among these, environmental contaminants feature prominently. For example, allergic inflammation has a strong environmental component in its etiology (Yang et al., 2014) and selected other examples are provided later in the chapter.

2. Peripheral inflammation and its contribution to non-neurodegenerative diseases

Chronic peripheral inflammation is associated with the etiology of multiple peripheral diseases. One example is cancer. Increased inflammation in the tumor microenvironment enhances the proliferation and survival of malignant cells, facilitates angiogenesis and metastasis, and it decreases the efficacy of both chemo- and immunotherapies (Mantovani et al., 2010). Multiple inflammatory mediators, including cytokines, chemokines, prostaglandins, proteases and Danger-Associated-Molecular Patterns (DAMPs), participate in the cellular communication in the tumor microenvironment and in the systemic response to tumor presence and/or status (Candido and Hagemann, 2013). TNF-alpha, IL-6, and IL-17 are examples of inflammatory cytokines linked clinically and experimentally to multiple cancers (Candido and Hagemann, 2013); the degree and/or type of inflammatory profile in many malignancies is associated with severity, progression and, importantly, cancer prognosis (Guthrie et al., 2013).

Immune cells are recruited at the tumor site both directly and indirectly, i.e., via chemokines (Balkwill and Mantovani, 2012; Mantovani et al., 2010), and profiles of circulating immune cells, acute phase proteins, or cytokines are associated with cancer type, metastasis, and treatment efficacy (Sylman et al., 2018). In this regard, the utility of using circulating inflammatory markers, such as neutrophil-to-lymphocyte ratio or C-reactive protein, for cancer tracking, staging and outcome prediction, has been evaluated (Sylman et al., 2018). Among the multiple immune cells associated with cancer inflammation, macrophages feature prominently (Balkwill and Mantovani, 2012). As highlighted earlier, inflammation has been associated with many different cancers, but the knowledge about the inflammation-cancer relationship is perhaps more advanced for liver cancer and cancers of the gastro-intestinal tract (Cani and Jordan, 2018; Ray, 2018) where inflammatory dysfunction is an indicative of a poor prognosis (Margetts et al., 2018; Shawki et al., 2018).

Cardiovascular disease is another prominent example of inflammation associated peripheral disease (Fort-Gallifa et al., 2016; Martinez and White, 2018; Oikonomou and Tousoulis, 2018; Zelinskiy et al., 2015). Of the multitude of inflammatory mediators implicated as effector molecules in inflammation-induced cardiovascular damage, targeting of IL-1 beta is explored clinically (Martinez and White, 2018). The effects of unregulated inflammation on the cardiovascular system are not uniform and, at least when it comes to the arteries, inflammation is markedly higher in the context of peripheral artery disease compared to coronary artery disease (Rein et al., 2015).

Inflammation is also strongly implicated in the etiology of respiratory diseases, ranging from asthma (Tenero et al., 2018), infection-related respiratory dysfunction, i.e., one induced by *Mycobacterium tuberculosis* (Arora and Chopra, 2018), to chronic obstructive pulmonary disease and fibrosis (Racanelli et al., 2018). In fact, anti-inflammatory medications are an essential part of the treatment regimen for these diseases (Koster et al., 2009; Taylor-Cousar et al., 2010).

Metabolic disease, obesity and diabetes have chronic inflammation as a common denominator (Lim et al., 2018; Prattichizzo et al., 2018; San et al., 2018). Of note, these conditions are typically associated with a low-grade chronic inflammation that shares some, but not all, of the features of the chronic peripheral inflammation associated with other peripheral diseases and it has been recently referred to as "metaflammation" (Prattichizzo et al., 2018).

Finally, aging is associated with increased activation of the innate immune system and low-grade chronic inflammation that is not restricted to the periphery and features in the brain as well (discussed later). There is a substantial overlap between the "metaflammation" observed in metabolic disorders/diabetes and "inflammaging," which is the term coined for the aging-associated low-grade inflammation (Ferrucci and Fabbri, 2018; Muller-Werdan et al., 2017; Prattichizzo et al., 2018). This overlap suggests common treatment possibilities.

While in some cases locally initiated chronic inflammatory responses might be important for disease pathogenesis, an overarching and perhaps unifying denominator associated with chronic inflammation systemically and centrally (discussed later) is gut dysbiosis (Carding et al., 2015). Thus, dysfunctional gut microbiota-mediated inflammation has been associated with obesity and gastrointestinal cancers (Cani and Jordan, 2018). Microbiome dyshomeostasis is also associated with cardiovascular disease and microbiome-driven/modulated inflammation is considered the link between the gut microbiome and cardiovascular dysfunction (Jovanovich et al., 2018). Gut dysbiosis features in kidney disease (Ramezani and Raj, 2014) and, importantly, in metabolic disease and diabetes (Wilson et al., 2018). Dysfunctional gut microbiota is an aging phenomenon (O'Toole and Jeffery, 2015) and this, together with aging-related gut leakiness (Keebaugh and Ja, 2017), is considered a major contributor to the "inflammaging" inflammatory phenotype discussed earlier. It is because of the wide-spread associations of gut dysbiosis with chronic inflammation and a myriad of human diseases that microbiota-targeted therapeutic strategies are now at the research forefront (Collado et al., 2015).

3. Selected examples of toxicant or toxicant-pathogen exposures where peripheral inflammation is a major consequence of the exposure and a contributor to the toxicant-induced pathology in the periphery

How toxicants impact the innate immune system or interact with bacterial, viral, or parasitic pathogens to modulate the inflammatory response in the periphery has been a major subject of investigation for the last quarter century. For example, several chemicals have been demonstrated to function as co-factors in carcinogenesis by tipping the pro/anti-inflammatory balance. Bisphenol A (BPA) and phthalates are among the chemicals known to affect critical pathways responsible for inflammation-induced/modulated tumorigenesis (Thompson et al., 2015). Of specific interest is the case of non-asbestos mesothelioma where the industrial toxicant vinylidene chloride (VDC) markedly increased malignant mesotheliomas by providing chronic inflammatory environment for the tumor (Blackshear et al., 2015).

Multiple toxicants contribute to lung inflammation and respiratory disease (Wallaert et al., 2000). Thus, several endocrine disruptors and particulates, such as diesel exhaust particles, contribute to allergic inflammation (Yang et al., 2014). The respiratory injury caused by the potent chemical warfare agent sulfur mustard is inflammation-dependent (Gao et al., 2007). Toxicants also interact with pathogens to augment the pathogen-induced inflammation and tissue damage. One such example is acrolein, which exacerbates the lung inflammation and damage of mice infected with influenza A (Ong et al., 2012).

Metals, such as copper and arsenic, cause kidney damage by over-activating a key inflammatory pathway, NFκB (Wang et al., 2017). On the other hand, polychlorinated biphenyls (PCBs), among many other effects, act as inflammagens in the vasculature and cause vascular dysfunction that way (Petriello et al., 2014). Liver inflammation and damage is induced experimentally by acrylamide and glycidamide (Chen et al., 2018). Toxic overexposures also cause inflammatory activities in reproductive organs. For example, certain phthalates cause testicular inflammation (Voss et al., 2018), dioxins are implicated in inflammation-dependent development of endometriosis (Bruner-Tran et al., 2008), and one of the effects of excessive alcohol is an enhancement of peripheral inflammation (Orio et al., 2018).

As it was the case in the absence of toxicant exposures, gut dysbiosis might also play a major role in the chronic inflammation-dependent peripheral effects of many toxicants. For example, the development of liver cancer by excessive inorganic arsenic is associated with gut microbiota shifts of inflammatory nature (Choiniere and Wang, 2016). Certain polycyclic aromatic hydrocarbons (PAHs) lead to pro-inflammatory gut microbiota and lesions reminiscent of inflammatory bowel disease (IBD) (Ribiere et al., 2016). Of note, when the exposure to the already mentioned BPA occurs in the perinatal period, this exposure leads to gut dysbiosis and intestinal permeability that might be responsible for increase disease burden later in life (Reddivari et al., 2017).

4. Central inflammation, a.k.a., neuroinflammation

Inflammation in the CNS shares many of the inflammatory mediators (cytokines, chemokines, prostaglandins) that are used by the immune system for the inflammatory process in the periphery (Perry, 2000; Perry et al., 1996). While peripheral immune cells might be recruited into the brain and participate in the central inflammatory process, aberrant activation of the resident immune cells, the microglia, and eventual engagement of astrocytes are considered key characteristics of neuroinflammation (Gomes-Leal et al., 2004; Lunnon et al., 2008). It should also be noted that standard immunosuppressive treatments, which will tone down the peripheral inflammatory response, might not do the same in the brain (O'Callaghan et al., 2008).

Microglia-associated neuroinflammation, which is largely an innate immune system phenomenon, should be distinguished from the so-called autoimmune neuroinflammation, where the central inflammatory process is dependent on the adaptive immune response against self-antigens or brain origin. Multiple sclerosis is a classic example for autoimmune neuroinflammation and its signature, both histologically, and at the level of the transcriptome, differs markedly from the non-autoimmune, microglia-dependent neuroinflammation (Filiou et al., 2014). Because the latter is implicated in the etiology of multiple neurodegenerative diseases (Perry et al., 2010) and evidence for environmental influence on microglia status and microglia-associated inflammation is most abundant (Cope et al., 2018; Langley et al., 2018), here, the focus is on non-autoimmune neuroinflammation.

5. Overview of microglia

Microglia play an important role in maintaining normal brain activity: they are vital for brain development, homeostasis and function (Perry et al., 2010; Wolf et al., 2017). While upon activation they produce a host of inflammatory mediators and contribute/drive the neuroinflammatory process, they also produce trophic factors and exhibit phagocytic activity, which are essential for maintenance of normal neural networking and communication (Wolf et al., 2017). Compared to peripheral cells of the innate immune system, microglia are long-lived, but recent evidence suggests that in both mice and humans there is a substantial microglial turnover such that the whole population is renewed several times over the course of a lifetime (Askew et al., 2017). This substantial remodeling further supports the critical role microglia play in the maintenance of brain homeostasis, as well as in their ability to respond to diverse stimuli.

Microglia are not uniform morphologically, functionally, and regionally (Lawson et al., 1990, 1993; Wolf et al., 2017). Morphologically, microglia are traditionally divided into ramified, or resting microglia and amoeboid, or activated microglia, but multiple intermediate types exist (Boche et al., 2013; Lawson et al., 1993). Functionally, based on activation states, microglia are divided into M1 (classically activated), and M2 (alternatively activated) microglia (Boche et al., 2013). As is the case for microglia's morphology, sub-types of these two activation states have been characterized, i.e., M2A, M2B, and M2C subtypes of the M2 microglia (Moehle and West, 2015). In addition, neonatal microglia, typically used for experimentation, differ from adult brain microglia isolated from a healthy brain, arguing for a caution when extrapolating research findings (Grabert and McColl, 2018). Spatially, microglia are not uniform and areas of the brain, such as the hippocampus, striatum, and cortex, have disproportionately higher microglia numbers (Lawson et al., 1990; Perry et al., 1996). It is interesting to note that not only are microglia-rich brain regions of critical functional importance, but they are also the major targets of neurodegenerative diseases, such as Parkinson's and Alzheimer's (De Lucia et al., 2016; Moehle and West, 2015).

6. Role of aging, diet, and sex on microglia phenotype and (inflammatory) function

The fact that the aged brain has an increased inflammatory load is well documented. It is also recognized that much of this increased load is due to aged microglia being with dystrophic morphology, diminished neuroprotective capacity, and augmented ability for inflammatory mediator output (Niraula et al., 2017; Norden and Godbout, 2013). The heightened inflammatory profile of aged microglia is associated with a different phenotype; microglia in aged brains are referred to as being "sensitized" or "primed" (Fonken et al., 2016; Norden and Godbout, 2013). These microglia respond more robustly to a variety of stimuli and their hyperactivation is associated with, for example, cognitive deficits (Corona et al., 2012).

Mounting evidence points to a causal link between the primed aged microglia profile and increased vulnerability to secondary insults, including pathogens and stressors of the aged brain (Niraula et al., 2017). Thus, strategies aimed at decreasing microglia activation, such as exercise, not only reduce inflammation, but also improve hippocampal function (Kohman et al., 2013). Interestingly, age is a major factor for microglia diversity and the aging process affects microglia phenotype in a spatially restrictive manner (Grabert et al., 2016). Such spatial restriction of the impact of age on microglia phenotype, together with the known differences in microglia distribution, is viewed as a potential reason for the spatial restriction of neurodegenerative diseases where neuroinflammation has been implicated (Grabert et al., 2016).

Diet is a major immune system modulator and it affects the microglia function as well. Many dietary approaches aimed at decreasing neuroinflammation have been explored (i.e., Burton et al., 2016; Rangarajan et al., 2016), including some within the context of neurodegenerative diseases (Lee et al., 2013). The role of diet/dietary components in promoting neuroinflammation has been studied the most in high-fat (calorie) diet paradigms used to model diet-induced obesity. One interesting study reported that pharmacological inhibition of microglia activation prevented obesity-associated structural and cognitive deficits. Interestingly, this team also reported that partial knockdown of the receptor for the chemokine fractalkine (neuronally expressed) prevented microglial activation and cognitive decline induced by obesity (Cope et al., 2018). The latter suggests that initial neuronal distress might serve as a trigger for neuroinflammation in obesity. Interestingly, it has been suggested that the effects of high-fat diet on microglia and neuroinflammation are regional (Baufeld et al., 2016).

Sex also plays an important role in microglia biology. Sex differences might be responsible, in part, for the sex-specific contribution of microglia and neuroinflammation to neurodegeneration. Important differences in timing of colonization and of microglia phenotype exist between the developing male and female brains (Schwarz et al., 2012). Moreover, in addition to sex differences in disease incidence, microglia are considered critical for the establishment and maintenance of sex-specific brain circuits and behaviors (Lenz and McCarthy, 2015). Sex differences in microglia reactivity and ensuing neuroinflammation have been documented within the context of multiple stimuli ranging from brain injury (Acaz-Fonseca et al., 2015; Villapol et al., 2017) to stress, which, depending on the brain region and/or its chronicity, has differential effects (Bollinger et al., 2017).

7. Neuroinflammation as a feature of major neurodegenerative diseases

Neuroinflammation features prominently in the etiology of Alzheimer's disease (AD) (Heppner et al., 2015; McGeer et al., 2000; Shadfar et al., 2015). It is interesting to note that the inflammation in the Alzheimer's brain appears to be locally induced, chronic, and reliant on activated microglia that are in the vicinity of the amyloid plaques (Eikelenboom et al., 2000). Moreover, AD-relevant mutations, i.e., tau, amyloid, co-localize in the brain with key inflammatory mediators, such as TNF-alpha (Hung et al., 2016). In addition, the most profound neuropathology associated with central administration of the inflammagen LPS was AD-like and it tracked LPS-induced neuroinflammation (Deng et al., 2014).

Parkinson's disease (PD) is another major neurodegenerative disease where neuroinflammation is implicated in its etiology (Moehle and West, 2015; Thome et al., 2015; Tiwari and Pal, 2017). It is noteworthy that fractalkine signaling, already implicated in age-related neuroinflammation, is also involved in the PD-like pathology of animal models of the disease (Thome et al., 2015). This provides one possible explanation of the age dependency of sporadic PD. Because of the overwhelming evidence for an in situ neuroinflammation in affected brains, various anti-inflammatory treatment modalities with brain-penetrating molecules have/are being explored with some success. One example is eicosanoyl-5-hydroxytryptamide (EHT), a coffee ingredient, which was anti-inflammatory and neuroprotective in the MPTP model of PD (Lee et al., 2013). Selective inhibitors of the inducible form of nitric oxide synthase (iNOS), which prevents microglia from producing excessive nitric oxide (Broom et al., 2011), cyclooxygenase-2 (COX-2), which prevents excessive prostaglandins (Gupta et al., 2011), or manipulation of cannabinoid signaling (Javed et al., 2016; Mounsey et al., 2015), have all been shown to be anti-inflammatory and neuroprotective experimentally.

Besides AD and PD, neuroinflammation and microglia hyperactivation is implicated in the etiology of amyotrophic lateral sclerosis (ALS) (Brites and Vaz, 2014; Phani et al., 2012). In fact, targeting of a key inflammatory miRNA, miR-155, is ameliorative in the SOD1 mouse model of ALS (Butovsky et al., 2015). Extensive neuroinflammation is also featured in a neurodegenerative disease with a strong genetic predisposition, such as Huntington's disease (Labadorf et al., 2015). Another neuropathology, receiving an ever increasing attention, where neuroinflammation is likely at the core of its pathogenesis, is concussion injury (Patterson and Holahan, 2012; Steenerson and Starling, 2017).

8. Selected examples of toxicant or toxicant/pathogen exposures where neuroinflammation is a major consequence of the exposure and a contributor to the toxicant-induced pathology in the brain

Multiple toxicants are capable of activating the immune cells in the brain and causing central inflammation. Several pesticides, drugs of abuse, metals, as well as particulates, are among them (Anderson et al., 2018; Filipov and Dodd, 2012; Kraft and Harry, 2011; Mushtaq et al., 2015; Segura-Aguilar and Kostrzewa, 2006). Among the metals, the role of neuroinflammation in the neurotoxic mechanism of manganese (Mn) is perhaps the best studied. Neuroinflammation is a prominent feature of Mn neurotoxicity with both microglia and astrocytes being involved, but microglia activation is needed to amplify the output of neurotoxic factors from astrocytes (Filipov and Dodd, 2012; Kirkley et al., 2017). Mn neurotoxicity is also a good example for toxicant-pathogen and gene-environment interactions within the context of neuroinflammation. Thus, inflammatory mediator output from inflammagen-activated microglia is increased by Mn (Filipov and Dodd, 2012; Filipov et al., 2005), and Mn synergizes with the gram-negative bacteria component LPS to increase microglia activation and ensuing dopaminergic toxicity (Filipov et al., 2005; Zhang et al., 2010). Moreover, Mn pretreatment increases CNS viral load and neuroinflammation (Seth et al., 2003), and basal ganglia dysfunction and neuroinflammation are increased by Mn in a genetic model of PD (Langley et al., 2018).

Neuroinflammation features in the neurotoxic mechanism of several pesticides, including some that are associated with PD and/or AD-like pathology. For example, the dopaminergic toxicity caused by rotenone depends, at least in part, on neuroinflammation (Main et al., 2017). Chronic neuroinflammation, and so-called "non-cholinergic toxicity," is implicated in the cognitive and mood deficits associated with organophosphate (OP) insecticide exposures (Naughton and Terry, 2018). Interestingly, the neuroinflammatory response to organophosphates is enhanced by stress or modeling of stress by corticosterone supplementation (O'Callaghan et al., 2015). The latter indicates that stress, much like age, may serve as a priming factor for microglia in their subsequent response to toxic exposures.

Excessive exposures to particulates, such as diesel exhaust particles (DEP) or manufactured/engineered nanoparticles (NPs), can also lead to neuroinflammation (Levesque et al., 2011; Mushtaq et al., 2015; Win-Shwe and Fujimaki, 2011). In the case of DEPs, microglia priming to subsequent/concurrent pathogen insults are a major feature (Levesque et al., 2011). Traffic air pollution, which contains various (type and size) particulates, has been associated with neuroinflammation and the neuroinflammatory response/brain damage to this type of exposure is fairly focused, i.e., the hippocampus is a major target (Costa et al., 2017). Given the exponential increase of NPs in the daily lives and the environment, it is conceivable that NP-induced and/or amplified neuroinflammation would be a major health concern in the upcoming years.

Neuroinflammation is implicated in the neurotoxicity of multiple drugs of abuse. For example, both methamphetamine and MDMA cause neuroinflammation (Yamamoto et al., 2010). Even more important is the fact that abuse of these substances frequently co-insides with chronic infections, such as HIV, or stress; in both cases, amphetamines' neurotoxicity is enhanced by the infection and/or stress in a neuroinflammation-dependent manner (Jung et al., 2010; Northrop and Yamamoto, 2013; Silverstein et al., 2011; Yamamoto et al., 2010). Multiple synthetic cathinones most commonly found in "bath salts," such as methcathinone, mephedrone, methylone, or methylenedioxypyrovalerone (MDPV), in addition to being excitotoxic, are also neuroinflammatory (Angoa-Perez et al., 2017). Alcohol abuse is also associated with neuroinflammation; key inflammatory pathways in the brain, i.e., NFκB, are pivotal for the addictive, as well as in the neurotoxic properties, of alcohol (Cui et al., 2014). Major innate immune system receptor, TLR4, is used by ethanol for exerting its neuroinflammatory and neurotoxic effects. Interestingly, sex is a major factor determining both the degree of neuroinflammation and susceptibility to the deleterious effects of alcohol, with females being more susceptible than males (Alfonso-Loeches et al., 2013).

For many toxicants or drugs of abuse, neuroinflammation is present in the absence of overt infection, i.e., the so-called sterile neuroinflammation. In these cases, the key neurotoxic mechanism appears to be an activation of the inflammasome (Anderson et al., 2018; Xu et al., 2017). This activation and subsequent neurotoxicity is triggered by cellular stress in situ, which leads to selective neuroinflammation, inflammasome activation, and neurotoxicity (Anderson et al., 2018).

9. Contribution of peripherally initiated inflammation to brain dysfunction and pathology

While, as described earlier in the text, central inflammation does not necessarily need the peripheral innate immune system to cause brain dysfunction, in many instances, peripherally initiated inflammation contributes to neurodegeneration (Herrera et al., 2015; King et al., 2018). There are several ways that peripheral inflammation contributes to neuronal dysfunction and damage. First, inflammatory cytokines and other inflammatory mediators, produced and released in the periphery, are involved in peripheral immune system-to-brain communication by activating microglia in the brain (Chesnokova et al., 2016). Once in the brain, these peripherally derived inflammatory molecules contribute to direct neurotoxic insult as well as to decreased neurogenesis in areas of the brain where this is an important homeostatic mechanism, i.e., the hippocampus (Chesnokova et al., 2016; Herrera et al., 2015). Second, peripheral immune cells could infiltrate the brain, where they will interact with microglia and/or astrocytes and amplify neuroinflammation and neurotoxicity (Zindler and Zipp, 2010). In some cases, entry of cells from the periphery is requisite for a CNS inflammation and neuropathology to be observed; a good example for this is the alpha-synuclein model of PD (Harms et al., 2018).

Third, and or increasing importance way, the periphery transmits inflammation to the brain is via the microbiome-gut-brain axis (Caputi and Giron, 2018; D'Mello et al., 2015). Multiple neurological diseases have been associated with gut dysbiosis, inflammation, and aberrant communication with the brain. Thus, inflammatory microbiota phenotype and intestinal inflammation are associated with motor neuron diseases, such as ALS (Rowin et al., 2017). Multiple chronic systemic inflammatory diseases lead to neurological dysfunction, i.e., sickness behavior, and treatments aimed at re-balancing microbiota and restoring peripheral inflammation are effective in mitigating the neurological dysfunction (D'Mello et al., 2015).

Peripherally derived inflammation is a major contributor to demyelinating diseases (Murta and Ferrari, 2016). It is noteworthy that early in the development of AD and other dementias, the peripheral inflammatory cytokine profile is predictive for disease severity (King et al., 2018). Once peripheral inflammation is triggered, especially if the inflammation is chronic, the neurological dysfunction might be present long after any signs of inflammation in the periphery are present. An example of this is a study where we administered peripherally LPS repeatedly to model chronic inflammation: chronic LPS treatment induced depressive behavior that persisted long after LPS treatment termination when all measurable signs of peripheral innate immune system activation and inflammation were absent (Krishna et al., 2016).

As discussed earlier, inflammatory cytokines or infiltrating cells from the periphery are the major ways peripheral inflammation contributes to neuroinflammation and neurological dysfunction. However, an additional way that is only now beginning to be explored is via the transfer of DAMPs from the periphery to the brain. High-mobility group box 1 (HMGB1) is one DAMP molecule implicated in promoting spinal cord injury (Papatheodorou et al., 2017). Moreover, plasma HMGB1 levels appear to track the severity of the CNS injury (Papatheodorou et al., 2017), and this particular DAMP is a key microglia priming signal in age-associated microglia activation (Fonken et al., 2016). If the fact that multiple DAMPs are increased in the circulation of aged individuals is taken into consideration (Huang et al., 2015), it is conceivable that DAMPs generated peripherally, upon reaching the brain, can trigger or enhance central inflammation.

10. Contribution of centrally initiated inflammation to peripheral dysfunction and pathology

Peripheral innate immune system activation and subsequent transmission of the inflammatory signals to the brain are well established and studied extensively. On the other hand, brain-to-periphery communication within the inflammation domain is far less researched. Nevertheless, evidence mostly from the obesity field indicates that there is an active brain-to-periphery communication at the innate immunity level. For example, selective inhibition of microglia expansion and activation in the brains of HFD-fed mice also led to a marked decrease of peripheral inflammation (Andre et al., 2017). In addition, locally induced neuroinflammation in the nodose ganglion and hypothalamus of HFD-fed mice was responsible for ghrelin resistance and peripheral immune dysregulation in these mice (Naznin et al., 2015).

Interesting discoveries of brain-to-periphery cell trafficking by draining activated immune cells from brain to lymph nodes via the lymphatics further substantiate the ability of the central immune system to communicate and regulate peripheral immunity and inflammation not only by soluble mediators, but by using cells as well (Negi and Das, 2018).

11. Blood-brain barrier (BBB) dysfunction and its role in peripheral-central-peripheral immune communication

The existence of the BBB, viewed in the past as essential for the designation of the brain as an "immuno-privileged site," controls the flow of immune information into/out of the brain. For example, BBB is essential for maintenance of the chemical composition of the neuronal environment and for separation of circulating components from the brain parenchyma (Zlokovic, 2008). However, even in healthy conditions, there is a bi-directional immune communication between the periphery and the brain that takes place (Wohleb and Godbout, 2013; Zlokovic, 2008).

Many conditions, such as aging, inappropriate diet, infection, substance abuse, or toxicant exposures, have direct or indirect effects on the BBB integrity. As a result, the flux of cells and inflammatory and/or signaling mediators across the BBB is increased, the in situ homeostatic balance is perturbed, and uncontrolled inflammation followed/accompanied by cell/tissue damage takes place. For instance, the aged BBB is less tight and this is considered an important reason for CNS neuroinflammation and neurological effects of pathogen exposures in the elderly that would normally be contained in the periphery (Wohleb and Godbout, 2013). Certain infections, i.e., HIV, likely perturb the BBB to some extent in all infected individuals, but neuroinflammation, and the exhibition of the so-called "Neuro AIDS," is observed mostly in cases where the infection is compounded by other factors (Berger and Avison, 2004; Erickson et al., 2012).

Diet, in addition to affecting the central and peripheral inflammatory states, impacts the BBB and it contributes to neuroinflammation and neurological dysfunction that way (Grabowski, 2017; Snelson et al., 2017; Tucsek et al., 2014). As is frequently the case, metabolic dysregulation caused by diet or other factors interact with, for example, age and the result is an exacerbation of BBB disruption, degree of neuroinflammation, and neurological dysfunction (Tucsek et al., 2014).

Effects of psychostimulants on the BBB deserve special attention as BBB disruption is an integral part of their neuroinflammatory, psychostimulant, and, importantly, neurotoxic effects. Amphetamines are perhaps most studied in this regard and ample evidence points to structural and functional BBB disruption by these substances (Kousik et al., 2012; Northrop and Yamamoto, 2015; Yamamoto et al., 2010). Much as the diet-age interaction highlighted earlier, psychostimulants' effects on the BBB and subsequent inflammation are augmented by other factors, such as stress (Northrop and Yamamoto, 2012), infection (Silverstein et al., 2011), or multi-drug use (O'Shea et al., 2014).

Toxicants can also disrupt the BBB and gain entry into the brain and/or facilitate entry of peripheral cells and molecules, which, ultimately, will lead to neurotoxicity. Organotins (Ferraz da Silva et al., 2017) and nanoparticles (Mushtaq et al., 2015) are two examples where toxicant's effects on brain inflammation and ensuing toxicity are associated with BBB disruption.

While more of the emphasis on BBB disruption is on the subsequent increased periphery-to-brain flux of cells and inflammatory mediators, as stated earlier, this traffic is bi-directional. Because of "leakage" of centrally produced inflammatory molecules, the peripheral inflammatory process can be initiated or enhanced (Zlokovic, 2008). Moreover, presence of certain DAMPs, i.e., HMGB1, in the circulation of patients with Alzheimer's correlates with degree of BBB disruption, neuroinflammation, and disease symptom severity (Festoff et al., 2016). Neuronal antigens like beta amyloid also make their way into circulation when the BBB is disrupted (Chakraborty et al., 2017; Festoff et al., 2016); there, they can activate the peripheral immune system. Brain antigen leakage is not limited to beta amyloid, i.e., tau proteins are also found outside the BBB and in the vasculature (Michalicova et al., 2017) and can engage in peripheral immune activation.

12. Summary

Graphical summary of the major points related to central and peripheral inflammation discussed in the chapter is highlighted in Fig. 1. As illustrated in Fig. 1:

(1) Inflammatory status in both the periphery and the brain is graded in terms of degree and duration. Cells of the innate immune system in each compartment have important homeostatic functions, i.e., they operate

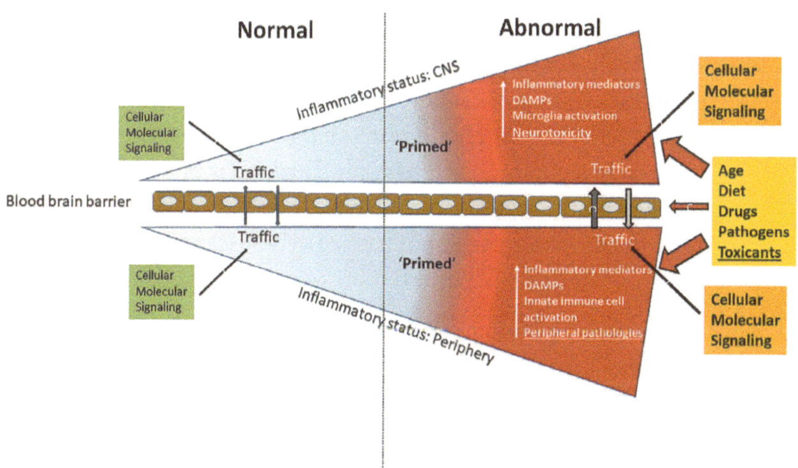

Fig. 1 Graphical summary of the major points related to central and peripheral inflammation discussed in the chapter.

under "Normal" conditions. The blood brain barrier (BBB) is intact, and the traffic, albeit bi-directional, is limited.

(2) Various factors, such as age, diet, drugs, pathogens, and toxic exposures, have negative consequences on central and/or peripheral inflammatory status; some of them may also disrupt the integrity of the BBB. These consequences would lead to the establishment of "Abnormal" inflammatory status in one or both compartments.

(3) In the brain, the "Abnormal" inflammatory status is associated with increased microglia activation and increased local levels of multiple inflammatory mediators (cytokines, prostaglandins) and DAMPs. The increased neuroinflammation would lead to, or, will be associated with, neurological dysfunction and neurotoxicity.

(4) In the periphery, the "Abnormal" inflammatory status is associated with increased innate immune cell, i.e., monocytes/macrophages activation and increased levels of inflammatory mediators and DAMPs. The increased peripheral inflammation would lead to, or, will be associated with, peripheral dysfunction and pathologies.

(5) The bi-directional traffic between the periphery and the CNS is substantially greater when one or both compartments are in the "Abnormal" inflammatory state. This is due, in part, to direct and indirect effects on the BBB by many of the factors that influence the central and/or peripheral inflammatory status. Because of this enhanced traffic, peripheral cells and molecules participate in the CNS inflammation and neurotoxicity, whereas CNS cell components and/or molecules activate/enhance the activation of the immune system in the periphery.

(6) As the inflammatory status is graded, inflammatory cells can be in a "primed" transition state, whereby they do not necessarily produce abnormal levels of inflammatory mediators. However, cells in this "primed" state respond to various stimuli much more robustly in terms of magnitude and duration, which makes the "prime" central and/or peripheral innate immune system particularly susceptible to inflammatory insults, including toxic ones.

References

Acaz-Fonseca, E., Duran, J.C., Carrero, P., Garcia-Segura, L.M., Arevalo, M.A., 2015. Sex differences in glia reactivity after cortical brain injury. Glia 63 (11), 1966–1981. https://doi.org/10.1002/glia.22867.

Alfonso-Loeches, S., Pascual, M., Guerri, C., 2013. Gender differences in alcohol-induced neurotoxicity and brain damage. Toxicology 311 (1–2), 27–34. https://doi.org/10.1016/j.tox.2013.03.001.

Anderson, F.L., Coffey, M.M., Berwin, B.L., Havrda, M.C., 2018. Inflammasomes: an emerging mechanism translating environmental toxicant exposure into neuroinflammation in Parkinson's disease. Toxicol. Sci. https://doi.org/10.1093/toxsci/kfy219. (Epub ahead of print).

Andre, C., Guzman-Quevedo, O., Rey, C., et al., 2017. Inhibiting microglia expansion prevents diet-induced hypothalamic and peripheral inflammation. Diabetes 66 (4), 908–919. https://doi.org/10.2337/db16-0586.

Angoa-Perez, M., Anneken, J.H., Kuhn, D.M., 2017. Neurotoxicology of synthetic cathinone analogs. Curr. Top. Behav. Neurosci. 32, 209–230. https://doi.org/10.1007/7854_2016_21.

Arora, V.K., Chopra, K.K., 2018. Inflammation plays a central role in respiratory diseases, including tuberculosis. Indian J. Tuberc. 65 (2), 103–105. https://doi.org/10.1016/j.ijtb.2018.03.001.

Askew, K., Li, K.Z., Olmos-Alonso, A., et al., 2017. Coupled proliferation and apoptosis maintain the rapid turnover of microglia in the adult brain. Cell Rep. 18 (2), 391–405. https://doi.org/10.1016/j.celrep.2016.12.041.

Balkwill, F.R., Mantovani, A., 2012. Cancer-related inflammation: common themes and therapeutic opportunities. Semin. Cancer Biol. 22 (1), 33–40. https://doi.org/10.1016/j.semcancer.2011.12.005.

Baufeld, C., Osterloh, A., Prokop, S., Miller, K.R., Heppner, F.L., 2016. High-fat diet-induced brain region-specific phenotypic spectrum of CNS resident microglia. Acta Neuropathol. 132 (3), 361–375. https://doi.org/10.1007/s00401-016-1595-4.

Berger, J.R., Avison, M., 2004. The blood brain barrier in HIV infection. Front. Biosci. 9, 2680–2685.

Blackshear, P.E., Pandiri, A.R., Nagai, H., et al., 2015. Gene expression of mesothelioma in vinylidene chloride-exposed F344/N rats reveal immune dysfunction, tissue damage, and inflammation pathways. Toxicol. Pathol. 43 (2), 171–185. https://doi.org/10.1177/0192623314537885.

Boche, D., Perry, V.H., Nicoll, J.A.R., 2013. Review: activation patterns of microglia and their identification in the human brain. Neuropathol. Appl. Neurobiol. 39 (1), 3–18. https://doi.org/10.1111/nan.12011.

Bollinger, J.L., Collins, K.E., Patel, R., Wellman, C.L., 2017. Behavioral stress alters corticolimbic microglia in a sex- and brain region-specific manner. PLoS One 12 (12), e0187631. https://doi.org/10.1371/journal.pone.0187631.

Brites, D., Vaz, A.R., 2014. Microglia centered pathogenesis in ALS: insights in cell interconnectivity. Front. Cell. Neurosci. 8, 117. https://doi.org/10.3389/fncel.2014.00117.

Broom, L., Marinova-Mutafchieva, L., Sadeghian, M., Davis, J.B., Medhurst, A.D., Dexter, D.T., 2011. Neuroprotection by the selective iNOS inhibitor GW274150 in a model of Parkinson disease. Free Radic. Biol. Med. 50 (5), 633–640. https://doi.org/10.1016/j.freeradbiomed.2010.12.026.

Bruner-Tran, K.L., Yeaman, G.R., Crispens, M.A., Igarashi, T.M., Osteen, K.G., 2008. Dioxin may promote inflammation-related development of endometriosis. Fertil. Steril. 89 (Suppl. 5), 1287–1298. https://doi.org/10.1016/j.fertnstert.2008.02.102.

Burton, M.D., Rytych, J.L., Amin, R., Johnson, R.W., 2016. Dietary luteolin reduces proinflammatory microglia in the brain of senescent mice. Rejuvenation Res. 19 (4), 286–292. https://doi.org/10.1089/rej.2015.1708.

Butovsky, O., Jedrychowski, M.P., Cialic, R., et al., 2015. Targeting miR-155 restores dysfunctional microglia and ameliorates disease in the SOD1 model of ALS. Int. J. Dev. Neurosci. 47, 5. https://doi.org/10.1016/j.ijdevneu.2015.04.023.

Candido, J., Hagemann, T., 2013. Cancer-related inflammation. J. Clin. Immunol. 33 (Suppl. 1), S79–S84. https://doi.org/10.1007/s10875-012-9847-0.

Cani, P.D., Jordan, B.F., 2018. Gut microbiota-mediated inflammation in obesity: a link with gastrointestinal cancer. Nat. Rev. Gastroenterol. Hepatol. 15 (11), 671–682. https://doi.org/10.1038/s41575-018-0025-6.

Caputi, V., Giron, M.C., 2018. Microbiome-gut-brain axis and toll-like receptors in Parkinson's disease. Int. J. Mol. Sci. 19 (6), 1689. https://doi.org/10.3390/ijms19061689.

Carding, S., Verbeke, K., Vipond, D.T., Corfe, B.M., Owen, L.J., 2015. Dysbiosis of the gut microbiota in disease. Microb. Ecol. Health Dis. 26, 26191. https://doi.org/10.3402/mehd.v26.26191.

Chakraborty, A., de Wit, N.M., van der Flier, W.M., de Vries, H.E., 2017. The blood brain barrier in Alzheimer's disease. Vascul. Pharmacol. 89, 12–18. https://doi.org/10.1016/j.vph.2016.11.008.

Chen, D., Liu, H., Wang, E., Yan, H., Ye, H., Yuan, Y., 2018. Toxicogenomic evaluation of liver responses induced by acrylamide and glycidamide in male mouse liver. Gen. Physiol. Biophys. 37 (2), 175–184. https://doi.org/10.4149/gpb_2017034.

Chesnokova, V., Pechnick, R.N., Wawrowsky, K., 2016. Chronic peripheral inflammation, hippocampal neurogenesis, and behavior. Brain Behav. Immun. 58, 1–8. https://doi.org/10.1016/j.bbi.2016.01.017.

Choiniere, J., Wang, L., 2016. Exposure to inorganic arsenic can lead to gut microbe perturbations and hepatocellular carcinoma. Acta Pharm. Sin. B 6 (5), 426–429. https://doi.org/10.1016/j.apsb.2016.07.011.

Collado, M.C., Rautava, S., Isolauri, E., Salminen, S., 2015. Gut microbiota: a source of novel tools to reduce the risk of human disease? Pediatr. Res. 77 (1–2), 182–188. https://doi.org/10.1038/pr.2014.173.

Cope, E.C., LaMarca, E.A., Monari, P.K., et al., 2018. Microglia play an active role in obesity-associated cognitive decline. J. Neurosci. 38 (41), 8889–8904. https://doi.org/10.1523/jneurosci.0789-18.2018.

Corona, A.W., Fenn, A.M., Godbout, J.P., 2012. Cognitive and behavioral consequences of impaired immunoregulation in aging. J. Neuroimmune Pharmacol. 7 (1), 7–23. https://doi.org/10.1007/s11481-011-9313-4.

Costa, L.G., Cole, T.B., Coburn, J., Chang, Y.C., Dao, K., Roque, P.J., 2017. Neurotoxicity of traffic-related air pollution. Neurotoxicology 59, 133–139. https://doi.org/10.1016/j.neuro.2015.11.008.

Cui, C., Shurtleff, D., Harris, R.A., 2014. Neuroimmune mechanisms of alcohol and drug addiction. Int. Rev. Neurobiol. 118, 1–12. https://doi.org/10.1016/B978-0-12-801284-0.00001-4.

De Lucia, C., Rinchon, A., Olmos-Alonso, A., et al., 2016. Microglia regulate hippocampal neurogenesis during chronic neurodegeneration. Brain Behav. Immun. 55, 179–190. https://doi.org/10.1016/j.bbi.2015.11.001.

Deng, X., Li, M., Ai, W., et al., 2014. Lipolysaccharide-induced neuroinflammation is associated with Alzheimer-like amyloidogenic axonal pathology and dendritic degeneration in rats. Adv. Alzheimer Dis. 3 (2), 78–93. https://doi.org/10.4236/aad.2014.32009.

D'Mello, C., Ronaghan, N., Zaheer, R., et al., 2015. Probiotics improve inflammation-associated sickness behavior by altering communication between the peripheral immune system and the brain. J. Neurosci. 35 (30), 10821–10830. https://doi.org/10.1523/Jneurosci.0575-15.2015.

Eikelenboom, P., Rozemuller, A.J., Hoozemans, J.J., Veerhuis, R., van Gool, W.A., 2000. Neuroinflammation and Alzheimer disease: clinical and therapeutic implications. Alzheimer Dis. Assoc. Disord. 14 (Suppl. 1), S54–S61.

Erickson, M.A., Dohi, K., Banks, W.A., 2012. Neuroinflammation: a common pathway in CNS diseases as mediated at the blood-brain barrier. Neuroimmunomodulation 19 (2), 121–130. https://doi.org/10.1159/000330247.

Ferraz da Silva, I., Freitas-Lima, L.C., Graceli, J.B., Rodrigues, L.C.M., 2017. Organotins in neuronal damage, brain function, and behavior: a short review. Front Endocrinol. (Lausanne) 8, 366. https://doi.org/10.3389/fendo.2017.00366.
Ferrucci, L., Fabbri, E., 2018. Inflammageing: chronic inflammation in ageing, cardiovascular disease, and frailty. Nat. Rev. Cardiol. 15 (9), 505–522. https://doi.org/10.1038/s41569-018-0064-2.
Festoff, B.W., Sajja, R.K., van Dreden, P., Cucullo, L., 2016. HMGB1 and thrombin mediate the blood-brain barrier dysfunction acting as biomarkers of neuroinflammation and progression to neurodegeneration in Alzheimer's disease. J. Neuroinflamm. 13 (1), 194. https://doi.org/10.1186/s12974-016-0670-z.
Filiou, M.D., Arefin, A.S., Moscato, P., Graeber, M.B., 2014. 'Neuroinflammation' differs categorically from inflammation: transcriptomes of Alzheimer's disease, Parkinson's disease, schizophrenia and inflammatory diseases compared. Neurogenetics 15 (3), 201–212. https://doi.org/10.1007/s10048-014-0409-x.
Filipov, N.M., Dodd, C.A., 2012. Role of glial cells in manganese neurotoxicity. J. Appl. Toxicol. 32 (5), 310–317. https://doi.org/10.1002/jat.1762.
Filipov, N.M., Seegal, R.F., Lawrence, D.A., 2005. Manganese potentiates in vitro production of proinflammatory cytokines and nitric oxide by microglia through a nuclear factor kappa B-dependent mechanism. Toxicol. Sci. 84 (1), 139–148. https://doi.org/10.1093/toxsci/kfi055.
Fonken, L.K., Frank, M.G., Kitt, M.M., et al., 2016. The alarmin HMGB1 mediates age-induced neuroinflammatory priming. J. Neurosci. 36 (30), 7946–7956. https://doi.org/10.1523/JNEUROSCI.1161-16.2016.
Fort-Gallifa, I., Garcia-Heredia, A., Hernandez-Aguilera, A., et al., 2016. Biochemical indices of oxidative stress and inflammation in the evaluation of peripheral artery disease. Free Radical Bio. Med. 97, 568–576. https://doi.org/10.1016/j.freeradbiomed.2016.07.011.
Gao, X., Ray, R., Xiao, Y., Barker, P.E., Ray, P., 2007. Inhibition of sulfur mustard-induced cytotoxicity and inflammation by the macrolide antibiotic roxithromycin in human respiratory epithelial cells. BMC Cell Biol. 8, 17. https://doi.org/10.1186/1471-2121-8-17.
Gilroy, D., De Maeyer, R., 2015. New insights into the resolution of inflammation. Semin. Immunol. 27 (3), 161–168. https://doi.org/10.1016/j.smim.2015.05.003.
Gomes-Leal, W., Corkill, D.J., Freire, M.A., Picanco-Diniz, C.W., Perry, V.H., 2004. Astrocytosis, microglia activation, oligodendrocyte degeneration, and pyknosis following acute spinal cord injury. Exp. Neurol. 190 (2), 456–467. https://doi.org/10.1016/j.expneurol.2004.06.028.
Grabert, K., McColl, B.W., 2018. Isolation and phenotyping of adult mouse microglial cells. Methods Mol. Biol. 1784, 77–86. https://doi.org/10.1007/978-1-4939-7837-3_7.
Grabert, K., Michoel, T., Karavolos, M.H., et al., 2016. Microglial brain region-dependent diversity and selective regional sensitivities to aging. Nat. Neurosci. 19 (3), 504–516. https://doi.org/10.1038/nn.4222.
Grabowski, G.A., 2017. Overview of inflammation in neurometabolic diseases. Semin. Pediatr. Neurol. 24 (3), 207–213. https://doi.org/10.1016/j.spen.2017.08.005.
Gupta, A., Kumar, A., Kulkarni, S.K., 2011. Targeting oxidative stress, mitochondrial dysfunction and neuroinflammatory signaling by selective cyclooxygenase (COX)-2 inhibitors mitigates MPTP-induced neurotoxicity in mice. Prog. Neuropsychopharmacol. Biol. Psychiatry 35 (4), 974–981. https://doi.org/10.1016/j.pnpbp.2011.01.017.
Guthrie, G.J., Roxburgh, C.S., Farhan-Alanie, O.M., Horgan, P.G., McMillan, D.C., 2013. Comparison of the prognostic value of longitudinal measurements of systemic inflammation in patients undergoing curative resection of colorectal cancer. Br. J. Cancer 109 (1), 24–28. https://doi.org/10.1038/bjc.2013.330.

Harms, A.S., Thome, A.D., Yan, Z.Q., et al., 2018. Peripheral monocyte entry is required for alpha-synuclein induced inflammation and neurodegeneration in a model of Parkinson disease. Exp. Neurol. 300, 179–187. https://doi.org/10.1016/j.expneurol.2017.11.010.

Heppner, F.L., Ransohoff, R.M., Becher, B., 2015. Immune attack: the role of inflammation in Alzheimer disease. Nat. Rev. Neurosci. 16 (6), 358–372. https://doi.org/10.1038/nrn3880.

Herrera, A.J., Espinosa-Oliva, A.M., Oliva-Martin, M.J., Carrillo-Jimenez, A., Venero, J.L., de Pablos, R.M., 2015. Collateral damage: contribution of peripheral inflammation to neurodegenerative diseases. Curr. Top. Med. Chem. 15 (21), 2193–2210. https://doi.org/10.2174/1568026615666150610142027.

Huang, J., Xie, Y., Sun, X., et al., 2015. DAMPs, ageing, and cancer: the 'DAMP hypothesis'. Ageing Res. Rev. 24 (Pt. A), 3–16. https://doi.org/10.1016/j.arr.2014.10.004.

Hung, A.S., Liang, Y., Chow, T.C., et al., 2016. Mutated tau, amyloid and neuroinflammation in Alzheimer disease—a brief review. Prog. Histochem. Cytochem. 51 (1), 1–8. https://doi.org/10.1016/j.proghi.2016.01.001.

Javed, H., Azimullah, S., Haque, M.E., Ojha, S.K., 2016. Cannabinoid type 2 (CB2) receptors activation protects against oxidative stress and neuroinflammation associated dopaminergic neurodegeneration in rotenone model of Parkinson's disease. Front. Neurosci. 10, 321https://doi.org/10.3389/fnins.2016.00321.

Jovanovich, A., Isakova, T., Stubbs, J., 2018. Microbiome and cardiovascular disease in CKD. Clin. J. Am. Soc. Nephrol. 13 (10), 1598–1604. https://doi.org/10.2215/CJN.12691117.

Jung, B.D., Shin, E.J., Nguyen, X.K., et al., 2010. Potentiation of methamphetamine neurotoxicity by intrastriatal lipopolysaccharide administration. Neurochem. Int. 56 (2), 229–244. https://doi.org/10.1016/j.neuint.2009.10.005.

Keebaugh, E.S., Ja, W.W., 2017. Breaking down walls: microbiota and the aging gut. Cell Host Microbe. 21 (4), 417–418. https://doi.org/10.1016/j.chom.2017.03.013.

King, E., O'Brien, J.T., Donaghy, P., et al., 2018. Peripheral inflammation in prodromal Alzheimer's and Lewy body dementias. J. Neurol. Neurosur. Psychiatry 89 (4), 339–345. https://doi.org/10.1136/jnnp-2017-317134.

Kirkley, K.S., Popichak, K.A., Afzali, M.F., Legare, M.E., Tjalkens, R.B., 2017. Microglia amplify inflammatory activation of astrocytes in manganese neurotoxicity. J. Neuroinflamm. 14 (1), 99. https://doi.org/10.1186/s12974-017-0871-0.

Kohman, R.A., Bhattacharya, T.K., Wojcik, E., Rhodes, J.S., 2013. Exercise reduces activation of microglia isolated from hippocampus and brain of aged mice. J. Neuroinflamm. 10, 114. https://doi.org/10.1186/1742-2094-10-114.

Koster, E.S., Raaijmakers, J.A., Koppelman, G.H., et al., 2009. Pharmacogenetics of anti-inflammatory treatment in children with asthma: rationale and design of the PACMAN cohort. Pharmacogenomics 10 (8), 1351–1361. https://doi.org/10.2217/pgs.09.79.

Kousik, S.M., Napier, T.C., Carvey, P.M., 2012. The effects of psychostimulant drugs on blood brain barrier function and neuroinflammation. Front. Pharmacol. 3, 121. https://doi.org/10.3389/fphar.2012.00121.

Kraft, A.D., Harry, G.J., 2011. Features of microglia and neuroinflammation relevant to environmental exposure and neurotoxicity. Int. J. Environ. Res. Public Health 8 (7), 2980–3018. https://doi.org/10.3390/ijerph8072980.

Krishna, S., Dodd, C.A., Filipov, N.M., 2016. Behavioral and monoamine perturbations in adult male mice with chronic inflammation induced by repeated peripheral lipopolysaccharide administration. Behav. Brain Res. 302, 279–290. https://doi.org/10.1016/j.bbr.2016.01.038.

Kumar, V., Robbins, S.L., 2007. Robbins Basic Pathology, eighth ed. Saunders/Elsevier, Philadelphia, PA.

Labadorf, A., Hoss, A.G., Lagomarsino, V., et al., 2015. RNA sequence analysis of human Huntington disease brain reveals an extensive increase in inflammatory and developmental gene expression. PLoS One 10 (12), e0143563. https://doi.org/10.1371/journal.pone.0143563.
Langley, M.R., Ghaisas, S., Ay, M., et al., 2018. Manganese exposure exacerbates progressive motor deficits and neurodegeneration in the MitoPark mouse model of Parkinson's disease: relevance to gene and environment interactions in metal neurotoxicity. Neurotoxicology 64, 240–255. https://doi.org/10.1016/j.neuro.2017.06.002.
Lawson, L.J., Perry, V.H., Dri, P., Gordon, S., 1990. Heterogeneity in the distribution and morphology of microglia in the normal adult-mouse brain. Neuroscience 39 (1), 151–170. https://doi.org/10.1016/0306-4522(90)90229-W.
Lawson, L.J., Matyszak, M.K., Perry, V.H., 1993. Lessons from microglia in special sites. Clin. Neuropathol. 12 (5), 310–313.
Lee, K.W., Im, J.Y., Woo, J.M., et al., 2013. Neuroprotective and anti-inflammatory properties of a coffee component in the MPTP model of Parkinson's disease. Neurotherapeutics 10 (1), 143–153. https://doi.org/10.1007/s13311-012-0165-2.
Lenz, K.M., McCarthy, M.M., 2015. A starring role for microglia in brain sex differences. Neuroscientist 21 (3), 306–321. https://doi.org/10.1177/1073858414536468.
Levesque, S., Taetzsch, T., Lull, M.E., et al., 2011. Diesel exhaust activates and primes microglia: air pollution, neuroinflammation, and regulation of dopaminergic neurotoxicity. Environ. Health Perspect. 119 (8), 1149–1155. https://doi.org/10.1289/ehp.1002986.
Lim, P.S., Chang, Y.K., Wu, T.K., 2018. Serum lipopolysaccharide-binding protein is associated with chronic inflammation and metabolic syndrome in hemodialysis patients. Blood Purif. 1–9. https://doi.org/10.1159/000492778.
Lunnon, K., Teeling, J., Szak, S., Relton, J., Glennic, M., Perry, V.H., 2008. Innate Inflammation's contribution to chronic neurodegeneration: microglia phenotype "switching" J. Neuroimmunol. 197 (2), 173.
Main, B.S., Zhang, M., Brody, K.M., Kirby, F.J., Crack, P.J., Taylor, J.M., 2017. Type-I interferons mediate the neuroinflammatory response and neurotoxicity induced by rotenone. J. Neurochem. 141 (1), 75–85. https://doi.org/10.1111/jnc.13940.
Mantovani, A., Garlanda, C., Allavena, P., 2010. Molecular pathways and targets in cancer-related inflammation. Ann. Med. 42 (3), 161–170. https://doi.org/10.3109/07853890903405753.
Margetts, J., Ogle, L.F., Chan, S.L., et al., 2018. Neutrophils: driving progression and poor prognosis in hepatocellular carcinoma? Br. J. Cancer 118 (2), 248–257. https://doi.org/10.1038/bjc.2017.386.
Martinez, B.K., White, C.M., 2018. The emerging role of inflammation in cardiovascular disease. Ann. Pharmacother. 52 (8), 801–809. https://doi.org/10.1177/1060028018765939.
McGeer, P.L., McGeer, E.G., Yasojima, K., 2000. Alzheimer disease and neuroinflammation. J. Neural Transm. Suppl. 59, 53–57.
Michalicova, A., Banks, W.A., Legath, J., Kovac, A., 2017. Tauopathies—focus on changes at the neurovascular unit. Curr. Alzheimer Res. 14 (7), 790–801. https://doi.org/10.2174/1567205014666170203143336.
Moehle, M.S., West, A.B., 2015. M1 and M2 immune activation in Parkinson's disease: foe and ally? Neuroscience 302, 59–73. https://doi.org/10.1016/j.neuroscience.2014.11.018.
Mounsey, R.B., Mustafa, S., Robinson, L., et al., 2015. Increasing levels of the endocannabinoid 2-AG is neuroprotective in the 1-methyl-4-phenyl-1,2,3,6-tetrahydropyridine mouse model of Parkinson's disease. Exp. Neurol. 273, 36–44. https://doi.org/10.1016/j.expneurol.2015.07.024.
Muller-Werdan, U., Nuding, S., Ost, M., 2017. Assessing inflammageing. Curr. Opin. Clin. Nutr. Metab. Care 20 (5), 346–348. https://doi.org/10.1097/MCO.0000000000000391.

Murta, V., Ferrari, C., 2016. Peripheral inflammation and demyelinating diseases. Adv. Exp. Med. Biol. 949, 263–285. https://doi.org/10.1007/978-3-319-40764-7_13.
Mushtaq, G., Khan, J.A., Joseph, E., Kamal, M.A., 2015. Nanoparticles, neurotoxicity and neurodegenerative diseases. Curr. Drug Metab. 16 (8), 676–684.
Naughton, S.X., Terry Jr., A.V., 2018. Neurotoxicity in acute and repeated organophosphate exposure. Toxicology 408, 101–112. https://doi.org/10.1016/j.tox.2018.08.011.
Naznin, F., Toshinai, K., Waise, T.M.Z., et al., 2015. Diet-induced obesity causes peripheral and central ghrelin resistance by promoting inflammation. J. Endocrinol. 226 (1), 81–92. https://doi.org/10.1530/Joe-15-0139.
Negi, N., Das, B.K., 2018. CNS: not an immunoprivilaged site anymore but a virtual secondary lymphoid organ. Int. Rev. Immunol. 37 (1), 57–68. https://doi.org/10.1080/08830185.2017.1357719.
Niraula, A., Sheridan, J.F., Godbout, J.P., 2017. Microglia priming with aging and stress. Neuropsychopharmacology 42 (1), 318–333. https://doi.org/10.1038/npp.2016.185.
Norden, D.M., Godbout, J.P., 2013. Review: microglia of the aged brain: primed to be activated and resistant to regulation. Neuropathol. Appl. Neurobiol. 39 (1), 19–34. https://doi.org/10.1111/j.1365-2990.2012.01306.x.
Northrop, N.A., Yamamoto, B.K., 2012. Persistent neuroinflammatory effects of serial exposure to stress and methamphetamine on the blood-brain barrier. J. Neuroimmune Pharmacol. 7 (4), 951–968. https://doi.org/10.1007/s11481-012-9391-y.
Northrop, N.A., Yamamoto, B.K., 2013. Cyclooxygenase activity contributes to the monoaminergic damage caused by serial exposure to stress and methamphetamine. Neuropharmacology 72, 96–105. https://doi.org/10.1016/j.neuropharm.2013.04.040.
Northrop, N.A., Yamamoto, B.K., 2015. Methamphetamine effects on blood-brain barrier structure and function. Front. Neurosci. 9, 69. https://doi.org/10.3389/fnins.2015.00069.
O'Callaghan, J.P., Sriram, K., Miller, D.B., 2008. Defining "neuroinflammation" Ann. N. Y. Acad. Sci. 1139, 318–330. https://doi.org/10.1196/annals.1432.032.
O'Callaghan, J.P., Kelly, K.A., Locker, A.R., Miller, D.B., Lasley, S.M., 2015. Corticosterone primes the neuroinflammatory response to DFP in mice: potential animal model of Gulf War Illness. J. Neurochem. 133 (5), 708–721. https://doi.org/10.1111/jnc.13088.
Oikonomou, E., Tousoulis, D., 2018. Inflammation: a pathogenetic mechanism or a mediator, linking risk factors and cardiovascular disease? Int. J. Cardiol. 264, 170–171. https://doi.org/10.1016/j.ijcard.2018.03.076.
Ong, F.H., Henry, P.J., Burcham, P.C., 2012. Prior exposure to acrolein accelerates pulmonary inflammation in influenza A-infected mice. Toxicol. Lett. 212 (3), 241–251. https://doi.org/10.1016/j.toxlet.2012.06.003.
Orio, L., Anton, M., Rodriguez-Rojo, I.C., et al., 2018. Young alcohol binge drinkers have elevated blood endotoxin, peripheral inflammation and low cortisol levels: neuropsychological correlations in women. Addict. Biol. 23 (5), 1130–1144. https://doi.org/10.1111/adb.12543.
O'Shea, E., Urrutia, A., Green, A.R., Colado, M.I., 2014. Current preclinical studies on neuroinflammation and changes in blood-brain barrier integrity by MDMA and methamphetamine. Neuropharmacology 87, 125–134. https://doi.org/10.1016/j.neuropharm.2014.02.015.
O'Toole, P.W., Jeffery, I.B., 2015. Gut microbiota and aging. Science 350 (6265), 1214–1215. https://doi.org/10.1126/science.aac8469.
Papatheodorou, A., Stein, A., Bank, M., et al., 2017. High-mobility group box 1 (HMGB1) is elevated systemically in persons with acute or chronic traumatic spinal cord injury. J. Neurotrauma 34 (3), 746–754. https://doi.org/10.1089/neu.2016.4596.
Patterson, Z.R., Holahan, M.R., 2012. Understanding the neuroinflammatory response following concussion to develop treatment strategies. Front. Cell. Neurosci. 6, 58. https://doi.org/10.3389/fncel.2012.00058.

Perry, V.H., 2000. Cytokines, macrophages and microglia. Brain Pathol. 10 (4), 794.

Perry, V.H., Anthony, D.C., Bell, M.D., Lawson, L.J., Reid, D.M., Gordon, S., 1996. Microglia activation and inflammation in the CNS. J. Neurochem. 66, S70.

Perry, V.H., Nicoll, J.A.R., Holmes, C., 2010. Microglia in neurodegenerative disease. Nat. Rev. Neurol. 6 (4), 193–201. https://doi.org/10.1038/nrneurol.2010.17.

Petriello, M.C., Newsome, B., Hennig, B., 2014. Influence of nutrition in PCB-induced vascular inflammation. Environ. Sci. Pollut. Res. Int. 21 (10), 6410–6418. https://doi.org/10.1007/s11356-013-1549-5.

Phani, S., Re, D.B., Przedborski, S., 2012. The role of the innate immune system in ALS. Front. Pharmacol. 3, 150. https://doi.org/10.3389/fphar.2012.00150.

Prattichizzo, F., De Nigris, V., Spiga, R., et al., 2018. Inflammageing and metaflammation: the yin and yang of type 2 diabetes. Ageing Res. Rev. 41, 1–17. https://doi.org/10.1016/j.arr.2017.10.003.

Racanelli, A.C., Kikkers, S.A., Choi, A.M.K., Cloonan, S.M., 2018. Autophagy and inflammation in chronic respiratory disease. Autophagy 14 (2), 221–232. https://doi.org/10.1080/15548627.2017.1389823.

Ramezani, A., Raj, D.S., 2014. The gut microbiome, kidney disease, and targeted interventions. J. Am. Soc. Nephrol. 25 (4), 657–670. https://doi.org/10.1681/ASN.2013080905.

Rangarajan, P., Karthikeyan, A., Dheen, S.T., 2016. Role of dietary phenols in mitigating microglia-mediated neuroinflammation. Neuromolecular Med. 18 (3), 453–464. https://doi.org/10.1007/s12017-016-8430-x.

Ray, K., 2018. Liver cancer: a complex interplay between inflammation and immunity in liver cancer. Nat. Rev. Gastroenterol. Hepatol. 15 (1), 3. https://doi.org/10.1038/nrgastro.2017.165.

Reddivari, L., Veeramachaneni, D.N.R., Walters, W.A., et al., 2017. Perinatal bisphenol a exposure induces chronic inflammation in rabbit offspring via modulation of gut bacteria and their metabolites. mSystems 2 (5), 1–16. https://doi.org/10.1128/mSystems.00093-17.

Rein, P., Saely, C.H., Silbernagel, G., et al., 2015. Systemic inflammation is higher in peripheral artery disease than in stable coronary artery disease. Atherosclerosis 239 (2), 299–303. https://doi.org/10.1016/j.atherosclerosis.2015.01.021.

Ribiere, C., Peyret, P., Parisot, N., et al., 2016. Oral exposure to environmental pollutant benzo[a]pyrene impacts the intestinal epithelium and induces gut microbial shifts in murine model. Sci. Rep. 6, 31027. https://doi.org/10.1038/srep31027.

Rowin, J., Xia, Y., Jung, B., Sun, J., 2017. Gut inflammation and dysbiosis in human motor neuron disease. Physiol. Rep. 5 (18). , e13443https://doi.org/10.14814/phy2.13443.

San, K.M.M., Fahmida, U., Wijaksono, F., Lin, H., Zaw, K.K., Htet, M.K., 2018. Chronic low grade inflammation measured by dietary inflammatory index and its association with obesity among school teachers in Yangon, Myanmar. Asia Pac. J. Clin. Nutr. 27 (1), 92–98. https://doi.org/10.6133/apjcn.042017.06.

Schwarz, J.M., Sholar, P.W., Bilbo, S.D., 2012. Sex differences in microglial colonization of the developing rat brain. J. Neurochem. 120 (6), 948–963. https://doi.org/10.1111/j.1471-4159.2011.07630.x.

Segura-Aguilar, J., Kostrzewa, R.M., 2006. Neurotoxins and neurotoxicity mechanisms. An overview. Neurotox. Res. 10 (3–4), 263–287.

Serhan, C.N., Ward, P.A., Gilroy, D.W., 2010. Fundamentals of Inflammation. Cambridge University Press, Cambridge, NY.

Seth, P., Husain, M.M., Gupta, P., et al., 2003. Early onset of virus infection and up-regulation of cytokines in mice treated with cadmium and manganese. Biometals 16 (2), 359–368.

Shadfar, S., Hwang, C.J., Lim, M.S., Choi, D.Y., Hong, J.T., 2015. Involvement of inflammation in Alzheimer's disease pathogenesis and therapeutic potential of anti-inflammatory agents. Arch. Pharm. Res. 38 (12), 2106–2119. https://doi.org/10.1007/s12272-015-0648-x.

Shawki, S., Ashburn, J., Signs, S.A., Huang, E., 2018. Colon cancer: inflammation-associated cancer. Surg. Oncol. Clin. N Am. 27 (2), 269–287. https://doi.org/10.1016/j.soc.2017.11.003.
Silverstein, P.S., Shah, A., Gupte, R., et al., 2011. Methamphetamine toxicity and its implications during HIV-1 infection. J. Neurovirol. 17 (5), 401–415. https://doi.org/10.1007/s13365-011-0043-4.
Snelson, M., Mamo, J.C.L., Lam, V., Giles, C., Takechi, R., 2017. Differential effects of high-protein diets derived from soy and casein on blood-brain barrier integrity in wild-type mice. Front Nutr. 4, 35. https://doi.org/10.3389/fnut.2017.00035.
Steenerson, K., Starling, A.J., 2017. Pathophysiology of sports-related concussion. Neurol. Clin. 35 (3), 403–408. https://doi.org/10.1016/j.ncl.2017.03.011.
Sylman, J.L., Mitrugno, A., Atallah, M., et al., 2018. The predictive value of inflammation-related peripheral blood measurements in cancer staging and prognosis. Front. Oncol. 8, 78. https://doi.org/10.3389/fonc.2018.00078.
Taylor-Cousar, J.L., Von Kessel, K.A., Young, R., Nichols, D.P., 2010. Potential of anti-inflammatory treatment for cystic fibrosis lung disease. J. Inflamm. Res. 3, 61–74.
Tenero, L., Zaffanello, M., Piazza, M., Piacentini, G., 2018. Measuring airway inflammation in asthmatic children. Front Pediatr. 6, 196. https://doi.org/10.3389/fped.2018.00196.
Thome, A.D., Standaert, D.G., Harms, A.S., 2015. Fractalkine signaling regulates the inflammatory response in an alpha-synuclein model of Parkinson disease. PLoS One 10 (10), e0140566. https://doi.org/10.1371/journal.pone.0140566.
Thompson, P.A., Khatami, M., Baglole, C.J., et al., 2015. Environmental immune disruptors, inflammation and cancer risk. Carcinogenesis 36 (Suppl. 1), S232–S253. https://doi.org/10.1093/carcin/bgv038.
Tiwari, P.C., Pal, R., 2017. The potential role of neuroinflammation and transcription factors in Parkinson disease. Dialogues Clin. Neurosci. 19 (1), 71–80.
Tucsek, Z., Toth, P., Sosnowska, D., et al., 2014. Obesity in aging exacerbates blood-brain barrier disruption, neuroinflammation, and oxidative stress in the mouse hippocampus: effects on expression of genes involved in beta-amyloid generation and Alzheimer's disease. J. Gerontol. A Biol. Sci. Med. Sci. 69 (10), 1212–1226. https://doi.org/10.1093/gerona/glt177.
Villapol, S., Loane, D.J., Burns, M.P., 2017. Sexual dimorphism in the inflammatory response to traumatic brain injury. Glia 65 (9), 1423–1438. https://doi.org/10.1002/glia.23171.
Voss, J., Stermer, A.R., Ghaffari, R., Tiwary, R., Richburg, J.H., 2018. MEHP-induced rat testicular inflammation does not exacerbate germ cell apoptosis. Reproduction 156 (1), 35–46. https://doi.org/10.1530/REP-18-0093.
Wallaert, B., Fahy, O., Tsicopoulos, A., Gosset, P., Tonnel, A.B., 2000. Experimental systems for mechanistic studies of toxicant induced lung inflammation. Toxicol. Lett. 112–113, 157–163.
Wang, Y., Zhao, H., Shao, Y., Liu, J., Li, J., Xing, M., 2017. Copper or/and arsenic induce oxidative stress-cascaded, nuclear factor kappa B-dependent inflammation and immune imbalance, trigging heat shock response in the kidney of chicken. Oncotarget 8 (58), 98103–98116. https://doi.org/10.18632/oncotarget.21463.
Wilson, Q.N., Wells, M., Davis, A.T., et al., 2018. Greater microbial translocation and vulnerability to metabolic disease in healthy aged female monkeys. Sci. Rep. 8 (1), 11373. https://doi.org/10.1038/s41598-018-29473-9.
Win-Shwe, T.T., Fujimaki, H., 2011. Nanoparticles and neurotoxicity. Int. J. Mol. Sci. 12 (9), 6267–6280. https://doi.org/10.3390/ijms12096267.
Wohleb, E.S., Godbout, J.P., 2013. Basic aspects of the immunology of neuroinflammation. Mod. Trends Pharmacopsychiatry 28, 1–19. https://doi.org/10.1159/000343964.

Wolf, S.A., Boddeke, H.W.G.M., Kettenmann, H., 2017. Microglia in physiology and disease. Annu. Rev. Physiol. 79, 619–643. https://doi.org/10.1146/annurev-physiol-022516-034406.
Xu, E., Liu, J., Wang, X., Xiong, H., 2017. Inflammasome in drug abuse. Int. J. Physiol. Pathophysiol. Pharmacol. 9 (6), 165–177.
Yamamoto, B.K., Moszczynska, A., Gudelsky, G.A., 2010. Amphetamine toxicities: classical and emerging mechanisms. Ann. N. Y. Acad. Sci. 1187, 101–121. https://doi.org/10.1111/j.1749-6632.2009.05141.x.
Yang, S.N., Hsieh, C.C., Kuo, H.F., et al., 2014. The effects of environmental toxins on allergic inflammation. Allergy Asthma Immunol. Res. 6 (6), 478–484. https://doi.org/10.4168/aair.2014.6.6.478.
Zelinskiy, V., Zhorina, A., Andreev, V., 2015. Association between inflammation markers and abdominal aortic calcification in patients with peripheral arterial disease. Atherosclerosis 241 (1), E35. https://doi.org/10.1016/j.atherosclerosis.2015.04.128.
Zhang, P., Lokuta, K.M., Turner, D.E., Liu, B., 2010. Synergistic dopaminergic neurotoxicity of manganese and lipopolysaccharide: differential involvement of microglia and astroglia. J. Neurochem. 112 (2), 434–443. https://doi.org/10.1111/j.1471-4159.2009.06477.x.
Zindler, E., Zipp, F., 2010. Neuronal injury in chronic CNS inflammation. Best Pract. Res. Clin. Anaesthesiol. 24 (4), 551–562. https://doi.org/10.1016/j.bpa.2010.11.001.
Zlokovic, B.V., 2008. The blood-brain barrier in health and chronic neurodegenerative disorders. Neuron 57 (2), 178–201. https://doi.org/10.1016/j.neuron.2008.01.003.

CHAPTER SEVEN

Neurotoxicity of air pollution: Role of neuroinflammation

Lucio G. Costa[a,b,*], Toby B. Cole[a,c], Khoi Dao[a], Yu-Chi Chang[a], Jacki Coburn[a], Jacqueline Garrick[a]
[a]Department of Environmental and Occupational Health Sciences, University of Washington, Seattle, WA, United States
[b]Department of Medicine & Surgery, University of Parma, Parma, Italy
[c]Center on Human Development and Disability, University of Washington, Seattle, WA, United States
*Corresponding author: e-mail address: lgcosta@uw.edu

Contents

1. Introduction

Air pollution comprises gases, metals, organic compounds, and particulate matter (PM), and all may contribute to its adverse health effects. PM, however, is the most implicated in disease, particularly $PM_{2.5}$ (i.e., particles $<2.5\,\mu m$ in diameter) and ultrafine PM (UFPM, with diameter $< 100\,nM$) (Costa et al., 2014; Møller et al., 2010). Millions of individuals especially in mega-cities in Asia (China, India) or Central America (e.g., Mexico City) are exposed to relatively high levels of air pollution ($PM_{2.5} > 100\,\mu g/m^3$) (Costa et al., 2017a; Van Donkelaar et al., 2015). Traffic-related air pollution (TRAP), and particularly its major component, diesel exhaust (DE), is a major contributor to air pollution and to urban ambient $PM_{2.5}$ (Karagulian et al., 2015). Furthermore, a wide range of vehicles, heavy equipment, and other

Advances in Neurotoxicology, Volume 3
ISSN 2468-7480
https://doi.org/10.1016/bs.ant.2018.10.007

machinery utilize diesel engines, thus causing high occupational exposures to millions of workers worldwide (Pronk et al., 2009).

The adverse effects of air pollution on the respiratory and cardiovascular systems are well established (Manzetti and Andersen, 2016; Rücker et al., 2011), and oxidative stress and inflammation are believed to represent the two principal underlying mechanisms. In recent years, accumulating evidence strongly suggests that air pollution may also negatively affect the central nervous system (CNS) and contribute to CNS diseases, possibly by similar oxidative and inflammatory mechanisms (Babadjouni et al., 2017; Block and Calderon-Garciduenas, 2009; Genc et al., 2012; Lancet Neurology, 2018; Xu et al., 2016). $PM_{2.5}$ and UFPM enter the circulation and distribute to various organs, including the brain (Genc et al., 2012; Oberdoerster et al., 2004); additionally, they can gain direct access to the brain through the nasal olfactory mucosa (Lucchini et al., 2012; Oberdoerster et al., 2004; Peters et al., 2006). Several human epidemiological studies have shown that exposure to elevated air pollution is associated with decreased cognitive function, olfactory dysfunction, auditory deficits, depressive symptoms, and other adverse neuropsychological effects (Costa et al., 2014, 2017a; Freire et al., 2010; Guxens and Sunyer, 2012; Pun et al., 2017). Animal studies have similarly found significant alterations in motor activity, spatial learning and memory, novel object recognition ability, and emotional behavior upon exposure to DE (Gerlofs-Nijland et al., 2010; Levesque et al., 2011; MohanKumar et al., 2008; Win-Shwe and Fujimaki, 2011). Additional evidence suggests that the aging brain may be particularly susceptible to air pollution-induced neurotoxicity, as several of the epidemiological studies identifying adverse effects of air pollution on behavior, particularly cognitive behavior, have identified significant effects in the elderly (Chen et al., 2015; Power et al., 2011; Ranft et al., 2009; Weuve et al., 2012). Aging is often associated with a wide variety of clinical and pathological conditions which can be classified as neurodegenerative diseases such as Alzheimer's disease (AD) or Parkinson's disease (PD). Findings in animals and humans indicate that air pollution increases expression of markers of neurodegenerative disease pathologies in brain, and the strongest evidence so far relates to a possible contribution of air pollution to the etiology of Alzheimer's disease (see Costa, 2017 for a full review).

Epidemiological and animal studies also suggest that young individuals may be particularly susceptible to air pollution-induced neurotoxicity (Calderon-Garciduenas et al., 2008, 2012; Freire et al., 2010; Guxens et al., 2014; Guxens and Sunyer, 2012). For example, exposure to air pollution during pregnancy was found to be associated with delayed psychomotor development (Guxens et al., 2014) and to lower cognitive development in primary school children (Sunyer et al., 2015). Animal studies support human findings (Costa et al., 2014, 2017a), as in utero or postnatal exposure to DE or ambient PM was found to cause alterations in motor activity, spatial learning and memory, and impulsive behavior (Allen et al., 2014; Hougaard et al., 2008; Tsukue et al., 2009; Win-Shwe et al., 2014; Yokota et al., 2013). In recent years, several studies have found associations between exposures to traffic-related air pollution and autism spectrum disorders (see Costa et al., 2017b for a review). For example, Volk et al. (2011, 2013) found that residential proximity to freeways and gestational and early-life exposure to traffic-related air pollution were associated with autism, and similar results were obtained in several other epidemiological studies (Becerra et al., 2013; Kalkbrenner et al., 2015). Developmental exposure of rodents to DE also causes alterations in behavioral domains (persistent/repetitive behaviors, communication, and social interactions) typically affected in autism (Chang et al., 2018; Church et al., 2018; Thirtamara Rajamani et al., 2013).

As indicated earlier, the main effects of air pollution are inflammatory in nature. Peripheral inflammation and/or neuroinflammation are believed to be responsible for, or at least significantly implicated in the adverse effects of air pollution on the CNS. This chapter will highlight animal, in vitro, and human studies describing neuroinflammatory processes induced by air pollution, and how such effects may be relevant in neurodevelopmental and neurodegenerative disorders associated with air pollution exposure.

2. Neuroinflammation

The term "inflammation" derives from the Latin word *inflammatio* which suggests the concepts of flames and fire. Indeed, the classical signs of inflammation are heat, pain, redness, and swelling. These are due to a complex response by the organism to protect itself from harmful stimuli. Thus, though the term inflammation has acquired a certain negativity, the ultimate functions of this response are extremely useful, as too little inflammation may compromise the organism by not eliminating the cause of cellular injury and initiate proper tissue repair. Inflammation can be either acute or chronic, and it is usually the latter which is involved in several chronic diseases such as rheumatoid arthritis or atherosclerosis. Inflammation traditionally involves the synthesis and release of pro-inflammatory mediators, such as cytokines and chemokines. Monocytes, neutrophils, and macrophages are the cells that in the periphery generate oxidative stress and pro-inflammatory responses (O'Callaghan et al., 2008).

Neuroinflammation refers to inflammatory responses in the CNS, which can also be acute or chronic. In recent years, neuroinflammation has become a dominant theme in neuroscience as inflammatory processes in the CNS have been associated with neurodevelopmental and neurodegenerative disorders (Harry and Kraft, 2008; Kraft and Harry, 2011; O'Callaghan et al., 2008). However, whether neuroinflammation represents a cause or a consequence of neurologic disease is still not fully understood, as is the relevance of inflammatory processes in toxicant-induced neurotoxicity (O'Callaghan et al., 2008). In the CNS, glial cells, particularly activated microglia and astrocytes, serve as source and targets of pro-inflammatory mediators. Indeed, a dominant response of the CNS to injuries (caused by disease, trauma, chemicals, or drugs) is the activation of these two types of glial cells. Thus, a primary event in neuroinflammation appears to be the change of microglia and astrocytes from a quiescent state to an activated state, though even this established view of neuroinflammation has been recently challenged (Masgrau et al., 2017).

Microglia, long considered the least important glial cells in the CNS, have now taken the center stage, given the increasing evidence of their role in maintaining normal brain homeostasis, and particularly their role in mediating CNS damage (Aguzzi et al., 2013; Block and Hong, 2005; Block et al., 2007; Chen and Trapp, 2016; Cherry et al., 2014; Cho and Choi, 2017; Hanamsagar and Bilbo, 2017; Harry and Kraft, 2008; Kraft and Harry, 2011; Moore et al., 2014; Salter and Stevens, 2017). Microglia are specialized macrophages originating from the yolk sac during development; early in development they populate the CNS parenchyma, have an extended lifespan, and little turnover (Moore et al., 2014). First described in 1932 by Pio de Rio-Hortega, they have a typical ramified morphology, which allows them to sense the local environment, communicate with other cells, and quickly respond to infection and injury (Moore et al., 2014). In the normal brain, microglia are essential for clearing cellular debris during development, and together with astrocytes, support axon development and modeling of synapses. Later in brain development they also participate in synaptic pruning (Moore et al., 2014). Upon activation, microglia adopt a characteristic amoeboid morphology, and release reactive oxygen species (ROS), cytokines, chemokines, and other pro-inflammatory mediators.

However, microglia are a "double-edged sword," as they may also possess anti-inflammatory/neuroprotective roles, by secreting various growth and trophic factors and anti-inflammatory cytokines. Peripheral macrophages are known to adopt different phenotypes in the activated state: the M1, pro-inflammatory or classically activated cells, and the M2, anti-inflammatory, or alternatively activated cells (Moore et al., 2014). By analogy with macrophages, microglia have also been described as polarized to both M1 and M2 phenotypes (Block et al., 2007; Chen and Trapp, 2016; Cherry et al., 2014; Cho and Choi, 2017), with the M1 phenotype being pro-inflammatory and the M2 phenotype being immunosuppressive, though this classification has been judged to be too simplified (Tang, 2018).

Astrocytes also play a dual role in neuroinflammation, as they can either promote inflammation or provide protection against it (Colombo and Farina, 2016; Keil et al., 2018; Li et al., 2011). Astrocytes are known to respond to CNS injury through a process known as astrogliosis, which comprises several structural and functional changes of these cells. Activated astrocytes proliferate and can form a sort of a protective barrier around the inflamed tissue. In addition, hypertrophic astrocytes do not proliferate but secrete pro- and anti-inflammatory mediators, as well as other oxidative stressors (Keil et al., 2018). Astrocytes are believed to be critical regulators of the neuroinflammatory process, as different stimuli may confer on them a tissue-protective or -detrimental phenotype, and the astrocytic response should be seen as the net result of a complex and variegated network of activated intracellular pathways (Colombo and Farina, 2016).

3. Air pollution and neuroinflammation: In vitro and animal studies and human observations

As indicated earlier, inflammation is a main mechanism by which air pollution causes pulmonary and cardiovascular toxicity (Akopian et al., 2016; Fiordelisi et al., 2017). Inflammation and oxidative stress also occur in the CNS upon exposure to high levels of air pollution and are believed to be causally associated with air pollution-induced neurotoxicity, including neurodevelopmental and neurodegenerative diseases. In this section, we will briefly summarize some evidence, obtained in cellular systems in vitro, in animals in vivo as well as in humans, of the ability of air pollution to induce neuroinflammation.

In vitro studies have shown that PM can cause neuroinflammatory responses and that the main cell type involved in this process are microglia.

Indeed, microglia play an important role in mediating the neurotoxicity of DE particles in vitro (Block et al., 2004). A recent study from our laboratory is described in more detail to illustrate key findings related to the role of microglia (Roqué et al., 2016). The neuronal toxicity of DE particles was investigated using a neuron-microglia co-culture system, consisting of mouse cerebellar granule neurons and microglia isolated from mouse cerebral cortex, in a 9:1 ratio (Roqué et al., 2016). DE particles were only toxic to neurons in the presence of microglia, indicating a central role of these cells in mediating DE neurotoxicity. DE particles can activate microglia, both in vitro (Block et al., 2004; Levesque et al., 2013; Roqué et al., 2016) as well as in vivo (Cole et al., 2016; Durga et al., 2015; Levesque et al., 2011). Activated microglia are known to release both cytotoxic and protective factors, which may influence neuronal viability (Block et al., 2007; Luo and Chen, 2012; Ransohoff and Perry, 2009). Activated microglia contribute to oxidative stress and to neuroinflammation, and both may contribute to DE neurotoxicity (Block et al., 2004; MohanKumar et al., 2008; Roqué et al., 2016). In our neurons-microglia co-culture system, DE particles increased microglial reactive oxygen species, but two antioxidants (PBN and melatonin) failed to protect neurons from DE particle-induced, microglia-mediated, toxicity (Roqué et al., 2016). In the same system, DE particles also caused a significant increase of IL-6 and IL-1β mRNA in microglia and an increased release to the medium of IL-6 (Table 1). Most interestingly, conditioned medium from DE particle-treated microglia was

Table 1 Effect of diesel exhaust on levels of the pro-inflammatory cytokine IL-6.

Experimental condition	Effect (% increase)	Reference
Acute exposure (6h, 250 μg/m^3), hippocampus, Gclm$^{+/+}$ male mice	+1782 (protein)	Cole et al. (2016)
Acute exposure (6h, 250 μg/m^3), hippocampus, Gclm$^{+/-}$ male mice	+4688 (protein)	Cole et al. (2016)
Three-week exposure (250 μg/m^3), hippocampus, Gclm$^{+/+}$ male mice	+330 (mRNA)	Unpublished results
Developmental exposure (GD0-PND3, 250 μg/m^3, whole brain, male mice)	+105 (mRNA)	Unpublished results
In vitro DE particles (24h, 50 μg/2cm^2), mouse cortical microglia	+1600 (mRNA)	Roqué et al. (2016)
In vitro DE particles (24h, 50 μg/2cm^2), medium from mouse microglia	+650 (protein)	Roqué et al. (2016)

highly toxic to neurons, suggesting that pro-inflammatory cytokines and other factors secreted by activated microglia play a central role in DE neurotoxicity. Furthermore, two compounds known to inhibit microglia activation, i.e., the antibiotic minocycline and the peroxisome proliferator-activated receptor-γ agonist pioglitazone, attenuated DE particle-induced neuronal death in the co-culture system (Roqué et al., 2016). A recent study by Duffy et al. (2018) similarly showed that low concentrations of combustion-generated nanoparticles collected from a diesel engine generated a pro-inflammatory response in immortalized murine microglial cells (BV2) (i.e., an increase in TNF-α), and that conditioned medium from nanoparticle-stimulated microglia reduced mouse hypothalamic neurons survival. Rat mixed glial cultures (astrocytes and microglia) exposed in vitro to nanoparticles (diameter $< 0.2\,\mu m$) from TRAP collected from a California freeway caused a general increase of pro-inflammatory cytokine mRNAs, like that induced by LPS (Woodward et al., 2017). The response was also observed in vivo and was mediated by activation of the Toll-like receptor 4 (TLR4) (Woodward et al., 2017). In cultured mouse hippocampal slices, $PM_{2.5}$ increased the expression of cyclooxygenease-2 (COX-2), a key player in neuroinflammation, and of reactive oxygen species, and disrupted synaptic transmission (Li et al., 2018).

As already indicated by some examples, animal studies corroborate the in vitro observations (Costa et al., 2014). For example, dogs exposed to heavy air pollution presented evidence of chronic inflammation and neurodegeneration in various brain regions (Calderon-Garciduenas et al., 2003), and mice exposed to traffic in a highway tunnel had higher levels of pro-inflammatory cytokines in brain (Bos et al., 2012). Even an acute exposure of C57BL/6 mice to DE (250–300 $\mu g/m^3$ for 6 h) caused significant increases in lipid peroxidation and of pro-inflammatory cytokines (IL-1α, IL-1β, IL-3, IL-6, TNF-α) in various brain regions (particularly olfactory bulb and hippocampus), and in several cases the observed effects were more pronounced in male than in female mice. DE exposure also caused microglia activation, as measured by increased expression of Iba1 (ionized calcium-binding adapter molecule 1), and an increase in TSPO (translocator protein) binding (Cole et al., 2016). Mice heterozygotes for the modifier subunit of glutamate cysteine ligase (the limiting enzyme in glutathione biosynthesis; *Gclm*$^{+/-}$ mice) appeared to be significantly more susceptible to DE-induced neuroinflammation than wild-type mice (Table 1; Cole et al., 2016). These findings indicate that even an acute exposure to DE causes neuroinflammation and oxidative stress in brain and suggests that sex and genetic background may play important roles in modulating susceptibility to DE neurotoxicity. In mice exposed to concentrated PM, neuroinflammation was seen in the brain, as evidenced by increased levels of various cytokines and of NF-κB (Campbell et al., 2005). The latter result was also found in ApoE$^{-/-}$ mice exposed to UFPM, together with increases in mitogen-activated kinase pathways and of GFAP (glial fibrillary acidic protein) (Kleinman et al., 2008). Various markers of neuroinflammation have been found to be increased in rodents following DE exposure (Gerlofs-Nijland et al., 2010; Levesque et al., 2011). In addition, alterations in expression of some oxidative stress-related genes and other markers of oxidative stress have also been shown in rodents following DE exposures (Cole et al., 2016; Lodovici and Bigagli, 2011; Manzetti and Andersen, 2016). While air pollution components may exert their deleterious effects directly on the CNS, the possibility and the extent of a peripheral contribution to the central effects should be further explored. High levels of circulating pro-inflammatory cytokines may negatively affect the CNS (Calderon-Garciduenas et al., 2013; Hernandez-Romero et al., 2012; Jayaraj et al., 2017; Kempuraj et al., 2016), and the blood-brain barrier may represent an important site for air pollution neurotoxicity, that has been little studied so far (Varatharaj and Galea, 2017).

Limited information available from humans confirms the experimental findings. In highly air pollution-exposed individuals in Mexico City, a series of studies in post-mortem samples have revealed increased markers of oxidative stress and neuroinflammation in adults, young adults, and children (Calderon-Garciduenas et al., 2008, 2012, 2013). In addition, markers of peripheral inflammation (e.g., C-reactive protein, fibrinogen) are also increased in humans exposed to high levels of air pollution (Fiordelisi et al., 2017).

4. Air pollution, neuroinflammation, and developmental disorders

Epidemiological and animal studies suggest that young individuals may be particularly susceptible to air pollution-induced neurotoxicity (Calderon-Garciduenas et al., 2008, 2012, 2015; Freire et al., 2010; Guxens et al., 2014; Suades-Gonzalez et al., 2015; Vrijheid et al., 2016). Human studies have revealed biochemical and behavioral alterations in children exposed pre- and/or postnatally to elevated air pollution. A series of studies in Mexico City in children exposed to high air pollution have found elevated levels of neuroinflammatory markers in brain, and cognitive deficits (Calderon-Garciduenas et al., 2008, 2015). Newman et al. (2013) reported hyperactivity in 7-year-old children associated with early-life exposure to TRAP and in six European cohorts, exposure to air pollution during pregnancy was associated with delayed psychomotor development (Guxens et al., 2014). Similar results were found in Japan, where gestational air pollution exposure was associated with delays in developmental milestones in young children (Yorifuji et al., 2016). Further studies reported that exposure to TRAP was inversely associated with sustained attention in adolescents (Kicinski et al., 2015), and to lower cognitive development in primary school children (Basagaña et al., 2016; Sunyer et al., 2015). Developmental exposure to air pollution was also associated with diminished executive functions at 6–10 years of age (Harris et al., 2016), and with several behavioral alterations in children, mostly in boys, including lower IQ, diminished attention, and deficits in reaction time and memory (Chiu et al., 2016). Experimental studies also indicate that developmental exposure to DE causes neurotoxicity (Ema et al., 2013). In utero exposure to high levels of DE ($1.0\,mg/m^3$) caused alterations in motor activity, motor coordination, and impulsive behavior in male mice (Yokota et al., 2013). Early postnatal exposure of mice to concentrated ambient PM was reported to cause various behavioral changes, including long-term impairment of short-term memory, and impulsivity-like behavior (Allen et al., 2013, 2014). Depression-like responses were found in mice exposed prenatally to urban air nanoparticles (Davis et al., 2013). Additional studies have shown that developmental DE exposure of mice alters motor activity, spatial learning and memory, and novel object recognition ability, and causes changes in gene expression, neuroinflammation, and oxidative damage (Hougaard et al., 2008; Tsukue et al., 2009; Win-Shwe et al., 2014).

There is substantial evidence demonstrating an association between inflammation and neurodevelopmental disorders, particularly autism spectrum disorders (ASD; Davis, 2018; Brockmeyer and D'Angiulli, 2016; Mottahedin et al., 2017; Prata et al., 2017). In recent years, evidence has been accumulating suggesting that developmental exposure to air pollution, and particularly TRAP, may be involved in the etiology of autism spectrum disorders (ASD). Autism is a neurodevelopmental disorder characterized by marked reduction of social and communicative skills, and by the presence of stereotyped behaviors (Levy et al., 2009). The symptoms of ASD are typically present before the age of three, and are often accompanied by abnormalities of cognitive functioning, learning, attention, and sensory processing

(Levy et al., 2009). The incidence of ASD appears to have increased in the past few decades, and it is now estimated at about 16.8/1000, representing 1 every 59 children (Baio et al., 2018). ASD is more common in males than in females (ratio 4:1), and represents an important societal problem, as the economic burden of caring for an individual with ASD and intellectual disability during his/her lifespan has been estimated at $2.4 million (see references in Costa et al., 2017b). Children with ASD present morphological abnormalities in the brain (Levy et al., 2009; Stoner et al., 2014), a higher level of oxidative stress (Rose et al., 2012), as well as neuroinflammation and increased systemic inflammation (Depino, 2013; El-Ansary and Al-Ayadhi, 2012; Theoharides et al., 2013). The etiological basis of ASD is unknown, and susceptibility is attributable to both genetic and environmental factors (Levy et al., 2009; Rossignol et al., 2014; Sandin et al., 2014). Several candidate susceptibility genes for ASD have been identified, but no single anomaly appears to predominate, and the total fraction of ASD attributable to genetic inheritance may be only about 30–50% (Sandin et al., 2014). DNA methylation is also altered in the autistic brain, suggesting that epigenetic dysregulation may also contribute to ASD (Berko and Greally, 2015). Current evidence thus suggests that ASD likely results from the complex interactions between genes conferring vulnerability and diverse environmental factors.

Several studies have found associations between exposures to traffic-related air pollution and ASD. Two studies in California by Volk et al. (2011, 2013) found that residential proximity to freeways and gestational and early-life exposure to TRAP were associated with autism. Similar results were obtained in a separate epidemiological study in California (Becerra et al., 2013), and in another one, part of the Nurses' Health Study II, in which perinatal DE exposure was significantly associated with ASD, particularly in boys (Roberts et al., 2013). Several other studies in different countries support the finding that air pollution exposure is associated with an increased risk of ASD, particularly if exposure occurs during the third trimester of pregnancy (reviewed in Costa et al., 2017b). Animal studies (Allen et al., 2014; Chang et al., 2018; Church et al., 2018; Thirtamara Rajamani et al., 2013) agree with the human observations. For example, C57Bl/6J mice were exposed from embryonic day 0 to postnatal day 21 to 250–300 $\mu g/m^3$ DE or filtered air (FA) as control. Mice exposed developmentally to DE exhibited deficits in all three of the hallmark categories of ASD behavior: reduced social interaction in the reciprocal interaction and social preference tests, increased repetitive behavior in the T-maze and marble-burying test, and reduced or altered communication as assessed by measuring isolation-induced ultrasonic vocalizations and responses to social odors (Chang et al., 2018). These findings, and similar ones by other investigators, show that exposure to TRAP, in particular that associated with diesel-fuel combustion, can cause ASD-related behavioral changes in mice.

So far, the strongest association between an environmental factor and ASD has been found with maternal infection (maternal immune activation—MIA) (Bilbo et al., 2018; Estes and McAllister, 2016; Patterson, 2011). Supportive studies in various animal species have evidenced that MIA increases neuroinflammation in the placenta and in the fetal brain, leading to offspring which display ASD-like behaviors (Estes and McAllister, 2016; Malkova et al., 2012). An important consideration is that developmental DE exposure causes neuroinflammatory biochemical changes in the brain (e.g., increase in IL-6 levels; Table 1) mimicking those observed in MIA, which represent the strongest model for a developmental etiology of ASD.

Activation of microglia and subsequent neuroinflammation caused by air pollution may explain the effects seen in brain of rodents following developmental air pollution exposure. For example, microglia-generated pro-inflammatory cytokines could lead to the observed hypomyelination and ventriculomegaly via toxicity to oligodendrocytes. Closely related to ASD is also the hypothesis of a possible impairment by air pollution of the reelin signaling system. Reelin is a signaling glycoprotein, secreted in the marginal zone of the developing cerebral cortex by Cajal-Retzius cells, which plays a most relevant role in neuronal migration and establishment of neuronal polarity (Folsom and Fatemi, 2013; Förster, 2014). The canonical reelin signaling pathway is activated upon binding of reelin to VLDL receptor and APoE receptor 2, which triggers tyrosine phosphorylation of the intracellular adaptor protein disabled-1 (Dab1). Phosphorylated Dab1 then activates a kinase cascade involving PI-3 kinase, LIM kinase-1, and several others (Fatemi et al., 2005). Such complex network of signaling pathway mediates the ultimate effects of reelin on neuronal migration and polarity in the developing brain. Strong evidence exists for an involvement of reelin in ASD (reviewed in Costa et al., 2017b) as indicated by decreased reelin expression in brain from ASD subjects, several epigenetic and genetic abnormalities of the reelin gene in ASD brain, and the fact that the reeler ($rl^{-/-}$) mouse displays several ASD-like morphological and behavioral traits (Fatemi et al., 2005; Folsom and Fatemi, 2013; Serajee et al., 2006). In addition, MIA which—as said—leads to offspring which display neuroinflammation and ASD-like behaviors, has been shown to decrease levels of reelin protein and mRNA in brain of offspring (Meyer et al., 2006). The notion that oxidative stress and neuroinflammation may play an important role in modulating reelin expression is also supported by studies showing that *N*-acetylcysteine completely prevents lipopolysaccharide (LPS)-induced decreases of reelin (Novais et al., 2013). In our laboratory, we have found that developmental DE exposure (250–300 $\mu g/m^3$ from GD0 to PND21) causes neuroinflammation, a long-lasting decrease of reelin expression, and disorganization of cortical layering, all highly relevant for ASD (Chang et al., submitted for publication).

5. Air pollution, neuroinflammation, and neurodegenerative disorders

As indicated earlier, there is increasing evidence suggesting that the aging brain may be particularly susceptible to air pollution-induced neurotoxicity, and several epidemiological studies have identified significant associations between air pollution and behavioral and cognitive effects in the elderly (Chen et al., 2015; Power et al., 2011; Ranft et al., 2009; Weuve et al., 2012). Aging is also often associated with a variety of clinical and pathological conditions which can be classified as neurodegenerative diseases, such as Alzheimer's disease (AD) or Parkinson's disease (PD). Aging indeed remains the main risk factor for neurodegenerative diseases, and in the European Union and in the United States, the number of people with dementia is expected to double or triple in the next 30 years, causing a significant medical, societal, and economic burden (Hebert et al., 2013; Maresova et al., 2016).

Neurodegenerative diseases are heterogeneous disorders and have complex clinical and pathological pictures (Musgrove et al., 2015). While susceptibility genes for several disorders have been identified, environmental factors or gene-environment interactions are believed to play the most important role in most neurodegenerative diseases. PD is a neurodegenerative disorder characterized by a slow and progressive degeneration of dopaminergic neurons in the substantia nigra, with degeneration of nerve terminals in the striatum. Once loss of dopaminergic neurons has reached about 80%, clinical signs appear which include resting tremor, rigidity, bradykinesia, and gait disturbances (Cubo and Goetz, 2014). Olfactory dysfunction is an important early symptom of PD (Doty, 2012) and damage to the olfactory bulb precedes neuropathology in the motor areas (Braak et al., 2004). The prevalence of PD in the general population is about 0.3%, but increases to 1% in people above 60, and to 3–5% in those over 85 years (Lema-Tomè et al., 2013). While genetic forms of PD have been associated with specific mutations in a number of genes (PARK1-PARK13), the great majority of PD cases is sporadic, and may be due to environmental factors or to gene-environment interactions, i.e., neurotoxic exposures in susceptible individuals. Among environmental factors believed to be associated with PD there are certain pesticides (e.g., the herbicide paraquat or the insecticide rotenone), metals (e.g., manganese), or air pollution. Protein aggregations in the form of Lewy bodies in surviving neurons are a hallmark of PD (Beier and Richardson, 2015). These Lewy bodies are a dense core inclusion encircled by a halo radiating fibrils composed of misfolded α-synuclein. Levels of α-synuclein are higher in olfactory bulb and striatum, two brain regions affected in PD, and are higher in PD brain than in normal aging (Beier and Richardson, 2015; Ulusoy and Di Monte, 2013).

The term dementia refers to an acquired cognitive impairment involving multiple domains and is used as an umbrella term for several distinctive diseases (Perry, 2014). Dementia is predominantly a disease of later life, and after 65 years of age its prevalence doubles every few years. As populations are aging at an increasing rate, the number of cases in the world is predicted to reach 115 million by 2050 (Prince et al., 2013). The main type of dementia in individuals of >65 years is AD, which accounts for 55–75% cases (Scheltens et al., 2016). Other dementias are dementia with Lewy bodies, vascular dementia, and frontotemporal dementia (Perry, 2014). The most common symptom of AD is memory loss for recent events, while its gross pathology is represented by diffuse cortical and hippocampal atrophy. Accumulation of abnormally folded amyloid beta (Aβ) and of tau proteins in amyloid plaques and neuronal tangles are the neuropathological hallmarks of AD (Khan and Bloom, 2016; Selkoe and Hardy, 2016). The amyloid precursor protein (APP) is cleaved by secretases to generate an Aβ polypeptide, of which Aβ42 is the major form found in amyloid plaques (Pressman and Rabinovici, 2014). Neurofibrillary tangles are composed of hyperphosphorylated tau protein, which cause disassembly of microtubules. A few genes have been associated with AD (e.g., presenilins), but the major genetic risk factor for AD is apolipoprotein E (APOE), with APOEε4 predisposing the carrier to AD (Riedel et al., 2016).

The fact that air pollution causes systemic inflammation, microglia activation, oxidative stress, and neuroinflammation provides biological plausibility and potential underlying mechanisms for the observed association between exposures and ensuing risk of neurodegenerative diseases (Kraft and Harry, 2011). There is ample evidence that similar inflammatory processes occur in various neurodegenerative diseases and contribute to their etiopathology (Manoharan et al., 2016; Ransohoff, 2016). Microglia activation plays an important role in PD and has been strongly linked to its pathology (Lull and Block, 2010). Activated microglia have been found by positron emission tomography in the substantia nigra of living PD patients, in human post-mortem PD brains, and in animal models of PD (Lull and Block, 2010). Activation of microglia causes an increase in oxidative stress and in pro-inflammatory cytokines. Oxidative stress is believed to play a role in PD pathogenesis and has been shown to cause α-synuclein aggregation (Takahashi et al., 2007). There is also a growing recognition of the central role of neuroinflammation in the pathogenesis of PD (Hirsch et al., 2012; Qian et al., 2010), and peripheral inflammation may initiate or contribute to the neuroinflammation and dopaminergic degeneration in the CNS. Oxidative stress and neuroinflammation also play a cardinal role in AD (Heneka et al., 2015; Huang et al., 2016; Lee et al., 2010; Moulton and Yang, 2012; Rubio-Perez and Morillas-Ruiz, 2012). Important roles are played by pro-inflammatory cytokines such as IL-1, IL-6, and TNF-α originating from microglia and astrocytes (Leyns and Holtzman, 2017; Rubio-Perez and Morillas-Ruiz, 2012), which may lead to inhibition of adult neurogenesis (Fuster-Matanzo et al., 2013). Neuroinflammation can contribute to amyloid toxicity (Minter et al., 2016), and APOEε4 (a strong genetic risk factor for AD) is less protective against neuroinflammation than APOEε3 (Tai et al., 2015). As indicated earlier, the contribution of peripheral inflammation to neuroinflammation also needs to be considered. Though neuroinflammation and inflammation present significant differences (Filiou et al., 2014), systemic inflammation can affect inflammatory processes in the CNS (Hopkins, 2007; Lema-Tomè et al., 2013). As peripheral inflammation is a most relevant effect of exposure to air pollution, a possible contribution of peripheral inflammatory processes to adverse CNS effects needs to be further investigated.

Several studies have investigated the possible association between air pollution and PD, AD, or other neurodegenerative disease, and the strongest evidence so far is for an association with AD and other dementias (reviewed by Costa, 2017). The evidence for an association with PD is still limited, though observations in humans and in controlled animal studies also suggest that air pollution may increase the expression of the PD marker α-synuclein. Nevertheless, oxidative stress and neuroinflammation, two cardinal effects of air pollution, are also believed to play most relevant roles in PD, and an in vitro study showed that diesel exhaust particles could activate microglia leading to the demise of dopaminergic neurons (Block et al., 2004).

In recent years, air pollution has been emerging as an important etiological factor for AD (Killin et al., 2016; Power et al., 2016). As cognitive impairment is an initial important aspect of AD, several studies have focused on various assessments of cognitive level (e.g., performance on cognitive tests) in relationship to air pollution (Clifford et al., 2016; Cohen and Gerber, 2017; Peters et al., 2015), and most have found that increasing levels of air pollution, TRAP in particular, was associated with diminished cognitive abilities (reviewed by Costa, 2017). Additional studies have examined the association between air pollution and AD and other dementias. For example, three separate studies in Taiwan reported associations between air pollution and AD (Chang et al., 2014; Jung et al., 2015; Wu et al., 2015). Exposure to $PM_{2.5}$ was associated with increased risk of cognitive impairment and AD in a German population of elderly women (Schikowski et al., 2015). In Sweden, Oudin et al. (2016) found that TRAP was a risk factor for AD, and in Ontario, Canada, a large study involving about 2.2 million individuals aged 55–85 years, concluded that living <50 m from a major traffic road was a risk factor for AD (Chen et al., 2017). A few studies also examined the effects of high air pollution exposure on markers of AD (e.g., pTau, Ab42) in brain tissue of post-mortem humans or of experimental animals (Calderon-Garciduenas et al., 2004, 2008, 2012, 2016) or of experimental animals (Cacciottolo et al., 2017; Durga et al., 2015; Levesque et al., 2011), and found them to be elevated.

As said earlier, APOEε4 represents the strongest genetic risk factor for AD (Riedel et al., 2016), and a few studies have provided evidence that carriers of one or two of the APOEε4 alleles are more susceptible to the neurotoxic effects of air pollution (Calderon-Garciduenas et al., 2004, 2008, 2012; Schikowski et al., 2015). For example, the study by Cacciottolo et al. (2017) in older women found that the risk of cognitive decline and dementia was greater in homozygotes for the APOEε4 allele (HR = 3.95), than for those carrying the APOEε3 allele (HR = 1.65).

Neuroinflammation may "mechanistically" link air pollution and neurodegeneration, as suggested by various lines of evidence (Costa, 2017). Adult neurogenesis is impaired in humans with, or animal models of, neurodegenerative diseases (Horgusluoglu et al., 2017), including AD (Fuster-Matanzo et al., 2013) and PD (Marxreiter et al., 2013). Impairment of adult neurogenesis in the hippocampal region may be associated with decreased cognitive function in AD (Fuster-Matanzo et al., 2013; Horgusluoglu et al., 2017). Of relevance is that brain inflammation inhibits basal neurogenesis in the hippocampal subgranular zone (SGZ), an effect that is prevented by minocycline, an inhibitor of microglia activation (Carpentier and Palmer, 2009; Ekdahl et al., 2003). Evidence of decreased adult neurogenesis following early postnatal inflammation in rodents has also been reported (Dinel et al., 2014; Lajud and Torner, 2015); furthermore, exposure of adult mice to DE (250–300 μg/m^3 for 6 h) decreases neurogenesis in the SGZ in mice (Coburn et al., 2018), which is prevented by inhibiting microglia activation and ensuing neuroinflammation with pioglitazone (Coburn et al., 2018).

Reelin, already discussed in relationship to its possible role in ASD, may also be relevant in AD, and air pollution-induced alterations in reelin expression and/or signaling may play a role in the observed associations between exposure and AD. Altered reelin expression, and the ensuing alteration in reelin signaling, may contribute to AD (Yu et al., 2016), as suggested by the fact that brain reelin levels are decreased in natural aging (Stranahan et al., 2011), in transgenic animal models of AD (Chin et al., 2007; Kocherhans et al., 2010; Yu et al., 2014), and in brains of AD patients, particularly in the early stages (Chin et al., 2007; Herring et al., 2012). Reelin can inhibit Aβ generation, promote Aβ clearance, and prevent tau phosphorylation (Kocherhans et al., 2010; Pujadas et al., 2014; Yu et al., 2016). Indeed, the reeler mouse ($rl^{-/-}$) shows hyperphosphorylation of tau (Hiesberger et al., 1999; Ohkubo et al., 2003), and low levels of reelin are associated with higher tau phosphorylation (Yu et al., 2016). Decreased reelin levels also cause synaptic dysfunction in the hippocampus, leading to memory and cognitive deficits which may precede neuronal loss, as found in AD (Ma and Klann, 2012; Yu et al., 2016). Genetic polymorphisms of *RELN* are also associated with AD (Fehen et al., 2015). Neuroinflammation (and specifically an increase in IL-6) which follows microglia activation, is an initial event induced by air pollution, which can lead to a decrease of reelin, through an increased methylation of the reelin promoter (Noh et al., 2005; Palacios-Garcia et al., 2015).

6. Conclusions

Components of TRAP, particularly PM, can elicit oxidative stress and inflammation, and these effects are believed to be at the basis of their observed adverse actions. In the CNS oxidative stress and neuroinflammation are similarly observed upon exposure to air pollution. Significant increases in the expression of pro-inflammatory cytokines (e.g., IL-6; Table 1) can be found in various experimental protocols in vitro and in vivo. Furthermore, similar neuroinflammatory processes have also been reported in humans. Neuroinflammation is known to play a central role in developmental (e.g., ASD) and neurodegenerative (e.g., AD) disorders, and increasing evidence provided by experimental and epidemiological studies indicates strong causal association between exposure to high levels of air pollution and some developmental and neurodegenerative disorders. So far, traditional anti-inflammatory drugs have been found to be of limited effectiveness as pharmacological treatments for these diseases; nevertheless, alternative treatments, perhaps focusing on microglia activation, may be of more relevance. It should be noted however, that most neurodevelopmental and neurodegenerative diseases are of multiple etiologies, involving both genetic and environmental factors, and that neuroinflammation only represent a facet of the overall problem.

Acknowledgments

Research by the authors is supported by the grants from NIEHS (R01ES022949, R01ES028273, P30ES07033, and P42ES04696) and NICHD (U54HD08091) and by funds from the University of Washington and the Department of Environmental and Occupational Health Sciences at the University of Washington.

References

Aguzzi, A., Barres, B.A., Bennett, M.L., 2013. Microglia: scapegoat, saboteur, or something else? Science 339, 156–161.

Akopian, A.N., Fanick, E.R., Brooks, E.G., 2016. TRP channels and traffic-related environmental pollution-induced pulmonary disease. Semin. Immunopathol. 38, 331–338.

Allen, J.L., Conrad, K., Oberdorster, G., Johnston, C.J., Sleezer, B., Cory-Slechta, D.A., 2013. Developmental exposure to concentrated ambient particles and preference for immediate reward in mice. Environ. Health Perspect. 121, 32–38.

Allen, J.L., Liu, X., Weston, D., Prince, L., Oberdörster, G., Finkelstein, J.N., Johnston, C.J., Cory-Slechta, D.A., 2014. Developmental exposure to concentrated ambient ultrafine particulate matter air pollution in mice results in persistent and sex dependent behavioral neurotoxicity and glial activation. Toxicol. Sci. 140, 160–178.

Babadjouni, R.M., Hodis, D.M., Radwanski, R., Durazo, R., Patel, A., Liu, Q., Mack, W.J., 2017. Clinical effects of air pollution on the central nervous system: a review. J. Clin. Neurosci. 43, 16–24.

Baio, J., Wiggins, L., Christensen, D.L., Maenner, M.J., Daniels, J., Warren, Z., et al., 2018. Prevalence of autism spectrum disorders among children aged 8 years-Autism and Developmental Disabilities Monitoring Network, 11 Sites, United States, 2014. MMWR 67, 1–23.

Basagaña, X., Esnaola, M., Rivas, I., Amato, F., Alvarez-Pedrerol, M., Forns, J., Lopez-Vicente, M., Pujol, J., Nieuwenhuijsen, M., Querol, X., Sunyer, J., 2016. Neurodevelopmental deceleration by urban fine particles from different emission sources: a longitudinal observational study. Environ. Health Perspect. 124, 1630–1636.
Becerra, T.A., Wilhelm, M., Olsen, J., Cockburn, M., Ritz, B., 2013. Ambient air pollution and autism in Los Angeles County, California. Environ. Health Perspect. 121, 380–386.
Beier, E.E., Richardson, J.R., 2015. Parkinson's disease: mechanisms, models, and biological plausibility. In: Aschner, M., Costa, L.G. (Eds.), Environmental Factors in Neurodevelopmental and Neurodegenerative Disorders. Academic Press/Elsevier, New York, pp. 267–288.
Berko, E.R., Greally, J.M., 2015. How might epigenetic dysregulation in early embryonic life contribute to autism spectrum disorder? Epigenomics 7, 1–4.
Bilbo, S.D., Block, C.L., Bolton, J.L., Hanamsagar, R., Tran, P.K., 2018. Beyond infection—Maternal immune activation by environmental factors, microglial development, and relevance for autism spectrum disorders. Exp. Neurol. 299 (Pt. A), 241–251.
Block, M.L., Calderon-Garciduenas, L., 2009. Air pollution: mechanisms of neuroinflammation and CNS disease. Trends Neurosci. 32, 506–516.
Block, M.L., Hong, J.S., 2005. Microglia and inflammation-mediated neurodegeneration: multiple triggers with a common mechanism. Prog. Neurobiol. 76, 77–98.
Block, M.L., Wu, X., Pei, Z., Li, G., Wang, T., Qin, L., Wilson, B., Yang, J., Hong, J.S., Veronesi, B., 2004. Nanometer size diesel exhaust particles are selectively toxic to dopaminergic neurons: the role of microglia, phagocytosis, and NADPH oxidase. FASEB J. 18, 1618–1620.
Block, M.L., Zecca, L., Hong, J.S., 2007. Microglia-mediated neurotoxicity: uncovering the molecular mechanisms. Nat. Rev. Neurosci. 8, 57–69.
Bos, I., De Boever, P., Emmerechts, J., Buekers, J., Vanoirbeek, J., Meeusen, R., Van Poppel, M., Nemery, B., Nawrot, T., Panis, L., 2012. Changed gene expression in brains of mice exposed to traffic in a highway tunnel. Inhal. Toxicol. 24, 676–686.
Braak, H., Ghebremedhin, E., Rub, U., Bratzke, H., Del Tredici, K., 2004. Stages in the development of Parkinson's disease pathology. Cell Tissue Res. 318, 121–134.
Brockmeyer, S., D'Angiulli, A., 2016. How air pollution alters brain development: the role of neuroinflammation. Transl. Neurosci. 7, 24–30.
Cacciottolo, M., Wang, X., Driscoll, I., Woodward, N., Saffari, A., Reyes, J., Serre, M.L., Vizuete, W., Sioutas, C., Morgan, T.E., Gatz, M., Chui, H.C., Shumaker, S.A., Resnick, S.M., Espeland, M.A., Finch, C.E., Chen, J.C., 2017. Particulate air pollutants, APOE alleles and their contributions to cognitive impairment in older women and to the amyloidogenesis in experimental models. Transl. Psychiat. 7, e1022.
Calderon-Garciduenas, L., Jewells, V., Galaz-Montoya, C., van Zundert, B., Perez-Calatayud, A., Ascencio-Ferrel, E., Valencia-Salazar, G., Sandovan-Cano, M., Carlos, E., Solorio, E., Acuña-Ayala, H., Torres-Jardon, R., D-Angiulli, A., 2016. Interactive and additive influences of gender, BMI and apolipoprotein 4 on cognition in children chronically exposed to high concentrations of $PM_{2.5}$ and ozone. APOE 4 females are at highest risk in Mexico City. Environ. Res. 150, 411–422.
Calderon-Garciduenas, L., Maronpot, R.R., Torres-Jardon, R., Henriquez-Roldan, C., Schoonhoven, R., Acuña-Ayala, H., Villareal-Calderon, A., Nakamura, J., Fernando, R., Reed, W., Azzarelli, B., Swenberg, J.A., 2003. DNA damage in nasal and brain tissues of canines exposed to air pollutants is associated with evidence of chronic brain inflammation and neurodegeneration. Toxicol. Pathol. 31, 524–538.
Calderon-Garciduenas, L., Reed, W., Maronpot, R.R., Henriquez-Roldan, C., Delgado-Chavez, R., Calderon-Garciduenas, A., Dragustinovis, I., Franco-Lira, M., Aragon-Flores, M., Solt, A.C., Altenburg, M., Torres-Jardon, R., Swnberg, J.A., 2004. Brain inflammation and Alzheimer's-like pathology in individuals exposed to severe air pollution. Toxicol. Pathol. 32, 650–658.

Calderon-Garciduenas, L., Solt, A.C., Henriquez-Roldan, C., Torres-Jardon, R., Nuse, B., Herritt, L., Villareal-Calderon, R., Osnaya, N., Stone, I., Garcia, R., Brooks, D.M., Gonzalez-Maciel, A., Reynoso-Robles, R., Delgado-Chavez, R., Reed, W., 2008. Long-term air pollution exposure is associated with ultrafine particulate deposition, and accumulation of amyloid β-42 and α-synuclein in children and young adults. Toxicol. Pathol. 36, 289–310.

Calderon-Garciduenas, L., Kavanaugh, M., Block, M., D'Angiulli, A., Delgado-Chavez, R., Torres-Jardon, R., Gonzalez-Maciel, A., Reynoso-Robles, R., Osnaya, N., Villareal-Calderon, R., Guo, R., Hua, Z., Zhu, H., Perry, G., Diaz, P., 2012. Neuroinflammation, hyperphosphorylated tau, diffuse amyloid plaques, and down-regulation of the cellular prion protein in air pollution exposed children and young adults. J. Alzheimers Dis. 28, 93–107.

Calderon-Garciduenas, L., Cross, J.V., Franco-Lira, M., Aragon-Flores, M., Kavanaugh, M., Torres-Jardon, R., Chao, C.K., Thompson, C., Chang, J., Zhu, H., D'Angiulli, A., 2013. Brain immune interactions and air pollution: macrophage inhibitory factor (MIF), prion cellular protein (PrP^C), interleukin-6 (IL-6), interleukin 1 receptor antagonist (IL-1Ra), and interleukin-2 (IL-2) in cerebrospinal fluid and MIF in serum differentiate urban children exposed to severe vs. low air pollution. Front. Neurosci. 7, 183.

Calderon-Garciduenas, L., Kulesza, R.J., Doty, R.L., D'Angiulli, A., Torres-Jardon, R., 2015. Megacities air pollution problems: Mexico city Metropolitan Area critical issues on the central nervous system pediatric impact. Environ. Res. 137, 157–169.

Campbell, A., Oldham, M., Becaria, A., Bondy, S.C., Meacher, D., Sioutas, C., Misra, C., Mendez, L.B., Kleinman, M., 2005. Particulate matter in polluted air may increase biomarkers of inflammation in mouse brain. Neurotoxicology 26, 133–140.

Carpentier, P.A., Palmer, T.D., 2009. Immune influence on adult neural stem cell regulation and function. Neuron 64, 79–92.

Chang, K.H., Chang, M.Y., Muo, C.H., Wu, T.N., Chen, C.Y., Kao, C.H., 2014. Increased risk of dementia in patients exposed to nitrogen dioxide and carbon monoxide: a population-based retrospective cohort study. PLoS One 9, e103078.

Chang, Y.C., Cole, T.B., Costa, L.G., 2018. Prenatal and early-life diesel exhaust exposure causes autism-like behavioral changes in mice. Part. Fibre Toxicol. 15, 18.

Chen, Z., Trapp, B.D., 2016. Microglia and neuroprotection. J. Neurochem. 136, 10–17. Suppl.

Chen, J.C., Wang, X., Wellenius, G.A., Serre, M.L., Driscoll, I., Casanova, R., McArdle, J.J., Manson, J.E., Chiu, H.C., Espeland, M.A., 2015. Ambient air pollution and neurotoxicity on brain structure: evidence from women's health initiative memory study. Ann. Neurol. 78, 466–476.

Chen, H., Kwong, J.C., Copes, R., Tu, K., Villeneuve, P.J., van Donkelaar, A., Hystad, P., Martin, R.V., Murray, B.J., Jessiman, B., Wilton, A.S., Kapp, A., Burnett, R.T., 2017. Living near major roads and incidence of dementia, Parkinson's disease, and multiple sclerosis: a population-based cohort study. Lancet 389, 718–726.

Cherry, J.D., Olschowka, J.A., O'Banion, M.K., 2014. Neuroinflammation and M2 microglia: the good, the bad, and the inflamed. J. Neuroinflammation 11, 98.

Chin, J., Massaro, C.M., Palop, J.J., Thwin, M.T., Yu, G.Q., Bien-Ly, N., Bender, A., Mucke, L., 2007. Reelin depletion in the enthorinal cortex of human amyloid precursor protein transgenic mice and humans with Alzheimer's disease. J. Neurosci. 27, 2727–2733.

Chiu, Y.H.M., Hsu, H.H.L., Coull, B.A., Bellinger, D.C., Kloog, I., Schwartz, J., Wright, R.O., Wright, R.J., 2016. Prenatal particulate air pollution and neuro-development in urban children: examining sensitive windows and sex-specific associations. Environ. Int. 87, 56–65.

Cho, K., Choi, G.E., 2017. Microglia: physiological functions revealed through morphological profiles. Folia Biol. (Praha) 63, 85–90.

Church, J.S., Tijerina, P.B., Emerson, F.J., Coburn, M.A., Blum, J.L., Zelikoff, J.T., Schwartzer, J.J., 2018. Perinatal exposure to concentrated ambient particulates results in autism-like behavioral deficits in adult mice. Neurotoxicology 65, 231–240.
Clifford, A., Lang, L., Chen, R., Anstey, K.J., Seaton, A., 2016. Exposure to air pollution and cognitive functioning across the life course—a systematic review. Environ. Res. 147, 383–398.
Coburn, J.L., Cole, T.B., Dao, K.T., Costa, L.G., 2018. Acute exposure to diesel exhaust impairs adult neurogenesis in mice: prominence in males and protective effect of pioglitazone. Arch. Toxicol. 92, 1815–1829.
Cohen, G., Gerber, Y., 2017. Air pollution and successful aging: recent evidence and new perspectives. Curr. Envir. Health Rep. 4, 1–11.
Cole, T.B., Coburn, J., Dao, K., Roque, P., Kalia, V., Guilarte, T.R., Dziedzic, J., Costa, L.G., 2016. Sex and genetic differences in the effects of acute diesel exhaust exposure on inflammation and oxidative stress in mouse brain. Toxicology 374, 1–9.
Colombo, E., Farina, C., 2016. Astrocytes: key regulators of neuroinflammation. Trends Immunol. 37, 608–620.
Costa, L.G., 2017. Traffic-related air pollution and neurodegenerative diseases: epidemiological and experimental evidence and potential underlying mechanisms. In: Aschner, M., Costa, L.G. (Eds.), Environmental Factors in Neurodegenerative Diseases. In: Advances in Neurotoxicology, vol. 1. Elsevier, New York, pp. 1–46.
Costa, L.G., Cole, T.B., Coburn, J., Chang, Y.C., Dao, K., Roque, P., 2014. Neurotoxicants are in the air: convergence of human and in vitro studies on the effects of air pollution on the brain. Biomed. Res. Int. 2014, 736385. (8 p).
Costa, L.G., Cole, T.B., Coburn, J., Chang, Y.C., Dao, K., Roque, P., 2017a. Neurotoxicity of traffic-related air pollution. Neurotoxicology 59, 133–139.
Costa, L.G., Chang, Y.C., Cole, T.B., 2017b. Developmental neurotoxicity of traffic-related air pollution: focus on autism. Curr. Environ. Health Rep. 4, 156–165.
Cubo, E., Goetz, C.G., 2014. Parkinson's disease. In: Aminoff, M.J., Daroff, R.B. (Eds.), The Encyclopedia of Neurological Sciences, second ed. In: vol. 3. Elsevier, New York, pp. 828–832.
Davis, R.L., 2018. Neurodevelopment: inflammation matters. In: Aschner, M., Costa, L.G. (Eds.), Linking Environmental Exposure to Neurodevelopmental Disorders. In: Advances in Neurotoxicology, vol. 2. Elsevier, New York, pp. 227–264.
Davis, D.A., Bortolato, M., Godar, S.C., Sander, T.K., Iwata, N., Pakbin, P., Shih, J.C., Berhane, K., McConnell, R., et al., 2013. Prenatal exposure to urban air nanoparticles in mice causes altered neuronal differentiation and depression like responses. PLoS One 8 (5), e64128.
Depino, A.M., 2013. Peripheral and central inflammation in autism spectrum disorders. Mol. Cell. Neurosci. 53, 69–76.
Dinel, A.L., Joffre, C., Trifilieff, P., Aubert, A., Foury, A., Le Ruyet, P., Laye, S., 2014. Inflammation early in life is a vulnerability factor for emotional behavior at adolescence and for lipopolysaccharide-induced spatial memory and neurogenesis alteration at adulthood. J. Neuroinflammation 11, 155.
Doty, R.L., 2012. Olfaction in Parkinson's disease and related disorders. Neurobiol. Dis. 46, 527–552.
Duffy, C.M., Swanson, J., Northrop, W., Nixon, J.P., Butterick, T.A., 2018. Microglial immune response to low concentrations of combustion-generated nanoparticles: an in vitro model of brain health. Nanomaterials 8, 155.
Durga, M., Devasena, T., Rajasekar, A., 2015. Determination of LC_{50} and sub-chronic neurotoxicity of diesel exhaust nanoparticles. Environ. Toxicol. Pharmacol. 40, 615–625.
Ekdahl, C.T., Claasen, J.H., Bonde, S., Kokaia, Z., Lindvall, O., 2003. Inflammation is detrimental for neurogenesis in adult brain. Proc. Natl. Acad. Sci. 100, 13632–13637.

El-Ansary, A., Al-Ayadhi, L., 2012. Neuroinflammation in autism spectrum disorders. J. Neuroinflammation 9, 265.
Ema, M., Naya, M., Horimoto, M., Kato, H., 2013. Developmental toxicity of diesel exhaust: a review of studies in experimental animals. Reprod. Toxicol. 42, 1–17.
Estes, M.L., McAllister, A.K., 2016. Maternal immune activation: implications for neuropsychiatric disorders. Science 353, 772–777.
Fatemi, S.H., Snow, A.V., Stary, J.M., Araghi-Niknam, M., Brooks, A.I., Pearce, D.A., Reutiman, T.J., Lee, S., 2005. Reelin signaling is impaired in autism. Biol. Psychiatry 57, 777–787.
Fehen, A., Juhasz, A., Pakaski, M., Kalman, J., Janka, Z., 2015. Genetic analysis of the RELN gene: gender specific association with Alzheimer's disease. Psychol. Res. 230, 716–718.
Filiou, M.D., Arefin, A.S., Moscato, P., Graeber, M.B., 2014. Neuroinflammation differs categorically from inflammation: transcriptomes of Alzheimer's disease, Parkinson's disease, schizophrenia and inflammatory diseases compared. Neurogenesis 15, 201–212.
Fiordelisi, A., Piscitelli, P., Trimarco, B., Coscioni, E., Iaccarino, G., Sorriento, D., 2017. The mechanisms of air pollution and particulate matter in cardiovascular diseases. Heart Fail. Rev. 22, 337–347.
Folsom, T.D., Fatemi, S.H., 2013. The involvement of reelin in neurodevelopmental disorders. Neuropharmacology 68, 122–135.
Förster, E., 2014. Reelin, neuronal polarity and process orientation of cortical neurons. Neuroscience 269, 102–111.
Freire, C., Ramos, R., Puertas, R., Lopez-Espinosa, M.J., Julvez, J., Aguilera, I., Cruz, F., Fernandez, M.F., Sunyer, J., Olea, N., 2010. Association of traffic-related air pollution with cognitive development in children. J. Epidemiol. Community Health 64, 223–228.
Fuster-Matanzo, A., Llorens-Martin, M., Hernandez, F., Avila, J., 2013. Role of neuroinflammation in adult neurogenesis and Alzheimer's disease: therapeutic approaches. Mediat. Inflamm. 2013, 26092. (9 p).
Genc, S., Zadeoglulari, Z., Fuss, S.H., Genc, K., 2012. The adverse effects of air pollution on the nervous system. J. Toxicol. 2012, 782462. 23 p.
Gerlofs-Nijland, M.E., van Berlo, D., Cassee, F.R., Schins, R.P.F., Wang, K., Campbell, A., 2010. Effect of prolonged exposure to diesel engine exhaust on proinflammatory markers in different regions of the rat brain. Part. Fibre Toxicol. 7, 12.
Guxens, M., Sunyer, J., 2012. A review of epidemiological studies on neuropsychological effects of air pollution. Swiss Med. Wkly. 141, w13322.
Guxens, M., Garcia-Esteban, R., Giorgis-Allemand, L., Forns, J., Badaloni, C., Ballester, F., Cesaroni, G., Chatzi, L., et al., 2014. Air pollution during pregnancy and childhood cognitive and psychomotor development. Epidemiology 25, 636–647.
Hanamsagar, R., Bilbo, S.D., 2017. Environment matters: microglia function and dysfunction in a changing world. Curr. Opin. Neurobiol. 47, 146–155.
Harris, M.H., Gold, D.R., Rifas-Shiman, S.L., Melly, S.J., Zanobetti, A., Coull, B.A., Schwartz, J.D., et al., 2016. Prenatal and childhood traffic-related air pollution exposure and childhood executive function and behavior. Neurotoxicol. Teratol. 57, 60–70.
Harry, G.J., Kraft, A.D., 2008. Neuroinflammation and microglia: considerations and approaches for neurotoxicity assessment. Exp. Opin. Drug Metab. Toxicol. 4, 1265–1277.
Hebert, L.E., Weuve, J., Scherr, P.A., Evans, D.A., 2013. Alzheimer disease in the United States (2010-2050) estimated using the 2010 census. Neurology 80, 1778–1783.
Heneka, M.T., Carson, M.J., El Khoury, J., et al., 2015. Neuroinflammation in Alzheimer's disease. Lancet Neurol. 14, 388–405.
Hernandez-Romero, M.C., Delgado-Cortes, M.J., Sarmiento, M., de Pablos, R.M., Espinosa-Oliva, A.M., Arguelles, S., Bandez, M.J., Villaran, R.F., Maurino, R., Santiago, M., Venero, J.L., Herrera, A.J., Cano, J., Machado, A., 2012. Peripheral inflammation increases the deleterious effects of CNS inflammation on the nigrostriatal dopaminergic system. Neurotoxicology 33, 347–360.

Herring, A., Donath, A., Steiner, K.M., Widera, M.P., Hamzehian, S., Kanakis, D., Kölble, K., ElAli, A., et al., 2012. Reelin depletion is an early phenomenon of Alzheimer's pathology. J. Alzheimers Dis. 30, 963–979.

Hiesberger, T., Trommsdorff, M., Howell, B.W., Goffinet, A., Mumby, M.C., Cooper, J.A., Herz, J., 1999. Direct binding of reelin to VLDL receptor and ApoE receptor 2 induces tyrosine phosphorylation of disabled-1 and modulates tau phosphorylation. Neuron 24, 481–489.

Hirsch, E.C., Vyas, S., Hunot, S., 2012. Neuroinflammation in Parkinson's disease. Parkinsonism Relat. Disord. 18S1, S210–S212.

Hopkins, S.J., 2007. Central nervous system recognition of peripheral inflammation: a neural, hormonal collaboration. Acta Biomed. 78 (Suppl. 1), 231–247.

Horgusluoglu, E., Nudelman, K., Nho, K., Saykin, A.J., 2017. Adult neurogenesis and neurodegenerative diseases: a systems biology perspective. Am. J. Med. Genet. 174, 93–112.

Hougaard, K.S., Jensen, K.A., Nordly, P., Taxvig, C., Vogel, U., Saber, A.T., Wallin, H., 2008. Effects of prenatal exposure to diesel exhaust particles on postnatal development, behavior, genotoxicity and inflammation in mice. Part. Fibre Toxicol. 5, 3.

Huang, W.J., Zhang, X., Chen, W.W., 2016. Role of oxidative stress in Alzheimer's disease (review). Biomed. Rep. 4, 519–522.

Jayaraj, R.L., Rodriguez, E.A., Wang, Y., Block, M.L., 2017. Outdoor ambient air pollution and neurodegenerative diseases: the neuroinflammation hypothesis. Curr. Environ. Health Rep. 4, 166–179.

Jung, C.R., Lin, Y.T., Hwang, B.F., 2015. Ozone, particulate matter, and newly diagnosed Alzheimer's disease: a population-based cohort study in Taiwan. J. Alzheimers Dis. 44, 573–584.

Kalkbrenner, A.E., Windham, G.C., Serre, M.L., Akita, Y., Wang, X., Hoffman, K., Thayer, B.P., Daniels, J.L., 2015. Particulate matter exposure, prenatal and postnatal windows of susceptibility, and autism spectrum disorders. Epidemiology 26, 30–42.

Karagulian, F., Belis, C.A., Dora, C.F.C., Prüss-Ustün, A.M., Bonjour, S., Adair-Rohani, H., Amann, M., 2015. Contributions to cities' ambient particulate matter: a systematic review of local source contributions at global levels. Atmos. Environ. 120, 475–483.

Keil, K.P., Stamou, M., Lein, P.J., 2018. Molecular mechanisms of neurotoxicity. In: Smart, R.C., Hodgson, E. (Eds.), Molecular and Biochemical Toxicology. John Wiley & Sons, Hoboken, NJ, pp. 709–771.

Kempuraj, D., Thangavel, R., Natteru, P.A., Selvakumar, G.P., Saeed, D., Zahoor, H., Zaheer, S., Iyer, S.S., Zaheer, A., 2016. Neuroinflammation induces neurodegeneration. J. Neurol. Neurosurg. Spine 1, 1.

Khan, S.S., Bloom, G.S., 2016. Tau: the center of signaling nexus in Alzheimer's disease. Front. Neurosci. 10, 31.

Kicinski, M., Vermeir, G., Van Larebeke, N., Den Hond, E., Schoeters, G., Bruckers, L., Sioen, I., et al., 2015. Neurobehavioral performance in adolescents is inversely associated with traffic exposure. Environ. Int. 75, 136–143.

Killin, L.O.J., Starr, J.M., Shiue, I.I., Russ, T.C., 2016. Environmental risk factors for dementia: a systematic review. BMC Geriatr. 16, 175.

Kleinman, M.T., Araujo, J.A., Nel, A., Sioutas, C., Campbell, A., Cong, P.Q., Li, H., Bondy, S.C., 2008. Inhaled ultrafine particulate matter affects CNS inflammatory processes and may act via MAP kinase signaling pathways. Toxicol. Lett. 178, 127–130.

Kocherhans, S., Madhusudan, A., Doehner, J., Breu, K.S., Nitsch, R.M., Fritschy, J.M., Knuesel, I., 2010. Reduced reelin expression accelerates amyloid-β plaque formation and tau pathology in transgenic Alzheimer's disease mice. J. Neurosci. 30, 9228–9240.

Kraft, A.D., Harry, G.J., 2011. Features of microglia and neuroinflammation relevent to environmental exposure and neurotoxicity. Int. J. Environ. Res. Public Health 8, 2980–3018.

Lajud, N., Torner, L., 2015. Early life stress and hippocampal neurogenesis in the neonate: sexual dimorphism, long term consequences and possible mediators. Front. Mol. Neurosci. 8, 3.
Lancet Neurology, 2018. Editorial: air pollution and brain health: an emerging issue. Lancet Neurol. 17, 103.
Lee, Y.J., Han, S.B., Nam, S.Y., Oh, K.W., Tong, J.T., 2010. Inflammation and Alzheimer's disease. Arch. Pharm. Res. 33, 1539–1556.
Lema-Tomè, C.M., Tyson, T., Rey, N.L., Grathwohl, S., Britschgi, M., Brundin, P., 2013. Inflammation and α-synuclein prion-like behavior in Parkinson's disease—is there a link? Mol. Neurobiol. 47, 561–574.
Levesque, S., Surace, M.J., McDonald, J., Block, M.L., 2011. Air pollution and the brain: subchronic diesel exhaust exposure causes neuroinflammation and elevates early markers of neurodegenerative disease. J. Neuroinflammation 8, 105.
Levesque, S., Taetzsch, T., Lull, M.E., Johnson, J.A., McGraw, C., Block, M.L., 2013. The role of MAC1 in diesel exhaust particle-induced microglial activation and loss of dopaminergic neuron function. J. Neurochem. 125, 756–765.
Levy, S.E., Mandell, D.S., Schultz, R.T., 2009. Autism. Lancet 374, 1627–1638.
Leyns, C.E.G., Holtzman, D.M., 2017. Glial contribution to neurodegeneration in tauopathies. Mol. Neurodegener. 12, 50.
Li, C., Zhao, R., Gao, K., Wei, Z., Yin, M.Y., Lau, L.T., Chui, D., Yu, A.C.H., 2011. Astrocytes: implications for neuroinflammatory pathogenesis of Alzheimer's disease. Curr. Alzheimer Res. 8, 67–80.
Li, B., Guo, L., Ku, T., Chen, M., Li, G., Sang, N., 2018. $PM_{2.5}$ exposure stimulates COX-2-mediated excitatory synaptic transmission via ROS-NFκB pathway. Chemosphere 190, 124–134.
Lodovici, M., Bigagli, E., 2011. Oxidative stress and air pollution exposure. J. Toxicol. 2011, 487074.
Lucchini, R.G., Dorman, D.C., Elder, A., Veronesi, B., 2012. Neurological impacts from inhalation of pollutants and the nose-brain connection. Neurotoxicology 33, 838–841.
Lull, M.E., Block, M.L., 2010. Microglial activation and chronic neurodegeneration. Neurotherapeutics 7, 354–365.
Luo, X.G., Chen, S.D., 2012. The changing phenotype of microglia from homeostasis to disease. Transl. Neurodegener. 1, 9.
Ma, T., Klann, E., 2012. Amyloid beta: linking synaptic plasticity failure to memory disruption in Alzheimer's disease. J. Neurochem. 120 (Suppl. 1), 140–148.
Malkova, N.V., Yu, C.Z., Hsiao, E.Y., Moore, M.J., Patterson, P.H., 2012. Maternal immune activation yields offspring displaying mouse versions of the three core symptoms of autism. Brain Behav. Immun. 26, 607–616.
Manoharan, S., Guillemein, G.J., Abiramasundari, R.S., Essa, M.M., Akbar, M., Akbar, M.D., 2016. The role of reactive oxygen species in the pathogenesis of Alzheimer's disease, Parkinson's disease, and Huntington's disease: a mini review. Oxidative Med. Cell. Longev. 2016, 8590578. 15 p.
Manzetti, S., Andersen, O., 2016. Biochemical and physiological effects from exhaust emissions. A review of the relevant literature. Pathophysiology 23, 285–293.
Maresova, P., Klimova, B., Novotny, M., Kuca, K., 2016. Alzheimer's and Parkinson's diseases: expected economic impact on Europe—a call for a uniform European strategy. J. Alzheimers Dis. 54, 1123–1133.
Marxreiter, F., Regensburger, M., Winkler, J., 2013. Adult neurogenesis in Parkinson's disease. Cell. Mol. Life Sci. 70, 459–473.
Masgrau, R., Guaza, C., Ransohoff, R.M., Galea, E., 2017. Should we stop saying "glia" and "neuroinflammation"? Trends Mol. Med. 23, 486–500.

Meyer, U., Nyffeler, M., Engler, A., Urwyler, A., Schedlowski, M., Knuesel, I., Yee, B.K., Feldon, J., 2006. The time of prenatal immune challenge determines the specificity of inflammation-mediated brain and behavioral pathology. J. Neurosci. 26, 4752–4762.

Minter, M.R., Taylor, J.M., Crack, P.J., 2016. The contribution of neuroinflammation to amyloid toxicity in Alzheimer's disease. J. Neurochem. 136, 457–474.

MohanKumar, S.M.J., Campbell, A., Block, M., Veronesi, B., 2008. Particulate matter, oxidative stress and neurotoxicity. Neurotoxicology 29, 479–488.

Møller, P., Jacobsen, N.R., Folkmann, J.K., Danielsen, P.H., Mikkelsen, L., Hemmingsen, J.G., Vesterdal, L.K., Forchhammer, L., Wallin, H., Loft, S., 2010. Role of oxidative damage in toxicity of particulates. Free Radic. Res. 44, 1–46.

Moore, C.S., Durafourt, B.A., Antel, J.P., 2014. Innate immunity in the CNS—a focus on the myeloid cells. In: Woodroofe, N., Amor, S. (Eds.), Neuroinflammation and CNS Disorders. John Wiley & Sons, Hoboken, NJ, pp. 9–36.

Mottahedin, A., Ardalan, M., Chumak, T., Riebe, I., Ek, J., Mallard, C., 2017. Effect of neuroinflammation on synaptic organization and function in the developing brain: implications for neurodevelopmental and neurodegenerative disorders. Front. Cell. Neurosci. 11, 190.

Moulton, P.V., Yang, W., 2012. Air pollution, oxidative stress, and Alzheimer's disease. J. Environ. Public Health 2012, 472751.

Musgrove, R.E., Jewell, S.A., Di Monte, D.A., 2015. Overview of neurodegenerative disorders and susceptibility factors in neurodegenerative processes. In: Aschner, M., Costa, L.G. (Eds.), Environmental Factors in Neurodevelopmental and Neurodegenerative Disorders. Academic Press/Elsevier, New York, pp. 197–210.

Newman, N.C., Ryan, P., LeMasters, G., Levin, L., Bernstein, D., Khurana Hershey, G.K., Lockey, J.E., et al., 2013. Traffic-related air pollution exposure in the first year of life and behavioral scores at 7 years of age. Environ. Health Perspect. 121, 731–736.

Noh, J.S., Sharma, R.P., Veldic, M., Salvacion, A.A., Jia, X., Chen, Y., Costa, E., Guidotti, A., Grayson, D.R., 2005. DNA methyltransferase 1 regulates reelin mRNA expression in mouse primary cortical cultures. Proc. Natl. Acad. Sci. 102, 1749–1754.

Novais, A.R.B., Guiramand, J., Cohen-Sola, C., Crouzin, N., de Jesus Ferreira, M.C., Vignes, M., Barbanel, G., Cambonie, G., 2013. N-acetyl-cysteine prevents pyramidal cell disarray and reelin-immunorecative neuron deficiency in CA3 after prenatal immune challenge in rats. Pediatr. Res. 73, 750–755.

Oberdoerster, G., Sharp, Z., Atudorei, V., Elder, A., Gelein, R., Kreyling, W., Cox, C., 2004. Translocation of inhaled ultrafine particles to the brain. Inhal. Toxicol. 16, 437–445.

O'Callaghan, J.P., Shiram, K., Miller, D.B., 2008. Defining "Neuroinflammation". Lessons from MPTP- and methamphetamine-induced neurotoxicity. Ann. N. Y. Acad. Sci. 1139, 318–330.

Ohkubo, N., Lee, Y.D., Morishima, A., Terashima, T., Kikkawa, S., Tohyama, M., Sakanaka, M., Tanaka, J., Maeda, N., Vitek, M.P., Mitsuda, N., 2003. Apolipoprotein E and reelin ligands modulate tau phosphorylation through apolipoprotein E receptor/disabled-1/glycogen synthase kinase-3β cascade. FASEB J. 17, 295–297.

Oudin, A., Forsberg, B., Adolfsson, A.N., lind, N., Modig, L., Nordin, M., Nordin, S., Adolfsson, R., Nilsson, L.G., 2016. Traffic-related air pollution and dementia incidence in Northern Sweden: a longitudinal study. Environ. Health Perspect. 124, 306–312.

Palacios-Garcia, I., Lara-Vasquez, A., Montiel, J.F., Diaz-Veliz, G.F., Sepulveda, H., Utreras, E., Montecino, M., Gonzalez-Billault, C., Aboitiz, F., 2015. Prenatal stress down-regulates reelin expression by methylation of its promoter and induces adult behavioral impairment in rats. PLoS One 10 (2), e0117680.

Patterson, P.H., 2011. Maternal infection and immune involvement in autism. Trends Mol. Med. 17, 389–394.

Perry, D.C., 2014. Dementia. In: Aminoff, M.J., Daroff, R.B. (Eds.), The Encyclopedia of Neurological Sciences, second ed. In: vol. 3. Elsevier, New York, pp. 962–969.
Peters, A., Veronesi, B., Calderon-Garciduenas, L., Gehr, P., Chen, L.C., Geiser, M., Reed, W., Rothen-Rutishauser, B., Schurch, S., Schulz, H., 2006. Translocation and potential neurological effects of fine and ultrafine particles a critical update. Part. Fibre Toxicol. 3, 13.
Peters, R., Peters, J., Booth, A., Mudway, I., 2015. Is air pollution associated with increased risk of cognitive decline? A systematic review. Age Ageing 44, 755–760.
Power, M.C., Weisskopf, M.G., Alexeeff, S.E., Coull, B.A., Spiro III, A., Schwartz, J., 2011. Traffic-related air pollution and cognitive function in a cohort of older men. Environ. Health Perspect. 119, 682–687.
Power, M.C., Adar, S.D., Yanosky, J.D., Weuve, J., 2016. Exposure to air pollution as a potential contributor to cognitive function, cognitive decline, brain imaging, and dementia: a systematic review of epidemiologic research. Neurotoxicology 56, 235–253.
Prata, J., Santos, S.G., Almeida, M.I., Coelho, R., Barbosa, M.A., 2017. Bridging autism spectrum disorders and schizophrenia through inflammation and biomarkers-pre-clinical and clinical investigations. J. Neuroinflammation 14, 179.
Pressman, P., Rabinovici, G.D., 2014. Alzheimer's disease. In: Aminoff, M.J., Daroff, R.B. (Eds.), The Encyclopedia of Neurological Sciences, second ed. In: vol. 3. Elsevier, New York, pp. 122–127.
Prince, M., Bryce, R., Albanese, E., Wimo, A., Ribeiro, W., Ferri, C.P., 2013. The global prevalence of dementia: a systematic review and meta-analysis. Alzheimer Dement. 9, 63–75.
Pronk, A., Coble, J., Stewart, P.A., 2009. Occupational exposure to diesel engine exhaust; a literature review. J. Exp. Sci. Environ. Epidemiol. 19, 443–457.
Pujadas, L., Rossi, D., Andres, R., Teixeira, C.M., Serra-Vidal, B., Parcerisa, A., Maldonado, R., Giralt, E., Carulla, N., Soriano, E., 2014. Reelin delays amyloid-beta fibril formation and rescues cognitive deficits in a model of Alzheimer's disease. Nat. Commun. 5, 3443.
Pun, V.C., Manjourides, J., Suh, H., 2017. Association of ambient air pollution with depressive and anxiety symptoms in older adults: results from the NSHAP study. Environ. Health Perspect. 125, 342–348.
Qian, L., Flood, P.M., Hong, J.S., 2010. Neuroinflammation is a key player in Parkinson's disease and a prime target for therapy. J. Neural Transm. 117, 971–979.
Ranft, U., Schikowski, T., Sugiri, D., Krutmann, J., Krämer, U., 2009. Long-term exposure to traffic-related particulate matter impairs cognitive function in the elderly. Environ. Res. 109, 1004–1011.
Ransohoff, R.M., 2016. How neuroinflammation contributes to neurodegeneration. Science 353, 777–782.
Ransohoff, R.M., Perry, V.H., 2009. Microglia physiology; unique stimuli, specialized responses. Annu. Rev. Immunol. 27, 119–145.
Riedel, B.C., Thompson, P.M., Brinton, R.D., 2016. Age, APOE and sex: triad of risk of Alzheimer's disease. J. Steroid Biochem. Mol. Biol. 160, 134–147.
Roberts, A.L., Lyall, K., Hart, J.E., Laden, F., Just, A.C., Bobb, J.F., Koenen, K.C., Ascherio, A., Weisskopf, M.G., 2013. Perinatal air pollutant exposures and autism spectrum disorder in the Children of Nurses's Health Study II participants. Environ. Health Perspect. 121, 978–984.
Roqué, P.J., Dao, K., Costa, L.G., 2016. Microglia mediate diesel exhaust particle-induced cerebellar neuronal toxicity through neuroinflammatory mechanisms. Neurotoxicology 56, 204–214.
Rose, S., Melnyk, S., Pavliv, O., Bai, S., Nick, T.G., Frye, R.E., James, S.J., 2012. Evidence of oxidative damage and inflammation associated with low glutathione redox status in the autism brain. Transl. Psychiatry 2, e134.

Rossignol, D.A., Genuis, S.J., Frye, R.E., 2014. Environmental toxicants and autism spectrum disorders: a systematic review. Transl. Psychiatry 4, e360.
Rubio-Perez, J.M., Morillas-Ruiz, J.M., 2012. A review: inflammatory process in Alzheimer's disease, role of cytokines. Sci. World J. 2012, 756357. (15 p).
Rücker, R., Schneider, A., Breitner, S., Cyrys, J., Peters, A., 2011. Health effects of particulate air pollution: a review of the epidemiological evidence. Inhal. Toxicol. 23, 555–592.
Salter, M.W., Stevens, B., 2017. Microglia emerges as central players in brain disease. Nat. Med. 23, 1018–1027.
Sandin, S., Lichtenstein, P., Kuja-Halkola, R., Larsson, H., Hultman, C.M., Reichenberg, A., 2014. The familial risk of autism. JAMA 311, 1770–1777.
Scheltens, P., Blennow, K., Breteler, M.B.B., de Strooper, B., Salloway, S., Van der Flier, W.M., 2016. Alzheimer's disease. Lancet 388, 505–517.
Schikowski, T., Vossoughi, M., Vierkötter, A., Schulte, T., Teichert, T., Sugiri, D., Fehsel, K., Tzivian, L., Bae, I.S., Ranft, U., Hoffman, B., Probst-Hensch, N., Herder, C., Krämer, U., Luckhaus, C., 2015. Association of air pollution with cognitive functions and its modification by APOE gene variants in elderly women. Environ. Res. 142, 10–16.
Selkoe, D.J., Hardy, J., 2016. The amyloid hypothesis of Alzheimer's disease at 25 years. EMBO Mol. Med. 8, 595–608.
Serajee, F.J., Zhong, H., Huq, A.H.M.M., 2006. Association of reelin polymorphysms with autism. Genomics 87, 75–83.
Stoner, R., Chow, M.L., Boyle, M.P., Sunkin, S.M., Mouton, P.R., Roy, S., Wynshaw-Boris, A., Colamarino, S.A., Lein, E.S., Courchesne, E., 2014. Patches of disorganization in the neocortex of children with autism. New Engl. J. Med. 370, 1209–1219.
Stranahan, A.M., Haberman, R.P., Gallagher, M., 2011. Cognitive decline is associated with reduced reelin expression in the entorhinal cortex of aged rats. Cereb. Cortex 21, 392–400.
Suades-Gonzalez, E., Gascon, M., Guxens, M., Sunyer, J., 2015. Air pollution and neuropsychological development: a review of the latest evidence. Endocrinology 156, 3473–3482.
Sunyer, J., Esnaola, M., Alvarez-Pedrerol, M., Forns, J., Rivas, I., Lopez-Vicente, M., Suades-Gonzales, E., Foraster, M., Garcia-Esteban, R., et al., 2015. Association between traffic-related air pollution in schools and cognitive development in primary school children: a prospective cohort study. PLoS Med. 12, e1001792.
Tai, L.M., Ghura, S., Koster, K.P., et al., 2015. *APOE*-modulated Aβ-induced neuroinflammation in Alzheimer's disease: current landscape, novel dta, and future perspective. J. Neurochem. 133, 465–488.
Takahashi, M., Ko, L.W., Kulathingal, J., Jiang, P., Sevlever, D., Yen, S.H.C., 2007. Oxidative stress-induced phosphorylation, degradation and aggregation of α-synuclein are linked to upregulated CK2 and cathepsin D. Eur. J. Neurosci. 26, 863–874.
Tang, Y., 2018. Editorial: microglia polarization in the pathogenesis and therapeutics of neurodegenerative diseases. Front. Aging Neurosci. 10, 154.
Theoharides, T.C., Asadi, S., Patel, A.B., 2013. Focal brain inflammation and autism. J. Neuroinflammation 10, 46.
Thirtamara Rajamani, K., Doherty-Lyons, S., Bolden, C., Willis, D., Hoffman, C., Zelikoff, J., Chen, L.C., Gu, H., 2013. Prenatal and early life exposure to high level diesel exhaust particles leads to increased locomotor activity and repetitive behaviors in mice. Autism Res. 6, 248–257.
Tsukue, N., Watanabe, M., Kumamoto, T., Takano, H., Takeda, K., 2009. Perinatal exposure to diesel exhaust affects gene expression in mouse cerebrum. Arch. Toxicol. 83, 985–1000.

Ulusoy, A., Di Monte, D.A., 2013. Alpha-synuclein elevation in human neurodegenerative diseases: experimental, pathogenetic, and therapeutic implications. Mol. Neurobiol. 47, 484–494.
Van Donkelaar, A., Martin, R.V., Brauer, M., Boys, B.L., 2015. Use of satellite observations for long-term exposure of global concentrations of fine particulate matter. Environ. Health Perspect. 123, 135–143.
Varatharaj, A., Galea, I., 2017. The blood-brain-barrier in systemic inflammation. Brain Behav. Immun. 60, 1–12.
Volk, H.E., Hertz-Picciotto, I., Delwiche, L., Lurmann, F., McConnell, R., 2011. Residential proximity to freeways and autism in the CHARGE study. Environ. Health Perspect. 119, 873–877.
Volk, H.E., Lurmann, F., Penfold, B., Hertz-Picciotto, I., McConnell, R., 2013. Traffic-related air pollution, particulate matter, and autism. JAMA Psychiat. 70, 71–77.
Vrijheid, M., Casas, M., Gascon, M., Valvi, D., Nieuwenhuijsen, M., 2016. Environmental pollutants and child health—a review of recent concerns. Int. J. Hyg. Environ. Health 219, 331–342.
Weuve, J., Puett, R.C., Schwartz, J., Yanosky, J.D., Laden, F., Grodstein, F., 2012. Exposure to particulate air pollution and cognitive decline in older women. Arch. Int. Med. 172, 219–227.
Win-Shwe, T.T., Fujimaki, H., 2011. Nanoparticles and neurotoxicity. Int. J. Mol. Sci. 12, 6267–6280.
Win-Shwe, T.T., Fujitani, Y., Kyi-Tha-Thu, C., Furuyama, A., Michikawa, T., Tsukahara, S., Nitta, H., Hirano, S., 2014. Effects of diesel engine exhaust origin secondary organic aerosols on novel object recognition ability and maternal behavior in BALB/C mice. Int. J. Environ. Res. Public Health 11, 11286–11307.
Woodward, N.C., Levine, M.C., Haghani, A., Shirmohammadi, F., Saffari, A., Sioutas, C., Morgan, T.E., Finch, C.E., 2017. Toll-like receptor 4 in glial inflammatory responses to air pollution in vitro and in vivo. J. Neuroinflammation 14, 84.
Wu, Y.C., Lin, Y.C., Yu, H.L., Chen, J.H., Chen, T.F., Sun, Y., Wen, L.L., Yip, P.K., Chu, Y.M., Chen, Y.C., 2015. Association between air pollutants and dementia risk in the elderly. Alzheimer Dement. 1, 220–228.
Xu, X., Ha, S.U., Basnet, R., 2016. A review of epidemiological research on adverse neurological effects of exposure to ambient air pollution. Front. Public Health 4, 157.
Yokota, S., Moriya, N., Iwata, M., Umezawa, M., Oshio, S., Takeda, K., 2013. Exposure to diesel exhaust during fetal period affects behavior and neurotransmitters in male offspring mice. J. Toxicol. Sci. 38, 13–23.
Yorifuji, T., Kashima, S., Higa Diez, M., Kado, Y., Sanada, S., Doi, H., 2016. Prenatal exposure to traffic-related air pollution and child behavioral development milestone delays in Japan. Epidemiology 27, 57–65.
Yu, D., Fan., W., Wu, P., Deng, J., Liu, J., Niu, Y., Li, M., Deng, J., 2014. Characterization of hippocampal Cajal-Retzius cells during development in a mouse model of Alzheimer's disease (Tg2576). Neural Regen. Res. 9, 394–401.
Yu, N.N., Tan, M.S., Yu, J.T., Xie, A.M., Tan, L., 2016. The role of reelin signaling in Alzheimer's disease. Mol. Neurobiol. 53, 5692–5700.

Further reading

Viviani, B., Boraso, M., Marchetti, N., Marinovich, M., 2014. Perspectives on neuroinflammation and excitotoxicity: a neurotoxic conspiracy? Neurotoxicology 43, 10–20.
Wallach, D., Kang, T.B., Dillon, C.P., Green, D.R., 2016. Programmed necrosis in inflammation: toward identification of the effector molecules. Science 352, 51.

CHAPTER EIGHT

The role of neuroinflammation in developmental neurotoxicity, tackling complexity in children's exposures and outcomes

Amedeo D'Angiulli[a,b,*]
[a]Department of Neuroscience, Carleton University, Ottawa, ON, Canada
[b]Institute of Interdisciplinary Studies, Carleton University, Ottawa, ON, Canada
*Corresponding author: e-mail address: amedeo.dangiulli@carleton.ca

Contents

1. Complex developmental neurotoxicity, neuroinflammation and children

Substances known to trigger specific developmental neurotoxic (DNT) effects—called throughout this chapter *developmental neurotoxicants*, or for short, DNTs—belong to different chemical types such as organic solvents or metals, or to different manmade product categories such as pharmaceuticals, industrial chemicals, biocides or pesticides. Approximately 218 chemicals have been identified as potential DNTs, of which 27 are metals or inorganic

Advances in Neurotoxicology, Volume 3
ISSN 2468-7480
https://doi.org/10.1016/bs.ant.2018.10.008

compounds, 41 are organic solvents, 48 are other organic substances and 102 are pesticides (Grandjean and Landrigan, 2014). Most recently, Maffini and Neltner (2015) extended this unofficial list to >300 chemicals as potential DNTs including chemicals belonging to categories related to food, such as pesticides, food contact materials and food additives including flavorings, colorings and preservatives. Only for a small percentage of such chemicals the official status by various regulatory governmental boards changed from potential to actual DNT. Yet publications related to DNT have increased exponentially and we have now a plethora of data and hypothesis but not a clear map to put everything together. This chapter is an initial attempt to build a conceptual map and understanding of complex DNTs in the initial key periods of the lifespan.

Historically, lead, methylmercury, toluene and PCB were initially implicated in neurobehavioral deficits in children following prenatal exposures at subclinical adult dosages (see Moore, 2009 for an exhaustive review on children). Successively, epidemiological studies associated more chemicals—manganese, fluoride, chlorpyrifos, dichlorodiphenyltrichloroethane, tetrachloroethylene, and PBDEs—with diminished intellectual functioning, learning disabilities, attention problems, aggressiveness, hyperactivity, and autism (see De Felice et al., 2015).

The link between CNS development and much more complex forms of exposures, notably, air pollution and the traffic-related pollution is much more recent. Particulate matter (PM), ultrafine particulate matter (UFPM), carbon monoxide, nitrogen oxides, and polycyclic aromatic hydrocarbons are recently reported to act as DNTs associated with cognitive and neurological impairment (Calderón-Garcidueñas et al., 2008; Perera et al., 2009), as well as ADS (autistic spectrum disorders) (Volk et al., 2013), and ADHD (attention deficit hyperactivity disorder) (Froehlich et al., 2011). Crucially, several multi-methods, transdisciplinary lines of research, suggest that air pollution exposure is strongly linked with neuroinflammation, and due to this link the study of air pollution has now become the most fertile frontier to understand DNT effects and mechanisms. So much so that we are now in a position to extend tests of the role of neuroinflammation as a general framework, not just for air pollution, but for many other complex DNTs behaving in similar ways. In its shortest formulation, this general hypothesis, and the central focus of the present chapter, proposes that elevation of cytokines and reactive oxygen species (ROS) in the brain are the most significant mediators of the deleterious effects for most of the complex DNTs (see Brockmeyer and D'Angiulli, 2016).

Understandably, so far the literature offers a piecemeal single-chemical exposure approach to defining and understanding how neuroinflammation acts on humans, and most of what we know is based on few single chemicals in animal studies and indirect associative relationships (correlation) involving extrapolations to humans, mostly in epidemiological and neuropsychological or behavioral studies. However, although animal models are essential to introduce the dimension of experimental manipulation and measurement of exposure and dose-response investigation, if matters are not thorny enough by virtue of the very issue of the complexity of the causal nature of determinism of the mechanisms involved, we still face the issue that the exposure condition studied by majority of the animal studies is not ecologically comparable and valid to human exposures, which is especially true for the population which is the focus of this chapter: children (considered here as the developmental period from fetus to adolescence).

Main knowledge limitations relate to the fact that animal studies and in vitro models point to multiple pathways and targets of toxicity for several DNTs: these involve formation and closure of the neural tube, cell proliferation, migration, death, or synapse formation (Lyall et al., 2014) inducing changes ranging from overt morphological alterations (Miller, 1986) to subtle dysfunction in synaptic connectivity (Neal et al., 2011). The same chemical compound may affect different developmental processes or different cell types, depending on the time window of exposure (Roy et al., 2004). Furthermore, compounds belonging to the same chemical class may have dissimilar mechanisms of action (for example, organophosphate insecticides, see Slotkin and Seidler, 2009). The same agent may have multiple independent mechanisms of action: for example, they may have an endocrine action on some parts of the CNS or PNS and concurrently exert an independent influence on synaptic connections in specific brain areas (Rubin, 2011).

Another issue is the measure of the individual exposure history. Different factors contribute to variability in the effects measured, the temporal dimension of the exposure, co-exposures, individual genetic vulnerability. In addition, in the case of children chemical exposure is associated with family socioeconomic background (SES) variables such as race/ethnicity, income, housing quality, parental education, neighborhood (Moore, 2009). These factors, for example, children's sex or age, are usually taken into account as confounders or effect modifiers (the covariates) but they could be the cause (predictor variables) of the results of chemical exposures or, as in case of socioeconomic status (SES), have their direct independent influence on the neurocognitive outcomes *also* affected by the chemicals. For example, neurocognitive functions vary with SES (see Schibli et al., 2017), and the odds of diagnosis as opposed to risk for neurodevelopmental disorders such as ASD and ADHD change according to where the children are in the SES spectrum, diagnoses are higher for high SES families but risk is higher for low SES families (King and Bearman, 2011; Rai et al., 2012).

A third critical issue, which is of particular significance for risk assessment, is that of the validity of the outcome measurements. As stated by De Felice et al. (2015, p. 93) "Most of prospective epidemiologic studies on neurotoxicity do not report clinically defined conditions (i.e., autism or learning disability) but rather atypical behavioral traits that range from increased/decreased anxiety and aggressiveness, to poorer motor or intellectual development in a significant proportion of exposed infants/children" [...]. "Whether these sub-clinical behavioral alterations may signal increased risk to develop a frank neuropsychiatric disorder at some point in the individual's life course remains to be determined." Furthermore, the interpretation of the findings has also generally focused on brain/mental disease rather than population-level health impacts, consequently, small increases of subclinical, but detrimental outcomes in a population that could have important impact on economic and health costs can remain overlooked (Bellinger, 2012).

Further to the issues already reviewed, children present special vulnerabilities. Children present unique risks of exposure which are additional to those experienced by adults. They consume more air and water per unit ofbody size compared to adults, and they are more active and spend more time outdoors during traffic peak times, for example, during school recessions. Children are further at risk because key organs and epithelial barriers are not fully developed (UNICEF, 2016). Furthermore, young children tend to play closer to the ground, where PM, such as dust and other pollutants, is found in higher concentrations (Vanos, 2015).

The complexity of the underlying mechanisms is a major challenge (Lopez Gonzalez et al., 2016). Pro-inflammatory responses should not be considered simply as single mutually exclusive causes, but rather part of a cascade system of interacting and hierarchical recursive mechanisms that may reflect the result of time-sensitive suppressed protective/repair processes (Singhal et al., 2014; Tansey and Goldberg, 2010). Such complexity may explain the likelihood of a proportion of immuno-resistant individuals in pediatric populations, but also why another proportion of apparently healthy children may only show subtle, preclinical short-term effects. The findings that young adults show early signs of neurodegeneration which are reviewed below may very well indicate a wear off of the neuroprotection/repair in place during childhood. And yet, in another proportion of children, subtle effects may go undetected, if participants appear "clinically" healthy. The consistent use of validated tests or screening batteries to compare outcomes across studies, and proper interpretation of individual background variables would improve the quality of children's data. However, cracking the complex relations between DNTs and the etiology of major chronic diseases including neurodevelopmental and neurodegenerative diseases will likely require a multidisciplinary synthesis of mechanistic biological validity provided by experimental data in tandem with the exposure and health mediating factors involving individual (i.e., genetic susceptibility) and ecological (i.e., SES) variables.

The remainder of this chapter first reviews the best evidence showing the most direct links between children's neurodevelopmental outcomes, air pollution and neuroinflammation. Here, air pollution is used as the paradigmatic test-bench in order to discuss an *exposome* perspective of neuroinflammation, focusing on the best evidence of pathways and mechanisms in the target population. An exposome framework of neuroinflammation action in DNT is then sketched generalizing the complex case of air pollution, by focusing on one of the fundamental neurodevelopmental mechanisms operating across the spectrum of DNTs, namely, *apoptosis*. Finally, promising approaches to identify biomarkers and promote developmental neuroprotection are discussed specifically in relation to air pollutants but generalizable to other DNTs.

2. The paradigmatic DNT test-bench: Air pollution

Outdoor air quality is often defined by indices reflecting the concentrations of criteria air pollutants. In the United States the 1970 amendments to the Clean Air Act required the Environmental Protection Agency (EPA) to set National Ambient Air Quality Standards (NAAQS) for certain pollutants known to be hazardous to human health. EPA identified six criteria pollutants: particulate matter (PM), ozone (O3), carbon monoxide (CO), sulfur dioxide (SO2), nitrogen oxide (NO), and lead (Pb) and set the standards as a function of the characteristics and their potential health and welfare effects. In the United States alone, >103 million people are exposed annually to PM concentrations above the standards, while 123 million are exposed to ozone.

Most monitoring systems across the globe measure two types of PM: PM_{10} (particles $<10\,\mu m$ diameter) and $PM_{2.5}$ (particles $<2.5\,\mu m$ diameter). The two fractions of PM which have been predominantly implicated in brain research are $PM_{2.5}$ and ultrafine PM (UFPM) (particles $<100\,nm$ diameter). Outdoor $PM_{2.5}$, mostly comes from tailpipe and brake emissions from mobile sources, residential fuel combustion, power plants, wildfires, oil refineries, and metal processing facilities. The primary contributors to UFPM are tailpipe emissions from mobile sources.

Indoor air pollutants, including tobacco smoke, emissions from cook stoves, mycotoxins, plasticizers, flame retardants, and pesticides, also represent a major source of harmful substances. Indoor air quality in schools is a major issue, the presence of mold, poor air quality, close proximity to major highways, and contaminated playgrounds can result in serious health problems (Everett-Jones et al., 2010; Kingsley et al., 2014; Sampson, 2012). Moreover, there are major disparities in indoor air pollution exposures related to socio-economic status (SES): the lower the SES, the higher indoor exposures (Adamkiewicz et al., 2011). Children are also exposed to manufactured nanoparticles (NPs) ($<100\,nm$) in many consumer products including food, toys, sunscreens and toothpaste (Linsinger et al., 2013).

Ground-level monitoring of $PM_{2.5}$ in nearly 3000 cities around the world by the World Health Organization (WHO, 2016) between 2008 and 2015 shows that mean annual levels tend to be highest in East Asia, South Asia and the Middle East and North African regions. They tend to be lower in North America, Europe and parts of the Western Pacific region. Approximately 98% of cities in low and middle income countries do not meet WHO guideline safety limits for $PM_{2.5}$ ($<10\,\mu g/m^3$). However, in high-income countries 56% of cities do not meet WHO limits. In these urban areas, 80% of people are exposed to air quality levels that exceed WHO limits, ranging between from >10 to $>100\,\mu g/m^3$ (10 times over the cut-off). Accordingly, the most recent estimated numbers of exposed children measured through satellite imagery by continent are: 520 million in Africa, 1.2 billion in Asia, 130 million in the Americas and 120 million in Europe (van Donkelaar et al., 2016). Between 2008 and 2015, there was an 8% increase in global levels of urban air pollution (WHO, 2016). Although, improvements were seen in some regions, most projections agree that by 2050 both urbanization and air pollution are projected to increase dramatically (Lelieveld et al., 2015; OECD, 2016).

Consequently, urban pollution seems an escalating and pressing global risk. Indeed, there is now a substantial extensive literature on brain functions and structure in children exposed to severe chronic urban air pollution. Next, a selected narrative review is presented from several cohorts of children followed in the Mexico City Metropolitan Area (MCMA), where $PM_{2.5}$ is estimated to fluctuate annually between 60 and 99 $\mu g/m^3$. Brain changes found in MCMA have been directly related to functional outcomes identical or similar to those reported in many of the largest cities of the globe; the consistent finding in all these studies is the pervasive role of neuroinflammation (Brockmeyer and D'Angiulli, 2016).

3. Animal models of air pollution components and DNT effects

As discussed earlier, the complexity of exposure in outdoor and indoor environments makes it very difficult to establish a direct association of CNS effects with specific air pollutants in humans. Fortunately, animal models exposed to air pollutant components such as ozone, PM, diesel NPs, and endotoxins have contributed a good deal to our understanding of the potential mechanisms acting upon the CNS. Depending on the pollutant component, doses, exposure protocol, age and gender, health status, etc., the detrimental effects range from endothelial dysfunction, breakdown of the blood-brain-barrier (BBB), neuroinflammation, formation of free radicals and oxidative stress, dopaminergic neuronal damage, and RNA, DNA damage, to the identification of early hallmarks of Alzheimer and Parkinson's diseases (Brun et al., 2012; Fonken et al., 2011; Guo et al., 2012; Levesque et al., 2011, 2013; Oppenheim et al., 2013). Notably, diesel exhaust particles (DEP), a major component of urban air pollution, have been linked in mice to neuroinflammation and the accumulation of Aβ42, tau (linked to Alzheimer's disease), along with alpha synuclein, microglial activation and Parkinson's disease-like pathology (Levesque et al., 2011, 2013). Inhalation exposure to traffic-generated air pollutants promotes increased activity of powerful matrix metalloproteinases and degradation of tight junction proteins in the cerebral microvasculature resulting in altered brain-blood-barrier permeability and expression of neuroinflammatory markers (Oppenheim et al., 2013).

Oxidative stress caused by low doses of ozone results in dysregulation of inflammatory responses, progressive neurodegeneration, altered brain repair in the hippocampus, and brain plasticity changes in the rat analogous to those seen in Alzheimer's disease (Rivas-Arancibia et al., 2010). In rats, cigarette smoking a powerful source of oxidative stress and particles reduces the expression of pre-synaptic proteins, impairs axonal transport and produces neurodegenerative changes as those seen in Alzheimer's disease (Ho et al., 2012).

In spite of the complexity, the evidence conclusively shows that in animals prenatal exposure to either one or a combination of criteria pollutants causes permanent changes in neurotransmitters and alters brain development, most commonly resulting in long-term deficits in functions associated with one or more memory systems (Fonken et al., 2011; Schröder et al., 2013; Takahashi et al., 2010; Umezawa et al., 2012).

4. Systemic DNT effects of air pollution in children

The developmental detrimental effects in animals are conceptually related to or even mirror the effects that might be expected and are actually observed in children. Consequently, it is reasonable and plausible to assume a fundamental continuity underlying the processes that impact the developing brain, human or animal. In this section, we selected plausible biological pathways shown to be in place in urban children and in experimental animals exposed to particulate matter and/or their components. Fig. 1 shows the pathways specifically linked to key evidence found in Mexico City children; the figure provides a hypothetical roadmap of the complex network of independent or linked concomitant and/or cascade processes associated with behavioral and cognitive outcomes in children.

Extensive data in the literature support the human and animal breakdown of critical barriers: nasal, olfactory, blood-brain-barrier (BBB) and alveolar-capillary barriers and the expression of detrimental genes associated to urban air pollution (Calderón-Garcidueñas et al., 1992, 2001; Oppenheim et al., 2013). Mexico City residents exhibit breakdown of the nasal epithelial barrier, thereby, facilitating the passage of fine and ultrafine PM and xenobiotics to the systemic circulation and the brain (Calderón-Garcidueñas et al., 1992). The integrity of the GI barrier is also compromised in urbanites, allowing for the entrance of swallowed PM. Nanoparticles are of great importance in terms of gastrointestinal toxicity (Bergin and Witzmann, 2013). Disruption of epithelial integrity with structural changes in tight junctions (TJ), the major determinant of paracellular permeability characterized the small bowel architecture of Mexico City residents. MCMA children have significantly higher antibodies for cell junction and neural

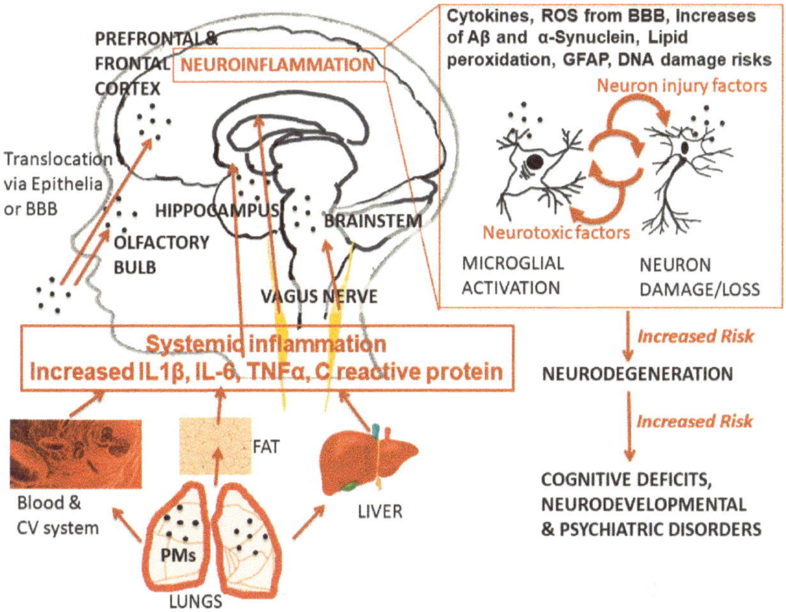

Fig. 1 Example of hierarchical cascade and recursive mechanisms of action leading to the different developmental neurotoxic effects from air pollution. *Note. IL-6*, interleukin 6; *IL-1β*, interleukin 1 beta; *TNF-α*, tumor necrosis factor alpha; *BBB*, brain blood barrier; *ROS*, reactive oxygen species; *GFAP*, astrogliosis; *Aβ*, beta amyloids. *Reused with author's permission, from D'Angiulli, A., 2018. (Severe urban outdoor air pollution and children's structural and functional brain development, from evidence to precautionary strategic action. Front. Public Health 6, 95).*

proteins raising the possibility the enteric nervous system plays a direct role to damage vulnerable brain regions in highly exposed children.

The pulmonary damage is equally severe (Calderón-Garcidueñas et al., 2003), and boys are more affected than girls, likely because of longer daily outdoor activities (Villarreal-Calderón et al., 2002). Systemic inflammation and endothelial dysfunction with high production of endothelin-1 (ET-1) are also important, high concentrations of powerful inflammatory mediators such as interleukin-1β, tumor necrosis factor alpha (TNF α), and interleukin 6 (IL6)—for which brain endothelial cells have receptors—are the rule in exposed children. High ET-1 plasma concentrations, a very potent vasoconstrictor, negatively impact the brain microvasculature, resulting in hypoperfusion and ET-1 concentrations are directly associated with $PM_{2.5}$ exposures (Calderón-Garcidueñas et al., 2007).

Neuropathology studies in children with accidental deaths revealed that 40% of urban children exhibited frontal tau hyperphosphorylation with pre-tangle material and 51% had Aβ42 diffuse plaques compared with 0% in controls (Calderón-Garcidueñas et al., 2012a). Hyperphosphorylated tau (HPτ) and Aβ42 plaques are hallmarks of Alzheimer's disease (AD) (Braak and Del Tredeci, 2011). Of utmost importance for this review, children with the apolipoprotein E allele 4 (a well-known risk factor for AD) had greater HPτ and diffuse Aβ plaques versus E3 carriers, suggesting genetic factors could make a significant portion of the exposed population more prone to accelerating their AD pathology (Calderón-Garcidueñas et al., 2012a). Arteriolar and capillary vascular changes with a diffuse breakdown of the BBB are at the core of the pathology in exposed children's brains. The changes are significant in the olfactory bulb and the prefrontal white matter, but can be found in every lobe and in the brainstem (Calderón-Garcidueñas et al., 2013b).

Clinically healthy children from Mexico City (MC) selected by stringent criteria including the absence of known risk factors for cognitive or neurological deficits, exhibited structural, neurophysiological and cognitive detrimental effects compared to matched SES, gender, age and mother's IQ low pollution exposed children (Calderón-Garcidueñas et al., 2008, 2011a). The cognitive deficits in MC children matched the MRI volumetric changes in their right parietal and bilateral temporal areas (Calderón-Garcidueñas et al., 2012b). Complex modulation of cytokines and chemokines influences children's CNS structural and volumetric responses and cognitive correlates resulting from environmental pollution exposures. MC children performed more poorly across a variety of cognitive tests, compared to control children (Calderón-Garcidueñas et al., 2008, 2011a, 2012b).

A number of abnormalities within the auditory brainstem nuclei have been identified in children exposed to severe air pollution. Specifically, the neuronal cell bodies within the medial superior olive (MSO) are significantly smaller and more round than those in age-matched control brains (Calderón-Garcidueñas et al., 2011b). This finding is important because the MSO has clear roles in localization of sound sources, encoding temporal features of sound and likely plays an important role in brainstem encoding of speech. Integrity of the auditory brainstem nuclei can be accessed through a number of noninvasive techniques, such as brainstem auditory evoked potentials (BAEPs), otoacoustic emissions, speech recognition tasks and listening in background noise. Incidentally, similar morphological alterations were observed in autistic children (Kulesza and Mangunay, 2008; Lukose et al., 2013) and correlated with abnormal brainstem reflexes (Kulesza et al., 2011). We will discuss the line of research on BAEP in much more detail later, but in interim, it is important to note the conclusion that urban children have delayed central conduction time of brainstem neural transmission, resulting in increased risk for auditory and vestibular impairment and altered speech recognition abilities (Calderón-Garcidueñas et al., 2011b) which can have very important implications in terms of risk of developing learning disabilities and life-long poor academic achievement.

Overall, the evidence would suggest that air pollution exposed children experience a chronic, intense state of oxidative stress and exhibit an early brain imbalance in genes involved in inflammation, innate and adaptive immune responses, cell proliferation and apoptosis (Calderón-Garcidueñas et al., 2012a, 2013a). Neuroinflammation, endothelial activation, the significant heterogeneity of endothelial cells in CNS microvessels, and the BBB breakdown contribute to cognitive impairment and pathogenesis and pathophysiology of neurodegenerative states (Jian et al., 2012; Paul et al., 2013; Roher et al., 2012).

5. Children's outcomes associated with the impacts of air pollution

The associations between cognition and urban pollution have been established not just in megacities like Mexico City but also in smaller cities like Boston, where black carbon—a marker for traffic PM—predicted decreased cognitive function across assessments of verbal and nonverbal intelligence and memory in 9 year olds (Suglia et al., 2008). Although genetic factors play a key role in CNS responses (as evidenced by the acceleration of neurodegenerative pathology in children carrying an APOE 4 allele), studies, such the abovementioned ones in Boston and others, sketch a complex scenario where air pollution and SES can influence neural development and cognition along with known factors such as psychosocial stress and poor nutrition, thereby influencing and determining mental health, academic achievements and overall life performance (Becerra et al., 2013; Siddique et al., 2011). It is critical to point out that although SES is an additive independent risk factor, in several of the studies conducted in megacities that we have reviewed earlier, the effects of outdoor air pollution on children's brain did not vary interactively with SES, thus outdoor air pollution effects are not a concern for just underprivileged populations although the fact that belonging to the lower end of the socioeconomic spectrum is very likely to aggravate detrimental health effects (D'Angiulli, 2015).

Related to the issue of cumulative risks, there is growing pediatric and public concern about the direct and indirect influences air pollution may have on several developmental outcomes such as school performance, behavioral changes and mood disorders in children and teens. A substantial body of research has shown that children attending schools near major roadways and freeways are exposed to severe traffic-related pollution, and several studies have documented the link between air pollution levels and lower academic performance, reduced standardized test scores or not meeting achievement standards elementary through high school (Clark-Reyna et al., 2016). Moreover, childhood aggression and teen delinquency are increasing in megacities, establishing early environmental health risk factors for violence prediction and prevention (Haynes et al., 2011; Liu, 2011) in populations at risk will be absolutely critical. In particular, psychiatrists, clinical psychologists and allied mental health and pediatric professionals have a critical role to play in identifying the potential associations between exposure and behavioral issues. An increasing body of evidence is showing associations between traffic-related exposure and diagnoses of psychiatric disorders in children and adolescents (Oudin et al., 2016).

Another intriguing association has been identified between autism spectrum disorders (see Costa et al., 2017) and attention deficit hyperactive disorders (ADHD) and indoor and outdoor pollution exposures have surged recently (Becerra et al., 2013; Larsson et al., 2009; Siddique et al., 2011; Volk et al., 2013; Zhang et al., 2010). Risk factors related to air pollution components include maternal second and third hand smoke exposure, residency during gestation at the highest quartile of exposure to traffic-related air pollution, condensation on windows (a proxy for low ventilation rate in the home) and polyvinyl chloride (i.e., airborne phthalates indoors) flooring, especially in the parents' bedroom. Interestingly, airway symptoms of wheezing and physician-diagnosed asthma were also associated with autism spectrum disorder 5 years later (Larsson et al., 2009). Because these associations are linking autism and ADHD with environmental variables, they warrant wider knowledge translation by and among the developmental, behavioral and clinical researchers and practitioners.

The reviewed evidence of brain, neurocognitive and behavioral outcomes associated with air pollution taken together suggests detrimental effects that may have long-term clinical repercussions in terms of degenerative diseases (Calderón-Garcidueñas et al., 2013a). Given the social and economic burden of accelerated aging in our society, whose far-reaching ramifications are simply incalculable, a multidisciplinary approach aiming at screening target school populations that are most at risk would seem a rather cost-effective and most beneficial public health strategy. Strong support for the need of neurocognitive and behavioral screening in target at risk populations of children comes from a growing psychological and epidemiological literature suggesting evidence of suboptimal cognitive functioning across the developmental span in clinically healthy children (Calderón-Garcidueñas et al., 2012b; D'Angiulli, 2018; Guxens and Sunyer, 2012). Importantly, a significant proportion of urban schools are situated near major traffic-related air pollution sources (Amram et al., 2011), and cognitive outcomes may be partly associated with air pollution levels around schools (Mohai et al., 2011).

6. Structure-function correspondences in the effects of air pollution

6.1 Magnetic resonance imaging (MRI) evidence of DNT effects on white matter connectivity

Clinically healthy children (i.e., appeared to be healthy and had no physical complaints), residents in a highly polluted atmosphere, exhibit selective impairment in attention, short term memory and risk for developing types of learning disability. The extent of neurocognitive deficits was found to correlate with developmental differences in barin volume (Calderón-Garcidueñas et al., 2011a). Children with white matter hyperintensities showed more pronounced cognitive deficits, while children with no white matter hyperintensities exhibited an additional persistent deficit in visual memory (picture completion). The cognitive deficits in these highly exposed children are compatible with the structural differences in temporal and parietal white matter regions observed over the 1 year follow-up. Mexico City children with no white matter hyperintensities (WMH) exhibited the most significant systemic inflammation as evidenced by the highest serum concentrations of MCP-1 and TNF α, and the lowest numbers of peripheral neutrophils reflecting endothelial activation and dysfunction.

Although there were no differences in the volume of subcortical structures, including hippocampus, caudate, putamen, globus pallidus and amygdala, we found altered pattern of brain development characterized by delayed age-appropriate increase in cortical white matter volume.

The cognitive changes are potentially related to the WM volumes through reduced brain connectivity (Gläscher et al., 2009, 2010; Woolgar et al., 2010) and neuroinflammation (Block and Calderón-Garcidueñas, 2009). The cognitive deficits present in this 6.9 ± 0.68 year old cohort follows the cognitive deficient profile of a similar but older cohort (9.85 ± 2.15 years) we reported previously (Calderón-Garcidueñas et al., 2011b). Older children exhibited significant cognitive deficits in areas of fluid cognition, memory, and executive functions, with 56% of MC children exhibiting WMH. Thus, cognition deficits are already present in first graders and persist in older cohorts.

General intelligence (g) reflects the performance variance shared across cognitive tasks (Gläscher et al., 2010). Statistically significant associations have been found between g and damage to a frontal and parietal cortex network, critically including white matter association tracts and frontopolar cortex (Gläscher et al., 2010). The authors suggest that g draw on connections between regions that integrate verbal, visuospatial, working memory and executive processes (Gläscher et al., 2010). Significant effects were located in the left hemisphere and included expected locations of major white matter tracts including the anterior and dorsal bundle of the superior longitudinal/arcuate fasciculus connecting temporal, parietal and inferior frontal regions. Right hemisphere critical regions were located at the occipitoparietal junction reaching into the postcentral sulcus and in the anterior bank of the central sulcus. In correlating with WAIS, the authors showed a modest significance for Picture Arrangement, Block Design and Picture Completion, indicating that visuospatial skills are vulnerable to damage in much larger areas of the right hemisphere. Working memory and verbal skills correlated substantially with the left hemisphere, most notably arithmetic and similarities.

In our studies (see reviews in Brockmeyer and D'Angiulli, 2016; D'Angiulli, 2018), information, arithmetic, vocabulary, digit span, and picture completion exhibited q values below 0.05, while significant q values were recorded bilaterally for temporal white matter and for the right parietal white matter. Thus, subtests that rely heavily on the capacity for complex reasoning and integration of diverse forms of knowledge, in addition to basic verbal and working memory skills, i.e., Arithmetic, are affected in highly exposed children.

In Gläscher et al. work, the largest overlap between WAIS and g was found for arithmetic, similarities, information and digit span. Relevant to our work, three of these four subtests were significantly involved in exposed children emphasizing the residency detrimental effect on verbal knowledge, verbal reasoning, abstraction, working memory capacity and their anatomical correlation with the left inferior frontal gyrus, the superior longitudinal/ arcuate fascicle and the parietal cortex (Gläscher et al., 2009).

Moreover, the same authors reported that in a cohort of 241 patients with focal brain lesions, impairments in verbal comprehension index using WAIS-III were associated with damage in the left hemisphere, specifically the left inferior frontal cortex. Impairments in perceptual organization index were associated with damage in right parietal, occipito-parietal and superior temporal cortex and finally impairments in working memory index correlated with left hemispheric lesions focused on the superior parietal cortex (Gläscher et al., 2009).

Jung and Haier reported that variations in the parietal-frontal/temporal occipital network (Broadman areas 6, 9, 10, 45–47, 39–40, 7, 32, 21, 37, 18, 19) with involvement of white matter regions, predict individual differences found on intelligence and reasoning tasks (Jung and Haier, 2007). White matter integrity of the target regions is key for a correlation to performance on the Wechsler Intelligence Scales, both in young and adult cohorts (Haier et al., 2004).

Very recently, Pujol et al. (2016) assessed the extent of potential effects of Barcelona's urban pollution on child brain maturation using general indicators of vehicle exhaust measured in the school environment and a comprehensive neuroimaging evaluation. A group of 263 children, aged 8–12 years, underwent MRI to quantify regional brain volumes, tissue composition, myelination, cortical thickness, neural tract architecture, membrane metabolites, functional connectivity in major neural networks and activation/deactivation dynamics during a sensory task. A combined measurement of elemental carbon and NO_2 was used as a putative marker of vehicle exhaust. Air pollution exposure was associated with brain changes of a functional nature, with no evident effect on brain anatomy, structure or membrane metabolites. Specifically, a higher content of pollutants was associated with lower functional integration and segregation in key brain networks relevant to both inner mental processes (the default mode network) and stimulus-driven mental operations. Age and performance (motor response speed) both showed the opposite effect to that of pollution, thus indicating that higher exposure is associated with slower brain maturation. Therefore the authors concluded that urban air pollution appears to adversely affect brain maturation in a critical age with changes specifically concerning the functional domain.

Consistent with the Barcelona findings, a preliminary diffusion tensor imaging analysis (Calderón-Garcidueñas et al., 2013b) of MRI scans from a small subgroup of children from the same cohorts we studied (Calderón-Garcidueñas et al., 2012b), 7 MC versus matched low-pollution controls, showed qualitative visual evidence of less white-matter connectivity in 15 out of 16 (on Binomial test with $P(H_0)$ at 50%, $P=0.0005$) fibers along the bilateral cerebral-spinal tracts including: bilateral prefrontal lobes, bilateral cingulum, superior longitudinal fasciculus, parahippocampal areas, olfactory tract except in one, the uncinate fasciculus, which has not been functionally related specifically to cognition and was included for contrast. In the analyses gender and ApoƐ status were controlled for. Fractional anisotropy (FA) values along the left cerebral-spinal tract in all 14 children were used to build the DTI atlas. From superior to inferior direction, lowest FA values were observed close to the cortex and high FA values within the internal capsule and brainstem regions. In addition, the FA values along the left cerebral-spinal tract two MC ApoƐ4 carriers were compared with the average control profile. Several regions from superior to inferior direction indicated potential differences in white matter properties in one ApoƐ4 carrier, whereas the other exhibits properties close to the mean control.

The cerebral white matter is critically involved in many bio-behavioral functions. Volume is an evolutionary and developmentally regulated property of tissue, which is sensitive to the regularities of normal histogenetic sequence and normal systems operations (Caviness et al., 1999). Temporal WM, carrying the principal association fiber tracts of the superior, middle and inferior temporal gyri and fusiform and parahippocampal gyri, supports basic analytical processing and cognitive processes such as language and categorization (Heilman et al., 1987). While prefrontal, medial and lateral parietal, and mesial temporal regions are associated with functions such as memory, attention, learning, pain perception, motivation, emotion and autonomic function (Mesulam, 2000). Prefrontal white matter hyperintensities (WMHs) were key parameters taken into account to study Mexico City children (Calderón-Garcidueñas et al., 2008), the hypothesis being that children with prefrontal WMH would have the most pronounced cognitive deficits. However, regardless of the presence of prefrontal WMH, Mexico City children performances in distinct cognition tests were significantly deficient compared to CTL.

In the adult literature, WMHs are usually regarded as surrogates of small vessel disease and represent a change of water content in brain white matter that appears as foci of high signal intensity on fluid attenuated inversion recovery (FLAIR) or T2 weighted magnetic resonance images (Gouw et al., 2011; Morris et al., 2009; Xu et al., 2010). Small vessel disease specifically in white matter arterioles with prominent endothelial hyperplasia, enlarged Virchow-Robin spaces, gliosis and ultrafine particle deposition were critical findings in healthy 1 year old dogs exposed to the same polluted atmosphere in MC (Calderón-Garcidueñas et al., 2008). WMHs are among the most replicated structural neuroimaging findings in cognitive impairment risk, cerebral ischemia, neurodegeneration, cardiovascular, metabolic, and demyelinating diseases and neurodevelopmental conditions (Brickman et al., 2009; Bunce et al., 2010; Carmichael et al., 2010; Gouw et al., 2011; Murray et al., 2010; Silbert et al., 2009). In elderly people, WMHs only partially identify underlying white matter pathology and are associated with lesions developing in surrounding tissues (Wallin and Fladby, 2010). Executive function is the primary cognitive domain affected by WMH burden (Murray et al., 2010). In middle age healthy subjects (44–48 years) frontal white matter lesions were significantly associated with greater intra-individual reaction time variability in women and temporal WMHs were associated with face recognition deficits in men (Bunce et al., 2010). Xu et al. (2010) reported a distinctive white blood cells RNA expression gene profile characterized by genes involved in oxidative stress, inflammation, oligodendrocyte proliferation, axonal repair and neurotransmission from subjects with WMH versus subjects with no white matter lesions. This gene profiling associated with WMH is very relevant to our brain findings in children, teens and young adult residents in Mexico City versus matched subjects from low-polluted areas, given the up-regulation of COX2, IL1β and CD14 in supra and infra tentorial regions, along with disruption of the blood-brain-barrier (BBB), endothelial activation, oxidative stress, and inflammatory cell trafficking (Calderón-Garcidueñas et al., 2008).

Several possibilities arise when discussing volumetric changes and cognition deficits in children highly exposed to pollution and lacking risk factors for cognitive or neurological deficits: i.e., neuroinflammation as a diffuse process involving supratentorial and infratentorial regions is present in brains of highly exposed children and young adults (Calderón-Garcidueñas et al., 2008). As part of the neuroinflammatory process, there is breakdown of the BBB as evidenced by the abnormal tight junctions, endothelial activation, i.e., immunoreactivity to vascular adhesion molecule-1, inflammatory cell trafficking around blood vessels, activated microglia, and oxidative stress markers are expressed in neuronal and glial cells (Calderón-Garcidueñas et al., 2008). Hypoperfusion could be the end result of the prevailing neuroinflammation. Moreover, highly exposed children have significant concentrations of endothelin-1 (ET-1), a powerful vasoconstrictor that exerts its effects on the cerebrovascular endothelium and is capable of mediating hypoperfusion or endothelial dysfunction (Huang et al., 2010; Moldes et al., 2008; Salonia et al., 2010). One year old healthy dogs exposed to the same atmosphere as the children in this study exhibit vascular subcortical pathology, significant neuroinflammation and ultrafine particles were visualized in endothelial brain capillaries (Calderón-Garcidueñas et al., 2008).

Neuroinflammation and alterations in blood flow could result in damage of subcortical neural networks and volumetric white matter changes. Thus, the possibility of chronic white matter hypoperfusion and disruption of the BBB, leading to chronic leakage of plasma into the white matter (Debette and Markus, 2010), has to be entertained. Inflammation and oxidative stress have been identified as common and basic mechanisms responsible of tissue damage related to air pollution exposure (Brook et al., 2010).

The systemic up-regulation of MCP-1 is of particular interest to us in view of the influence of MCP-1 on permeability of the BBB (Yadav et al., 2010). Specifically, MCP-1 is involved in the recruitment of both monocytes/macrophages and activated lymphocytes into the CNS and induces increase in brain endothelial permeability. Since an intact BBB is key for proper functioning of neuronal circuits, and synaptic transmission, the breakdown of the BBB in highly exposed children could account for regional hypoxic conditions (Zlokovic, 2008).

Likewise, increased serum concentrations of TNF α in the setting of air pollution relate to the systemic inflammation observed repeatedly in highly exposed children (Calderón-Garcidueñas et al., 2007, 2008, 2010), but more importantly point to the role of TNF as a marker of brain disease (Clark et al., 2010). In adult populations, inflammatory markers including TNF are associated with total brain volume (Jefferson et al., 2007).

6.2 Neurophysiological evidence and underlying neurofunctional model

Very recently we (Calderón-Garcidueñas et al., 2017) used an ecologically valid dog model to investigate the potential effects of air pollution on the function and morphology of the auditory brainstem found in MC children. Twenty-four dogs living in clean air versus MC with similar air pollution exposure to that of children, average age 37.1 ± 26.3 months, underwent brainstem auditory evoked potential (BAEP) measurements. Eight dogs

(4 MC, 4 Controls) were analyzed for auditory brainstem morphology and histopathology. MC dogs showed ventral cochlear nuclei hypotrophy and MSO dysmorphology with a significant decrease in cell body size, decreased neuronal packing density with regions in the nucleus devoid of neurons and marked gliosis. MC dogs showed significant delayed BAEP absolute wave I, III and V latencies compared to controls. Remarkably, the pattern of results in both children's and animal replication is similar to the ones found in previous research which has established a link between delayed BAEPs and exposure to methylmercury (Murata et al., 2004). Of particular importance is the observation that the prolongation of the interpeak latency I–III in children involves an effect size of much larger magnitude than the one observed for III–V (see Fig. 2).

Relations between BAEP waveforms and brainstem auditory structures or cell types were established in animal species in most cases, especially cat. However, to a certain extent, these same relations can be transposed to man/children when it is postulated that the spherical cell system has a major role in auditory pathways. The first three positive deflections of BAEP in cat and man (P1–P3 in cat and I–III man) have similar characteristics. In addition, P4 in cat which is predominantly generating by MSO has similar generators than V in man. With this in mind, waveform III can be assumed to be originated both from cochlear nucleus (CN) and contralateral superior olivar complex (SOC) cells. In the cochlear nucleus, spherical cells of the anterior part of the AVCN generate a part of wave III whereas in the contralateral SOC, principal cells of the medial nucleus of the trapezoid body (MNTB) contribute to Wave III generation. Additionally, neurons of the lateral superior olive (LSO) are also implicated. Ipsi and contralateral cells of the SOC participate in waveform V generation with MSO principal cells identified as generators. Additionally, spherical cells of the anterior part of the AVCN mainly contribute to this waveform, even if they are not cellular generators of this waveform, by driving MSO principal cells. Similarly, MNTB modulates the wave V generators as MNTB principal cells have an inhibitory effect on MSO principal cells. Other cellular generators of wave P5, most likely of wave V, are located in the lateral lemniscus and/or the inferior colliculus (for review, see Biacabe et al., 2001). However, direct effects on generators produce morphological abnormalities, for example, involvement of brainstem auditory structures by demyelinating diseases produces loss of later BAEP waveforms, due to loss of synchronous cells activity. For example, in multiple sclerosis, V can be absent, despite the presence of normal subjective audiometric thresholds (Voordecker et al., 1988).

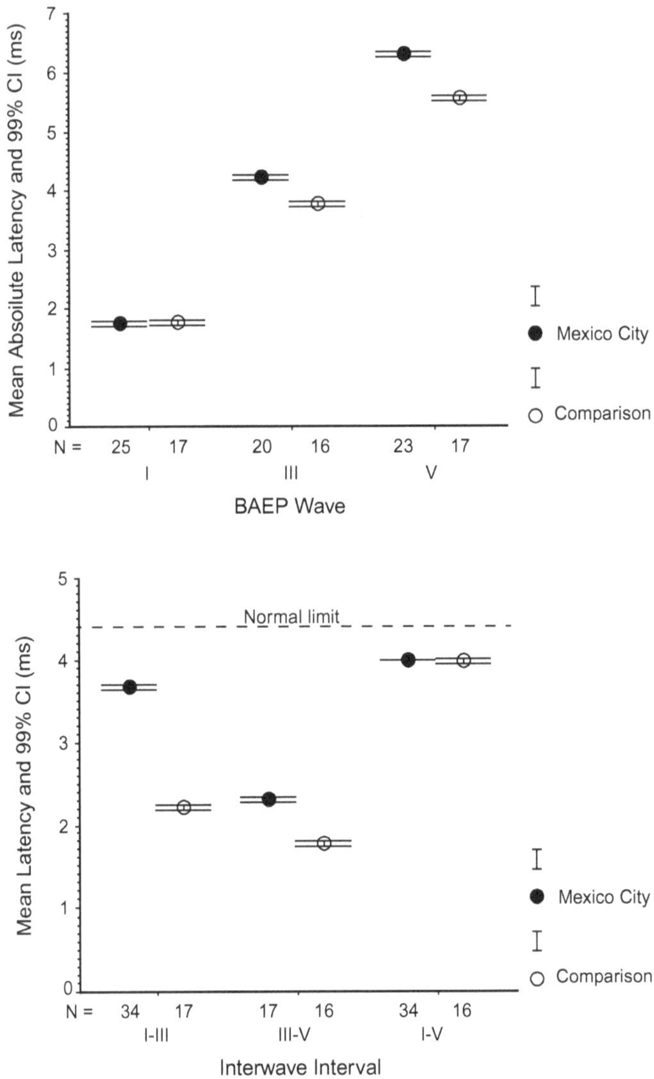

Fig. 2 Mean absolute latencies (in milliseconds) of the replicable waveform components of the brainstem auditory evoked potentials (top panel) measured in the MC children (filled circles) and the control group (open circles) and relative mean interwave differences (bottom panel). *(The plots are based on data from Calderón-Garcidueñas, L., D'Angiulli, A., Kulesza, R.J., Torres-Jardón, R., Romero, L., Keefe, S., Herritt, L., Brooks, D.M., Avila-Ramirez, J., Delgado-Chávez, R., Medina-Cortina, H., González-González, L.O., 2011b. Air pollution is associated with brainstem auditory nuclei pathology and delayed brainstem auditory evoked potentials. Int. J. Dev. Neurosci. 29, 365–75).*

In both children and dogs we observed α synuclein brainstem accumulation and medial superior olivary complex (MSO) dysmorphology. Thus, a variety of intracranial processes causing increased intracranial pressure can induce prolongation of BAEP waveforms, by interfering with typical functioning of the generators and/or the connections from and to these major nuclei. In addition, our findings were consistent with previous data indicating that delayed waves and interwaveform latencies may be linked with disturbed control of respiration. Patterns of results very similar to ours have been reported in adults and children with central alveolar hypoventilation syndrome (Beckerman et al., 1986; Litscher et al., 1996) and respiratory failure following encephalitis (Schwarz et al., 1990, 1994, 1996). One of the markers of this relationship may be the major effects in which wave III is involved (especially see bottom panel of Fig. 2).

Beckerman et al. (1986) have hypothesized a topographic-anatomic relation between the superior olivary complex which generates wave III on his opinion, and the medullary chemoreceptor zone, which is in the immediate vicinity. However, there are other possible candidates. Brainstem neural mechanisms are regarded as a complex neuronal system consisting of several functional subsystems. There is a major concentration of respiratory neurons—nucleus parabrachialis medialis, Kolliker Fuse nucleus (pontine respiratory group)—in the rostral pons. All these structures are assumed to be essential to the mechanisms controlling respiratory rhythm (Hukuhara, 1988). Litscher and colleagues (Litscher et al., 1996; Schwarz et al., 1996) have theorized that some of the BAEP patterns similar to the ones we report could help to document parallel dysfunction of the closely proximate neuronal structures generating BAEPs and of areas controlling respiration. Their model proposes topographic relations between different functional brainstem structures, which are very close to one another, and are involved in a common pathologic process.

The MRI findings of hyperintense white matter lesions found in frontal subcortical areas and in the prefrontal cortex (PFC) in MC children and in model comparison co-exposed dogs suggest that neuroinflammation may be widespread at many layers across the brainstem and could also implicate important structures in the pontine-mesencephalic area, in which cell bodies on the ascending monoaminergic brainstem systems have important reciprocal connections with the PFC and are known to modulate its functioning. Further converging evidence of the plausibility of such interpretation is provided by a longitudinal case study (Salgado et al., 2007) in which a persisting posttreatment scarified lesion in the right pontine tegmentum has been reported to be associated with PFC cognitive dysfunction leading to impairment in planning and behavioral inhibition. These may be the same underlying working memory deficits that presumably are responsible for the poor cognitive performance of Mexico City children on psychometric tests. Importantly, the PCF white matter lesions were most prevalent in the left side, coinciding with the prevalence in ipsilateral right side delays in BAEPs. The latter correlation may explain why the performance of the majority of the Mexico City children was particularly crippled in all the psychometric subtests tapping on verbal and linguistic working memory processes.

In conclusion, animal and pediatric BAEP data, together with other converging behavioral, imaging and anatomical evidence, show a link between exposure to air pollution and auditory dysfunctions. This association suggests damage to topographically proximal or even shared neural structures which are devoted, among other functions, to the central control of respiratory regulation.

7. Neuroinflammation and apoptosis

Although, for some aspects (discussed earlier) it is helpful to investigate the detailed and specific effects of single pollutants, this may have limited practical validity. Because of issues concerning the validity, reliability, and interpretation of DNT research, chemical-by-chemical and source-specific assessment of risk does not reflect the cumulative impacts of multiple toxic stress posed by NDTs (Corburn, 2017).

In particular, children move between indoor and outdoor environments seamlessly, and the many microenvironments between indoor and outdoor are hard to classify. Analysis of children's exposure in urban environments is much more complex than characterizations of "indoor" and "outdoor," and children do present unique exposure vulnerabilities. Thus, risk assessment cannot be narrowly focused on calculating the probability or significant effects of acceptable impact of independent single NDTs from single sources and through single exposure pathways, since the underlying (untested) assumption is that children's brains can tolerate an endless accumulation of single "acceptable" insults (NAS, 2009).

An alternative is needed to form the basis for a more comprehensive definition of risk for complex forms of DNT. To meet such need, a possible approach could be leveraged working from within the "exposome" framework. Such perspective permits to encompass the totality of human environmental exposures from conception onwards, and to address the need for more complete environmental exposure data, complementing the genome with a more comprehensive description of lifelong exposure history (Vrijheid, 2014; Wild, 2012).

The exposome paradigm analyzes both endogenous and exogenous sources of non-genetic exposure. In addition to DNTs, the exposome includes diet, life style factors, occupation, pre-existing diseases, infections, psychological stress, and socioeconomic status, together with several endogenous factors—inflammation, hormones, metabolism, and oxidative stress—potentially influencing the response of the organism to the external factors. The overlap and the interplay between the internal and the external domains, together with their relative temporal variation, represent the most challenging features for the exposome characterization. In keeping with such framework, DNT effects could be defined as total cumulative serious harm to brain development resulting from involuntary exposure to multiple hazardous or toxic pollutants over time, with serious threats of permanent or irreversible neurodegenerative disease in a clinically important fraction of children's population (D'Angiulli, 2018).

Following the exposome approach, one key effort relevant in the present context is to synthesize research in terms of formulation of complex and dynamic interactions between DNA sequence, epigenetic DNA modifications, gene expression and environmental factors which contribute to the identified preclinical, clinical or otherwise normative/protective neurodevelopmental trajectories.

In particular, neuronal and glial cell apoptosis is a form of programmed cell death crucial for the normal development of the CNS. Neuroapoptosis is involved in most regions of the brain during prenatal proliferation and postnatally in postmitotic neurons which are forming synaptic connections. Alteration of either timing or entity or both in the unfolding of such programs can result in neurodevelopmental deficits in neurocognitive and higher-order functions even when obvious pathology is not manifested (Lein et al., 2018), i.e., children appear clinically healthy. The latter aspect may be related to the fact that, differently than in necrosis, in apoptosis individual cells die without damaging neighboring healthy cells. In addition, while necrosis is induced by more severe and sustained insults, apoptosis is induced by less intense and more transient stresses. However, although per se apoptosis does not trigger inflammation, frequently necrosis and apoptosis coexist in the same population of cells undergoing neural injury. Chemicals which are associated with the promotion of apoptosis include some of the major NDTs cited or discussed in this chapter, such as lead, ethanol, methyl mercury, but also polychlorinated biphenyls (PCBs), dysregulated zinc, and pediatric anesthetics as implicated by developmental animal models (Notice that the role of the latter agents in infants and children is hotly debated and controversial right now, see O'Leary and Warner, 2017).

Thus, induction of apoptosis by NDTs during critical neurodevelopmental periods is deemed to be able to alter neurodevelopmental trajectories in important ways, resulting in persistent cognitive and behavioral deficits. Clear associations have been established between proinflammatory and oxidative stresses and neuroapoptosis at mitochondrial internal or external processes/sites and all major neurodevelopmental disorders such as fetal alcohol syndrome (i.e., exposure to ethanol, Komada et al., 2017), ASD (associations with mercury, Mostafa et al., 2016), ADHD, intellectual development, and learning disabilities (implication of Pb exposure, Zhang et al., 2015). There is also a growing support that chemical-induced apoptosis during brain development may increase the risk of neurodegenerative diseases and/or to susceptibility to stressors later in life that are detrimental to normal neurological functions (Farina et al., 2013). Although there are chemical-specific differences in the molecular processes that trigger apoptotic signaling, the processes by which apoptotic cell death is actually induced, albeit mechanistically not completely explained, involve a finite as well as delimited set of processes, which can be directly or indirectly linked with neuroinflammation and ROS (Lein et al., 2018).

However, one of the challenges in measuring this "neuroinflammasome" is establishing the time-dependent course of the causative or deterministic action in terms of which are the predisposing causes and which are the subsequent effects. Paradoxically, this becomes a "chicken and egg" type of problem if considered just within childhood. From the get go, neuroapoptosis and neuroinflammation may be linked by reciprocal circular causation, where one may escalate the other. Fortunately though, interdisciplinary data on chemical and physical external exposures complemented by the characterization of downstream biological events causally linked to exposures show potential "signatures" of the biological response to cumulative exposure, thus hopefully allowing a more holistic evaluation of exposure-related effects *before birth*.

Again, research on air pollution would seem to lead the way: A groundbreaking study by Martens et al. (2017) assessed the association of prenatal exposure to particulate matter (PM) with newborn telomere length as reflected by cord blood and placental telomere length. Maternal residential

$PM_{2.5}$ exposure during pregnancy was estimated using a high-resolution spatial-temporal interpolation method. In distributed lag models, both cord blood and placental telomere length were associated with average weekly exposures to $PM_{2.5}$ during pregnancy, allowing the identification of critical sensitive exposure windows. Results showed that, in 641 newborns, cord blood and placental telomere length were significantly and inversely associated with $PM_{2.5}$ exposure during midgestation (weeks 12–25 for cord blood and weeks 15–27 for placenta). A 5-$\mu g/m^3$ increment in $PM_{2.5}$ exposure during the entire pregnancy was associated with 8.8% (95% CI, −14.1% to −3.1%) shorter cord blood leukocyte telomeres and 13.2% (95% CI, −19.3% to −6.7%) shorter placental telomere length. A host of confounders were controlled for: date of delivery, gestational age, maternal body mass index, maternal age, paternal age, newborn sex, newborn ethnicity, season of delivery, parity, maternal smoking status, maternal educational level, pregnancy complications, and ambient temperature. Therefore, mothers who were exposed to higher levels of $PM_{2.5}$ gave birth to newborns with shorter telomere length. Telomere length is a marker of biological aging that may provide a cellular record of exposures to oxidative stress and inflammation. Critically, telomere shortening at birth has been related to postnatal apoptosis and life expectancy. Consequently, these findings offer evidence suggestive of the fundamental relationship between neuroinflammation and apoptosis before birth and the potential effects in terms of neurointegrity after birth. Likely, these networks of links will in the future be exploited as a model for investigating other complex DNTs. Indeed, a starting hypothesis for future research is that a core general fundamental mediational role of neuroinflammation on apoptosis (via a relatively small set of similar mechanisms) might nevertheless show a differentiation of outcomes depending on the specific chemicals, where the latter effects might shape the integrity of the brain development and its correlate behavioral, cognitive and higher-order phenotypes at specific times during the life course trajectory.

8. Conclusions

DNT effects on the developing brain may vary along a continuum from minor, subtle subclinical to irreversible serious cognitive and behavioral deficits that are identified readily by parents and/or teachers. The detrimental effects may also worsen with the age of the child, thus selected neurocognitive tools ought to be useful for longitudinal studies, across educational backgrounds and expecting overlaps in the functional areas and tests affected. As shown by reviewed evidence on IQ, MRI and neurophysiological data, complex cognitive responses that may be affected include: attention and short-term memory, information processing speed and executive function, verbal abstraction, and visuospatial and motor skills. Deficits in auditory and vestibular responses and sound localization could also be expected, along with olfaction deficits.

Most of the neuroimaging studies already mentioned (specifically, using techniques such as electroencephalography/event-related potentials, brainstem auditory evoked potentials, structural and functional magnetic resonance imaging, and magnetic resonance spectroscopy) conducted in clinical and preclinical settings have all been reported to show a gradient of effects. However, one gap in our knowledge and current research is how all the effects identified are different factors which are part of different disease processes and at the same time part of a complex systemic mechanism (again, consider the example of Fig. 1). Future studies need to be designed so that this limitation can be overcome. Consistent with these observations, the National Institute of Environmental Health Sciences/National Institute of Health panel on outdoor air pollution (Block et al., 2012) indicated cognitive, neuropsychological (and possibly neuroimaging) screening of children as one of the priority target areas for future research advocating a multidisciplinary collaborative approach wherein brain-related development testing would have a prominent role. Screening of children at the highest risk for example the 6.4 million US children attending schools within 250 m of a major roadway should be prioritized (Kingsley et al., 2014).

It would seem relatively straightforward that health professionals, behavioral scientists psychologists and psychiatrists should each have a carved responsibility to address the particular issues associated with DNTs influence in the measure and modality in which the individuals are impacted. However, the diffuse nature of the neuroinflammation and the evolving neurodegenerative changes observed in exposed children requires not to rely on a single study or measure but rather to employ a weight of evidence approach incorporating current clinical, neurophysiological, radiological and epidemiological research as well as the results of animal exposure studies to single pollutants/mixtures/or pollutant components. Inflammatory biomarkers play a key role in the identification of children with positive volumetric and cognitive responses to their lifelong pollutant exposures (Calderón-Garcidueñas et al., 2012b) and since neuroinflammation/vascular damage/neurodegeneration go hand in hand (Calderón-Garcidueñas et al.,

2013a), definition of inflammatory/endothelial dysfunction biomarkers establishing an association between brain growth and developmental behavioral and neuropsychological outcomes are urgently needed.

The evidence accumulated so far clearly indicates that for urban children the neurocognitive effects of air pollution and other complex DNTs are serious, they are apparent across all populations, not just disadvantaged ones, and most importantly the observed neurocognitive impairments are potentially clinically relevant as early evidence of evolving neurodegenerative changes. Our ultimate goal should be to protect severely exposed children through multidimensional interventions guided by the exposome framework, yielding both impact and reach (i.e., on cognitive/behavioral, family participation, and modifiable lifestyle factors such as diet and micronutrient supply). Accordingly, the potential contribution of high-throughput omics technologies and techniques does not just lie in measuring "signatures" of the biological response to cumulative exposure, the increased validity and reliability in the evaluation of exposure-related health effects that omics could provide would be essential to refine strategic neuroprotective interventions such as the introduction of public health targeted programs to push for anti-inflammatory healthy foods (cocoa, broccoli, etc., see), community-based initiatives to minimize children's exposure, and intensive school-based monitoring of children exposure and outcomes.

References

Adamkiewicz, G., Zota, A.R., Fabian, M.P., Chahine, T., Julien, R., Spengler, J.D., Levy, J.I., 2011. Moving environmental justice indoors: understanding structural influences on residential exposure patterns in low-income communities. Am. J. Public Health 101 (Suppl. 1), S238–S245.

Amram, O., Abernethy, R., Brauer, M., Davies, H., Allen, R.W., 2011. Proximity of public elementary schools to major roads in Canadian urban areas. Int. J. Health Geogr. 10, 68. https://doi.org/10.1186/1476-072X-10-68.

Becerra, T.A., Wilhelm, M., Olsen, J., Cockburn, M., Ritz, B., 2013. Ambient air pollution and autism in Los Angeles county, California. Environ. Health Perspect. 121, 380–386.

Beckerman, R., Meltzer, J., Sola, A., Dunn, D., Wegmann, M., 1986. Brain-stem auditory response in Ondine's syndrome. Arch. Neurol. 43, 698–701.

Bellinger, D.C., 2012. A strategy for comparing the contributions of environmental chemicals and other risk factors to neurodevelopment of children. Environ. Health Perspect. 120, 501–507.

Bergin, I.L., Witzmann, F.A., 2013. Nanoparticle toxicity by the gastrointestinal route: evidence and knowledge gaps. Int. J. Biomed. Nanosci. Nanotechnol. 3 (1–2).

Biacabe, B., Chevallier, J.M., Avan, P., Bonfils, P., 2001. Functional anatomy of auditory brainstem nuclei: application to the anatomical basis of brainstem auditory evoked potentials. Auris Nasus Larynx 28, 85–94.

Block, M., Calderón-Garcidueñas, L., 2009. Air pollution: mechanisms of neuroinflammation and CNS disease. Trends Neurosci. 32 (9), 506–516.

Block, M.L., Elder, A., Auten, R.L., Bilbo, S.D., Chen, H., Chen, J.C., Cory-Slechta, D.A., Costa, D., Diaz-Sanchez, D., Dorman, D.C., Gold, D.R., Gray, K., Jeng, H.A., Kaufman, J.D., Kleinman, M.T., Kirshner, A., Lawler, C., Miller, D.S., Nadadur, S.S., Ritz, B., Semmens, E.O., Tonelli, L.H., Veronesi, B., Wright, R.O., Wright, R.J., 2012. The outdoor air pollution and brain health workshop. Neurotoxicology 33, 972–984.

Braak, H., Del Tredeci, K., 2011. The pathological process underlying Alzheimer's disease in individuals under thirty. Acta Neuropathol. 121, 171–181.

Brickman, A.M., Zahra, A., Muraskin, J., Steffner, J., Holland, C.M., Habeck, C., Borogovac, A., Ramos, M.A., Brown, T.R., Asllani, I., Stern, Y., 2009. Reduction in cerebral flow in areas appearing as white matter hyperintensities on magnetic resonance imaging. Psychiatry Res. 172, 117–120.

Brockmeyer, S., D'Angiulli, A., 2016. How air pollution alters brain development: the role of neuroinflammation. Transl. Neurosci. 7 (1), 24–30.

Brook, R.D., Rajagopalan, S., Pope 3rd, C.A., Brook, J.R., Bhatnagar, A., Diez-Roux, A.V., Holguin, F., Hong, Y., Luepker, R.V., Mittleman, M.A., Peters, A., Siscovick, D., Smith Jr., S.C., Whitsel, L., Kaufman, J.D., American Heart Association Council on Epidemiology and Prevention, Council on the Kidney in Cardiovascular Disease, and Council on Nutrition, Physical Activity and Metabolism, 2010. Particulate matter air pollution and cardiovascular disease: an update to the scientific statement from the American Heart Association. Circulation 121, 2331–2378.

Brun, E., Carrière, M., Mabondzo, A., 2012. In vitro evidence of dysregulation of blood-brain barrier function after acute and repeated/long term exposure to TiO2 nanoparticles. Biomaterials 33, 886–896.

Bunce, D., Anstey, K.J., Cherbuin, N., Burns, R., Christensen, H., Wen, W., Sachdev, P.S., 2010. Cognitive deficits are associated with frontal and temporal lobe white matter lesions in middle age adults living in the community. PLoS One 5, e13567.

Calderón-Garcidueñas, L., Osornio-Velazquez, A., Bravo-Alvarez, H., Delgado-Chavez, R., Barrios-Marquez, R., 1992. Histopathological changes of the nasal mucosa in southwest Metropolitan Mexico City inhabitants. Am. J. Pathol. 140, 225–232.

Calderón-Garcidueñas, L., Valencia-Salazar, G., Rodríguez-Alcaraz, A., Gambling, T.M., García, R., Osnaya, N., Villarreal-Calderón, A., Devlin, R.B., Carson, J.L., 2001. Ultrastructural nasal pathology in children chronically and sequentially exposed to air pollutants. Am. J. Respir. Cell Mol. Biol. 24, 132–138.

Calderón-Garcidueñas, L., Mora-Tiscareño, A., Fordham, L.A., Valencia-Salazar, G., Chung, C.J., Rodriguez-Alcaraz, A., Paredes, R., Variakojis, D., Villarreal-Calderon, A., Flores-Camacho, L., Antunez-Solis, A., Henriquez-Roldan, C., Hazucha, M.J., 2003. Respiratory damage in children exposed to urban pollution. Pediatr. Pulmonol. 36, 148–161.

Calderón-Garcidueñas, L., Vincent, R., Mora-Tiscareño, A., Franco-Lira, M., Henríquez-Roldán, C., Barragán-Mejía, G., Garrido-García, L., Camacho-Reyes, L., Valencia-Salazar, G., Paredes, R., Romero, L., Osnaya, H., Villarreal-Calderón, R., Torres-Jardón, R., Hazucha, M.J., Reed, W., 2007. Elevated plasma endothelin-1 and pulmonary arterial pressure in children exposed to air pollution. Environ. Health Perspect. 115, 1248–1253.

Calderón-Garcidueñas, L., Mora-Tiscareño, A., Ontiveros, E., Gomez-Garza, G., Barragán-Mejía, G., Broadway, J., Chapman, S., Valencia-Salazar, G., Jewells, V., Maronpot, R.R., Henríquez-Roldán, C., Pérez-Guillé, B., Torres-Jardón, R., Herrit, L., Brooks, D., Monroy, M.E., González-Maciel, A., Reynoso-Robles, R., Villarreal-Calderón, R., Solt, A.C., Engle, R.W., 2008. Air pollution, cognitive deficits and brain abnormalities: a pilot study with children and dogs. Brain Cogn. 68, 117–127.

Calderón-Garcidueñas, L., Franco-Lira, M., Henriquez-Roldan, C., Osnaya, N., Gonzalez-Maciel, A., Reynoso-Robles, R., Villarreal-Calderon, R., Herritt, L., Brooks, D., Keefe, S., Palacios-Moreno, J., Villarreal-Calderon, R., Torres-Jardon, R., Medina-Cortina, H., Delgado Chavez, R., Aiello-Mora, M., Maronpot, R.R., Doty, R.L., 2010. Urban air pollution: influences on olfactory function and pathology in exposed children and young adults. Exp. Toxicol. Pathol. 62, 91–102.

Calderón-Garcidueñas, L., Engle, R., Mora-Tiscareño, A., Styner, M., Gómez-Garza, G., Zhu, H., Jewells, V., Torres-Jardón, R., Romero, L., Monroy-Acosta, M.E., Bryant, C., González-González, L.O., Medina-Cortina, H., D'Angiulli, A., 2011a. Exposure to severe urban air pollution influences cognitive outcomes, brain volume and systemic inflammation in clinically healthy children. Brain Cogn. 77, 345–355.

Calderón-Garcidueñas, L., D'Angiulli, A., Kulesza, R.J., Torres-Jardón, R., Romero, L., Keefe, S., Herritt, L., Brooks, D.M., Avila-Ramirez, J., Delgado-Chávez, R., Medina-Cortina, H., González-González, L.O., 2011b. Air pollution is associated with brainstem auditory nuclei pathology and delayed brainstem auditory evoked potentials. Int. J. Dev. Neurosci. 29, 365–375.

Calderón-Garcidueñas, L., Kavanaugh, M., Block, M., D'Angiulli, A., Delgado-Chávez, R., Torres-Jardón, R., González-Maciel, A., Reynoso-Robles, R., Villarreal-Calderon, R., Guo, R., Hua, Z., Zhu, H., Perry, G., Diaz, P., 2012a. Neuroinflammation, Alzheimer's disease-associated pathology and down regulation of the prion-related protein in air pollution exposed children and young adults. J. Alzheimers Dis. 28, 93–107.

Calderón-Garcidueñas, L., Mora-Tiscareño, A., Styner, M., Gómez-Garza, G., Zhu, H., Torres-Jardón, R., Carlos, E., Solorio-López, E., Medina-Cortina, H., Kavanaugh, M., D'Angiulli, A., 2012b. White matter hyperintensities, systemic inflammation, brain growth, and cognitive functions in children exposed to air pollution. J. Alzheimers Dis. 31, 183–191.

Calderón-Garcidueñas, L., Franco-Lira, M., Mora-Tiscareño, A., Medina-Cortina, H., Torres-Jardón, R., Kavanaugh, M., 2013a. Early Alzheimer's and Parkinson's disease pathology in urban children: friend versus foe responses—it is time to face the evidence. Biomed. Res. Int. 2013, 161687.

Calderón-Garcidueñas, L., Styner, M., D'Angiulli, A., 2013b. Diffusion Tensor Imaging MRI Analysis of Subgroups of Mexican Children. Department of Biomedical & Pharmaceutical Sciences. The University of Montana-Missoula, Montana, USA. Unpublished data.

Calderón-Garcidueñas, L., González-González, L.O., Kulesza, R.J., Fech, T.M., Pérez-Guillé, G., Jiménez-BravoLuna, M.A., Soriano-Rosales, R.E., Solorio, E., Miramontes-Higuer, J.J., Gómez-Maqueo Chew, A., Bernal-Morú, A.F., Mukherjee, P.S., Torres-Jardón, R., Mills, P.C., Wilson, W.J., Pérez-Guillé, B., D'Angiulli, A., 2017. Exposures to fine particulate matter (PM2.5) and ozone above USA standards are associated with auditory brainstem dysmorphology and abnormal auditory brainstem evoked potentials in healthy young dogs. Environ. Res. 158, 324–332.

Carmichael, O., Schwarz, C., Drucker, D., Fletcher, E., Harvey, D., Beckett, L., Jack Jr., C.R., Weiner, M., Decarli, C., 2010. Longitudinal changes in white matter disease and cognition in the first year of the Alzheimer disease neuroimaging initiative. Arch. Neurol. 67, 1370–1378.

Caviness Jr., V.S., Lange, N.T., Makris, N., Herbert, M.R., Kennedy, D.N., 1999. MRI-based brain volumetrics: emergence of a developmental brain science. Brain Dev. 21, 289–295.

Clark, I.A., Alleva, L.M., Vissel, B., 2010. The role of TNF in brain dysfunction and disease. Pharmacol. Ther. 128, 519–548.

Clark-Reyna, S.E., Grineski, S.E., Collins, T.W., 2016. Residential exposure to air toxics is linked to lower grade point averages among school children in El Paso, Texas, USA. Popul. Environ. 37 (3), 319–340.

Corburn, J., 2017. Concepts for studying urban environmental justice. Curr. Environ. Health Rep. 4, 61–67. https://doi.org/10.1007/s40572-017-0123-6.
Costa, L.G., Chang, Y.C., Cole, T.B., 2017. Developmental neurotoxicity of traffic related-air pollution: focus on autism. Curr. Environ. Health Rep. 4 (2), 156–165.
D'Angiulli, A., 2018. Severe urban outdoor air pollution and children's structural and functional brain development, from evidence to precautionary strategic action. Front. Public Health 6, 95.
D'Angiulli, A., 2015. Developmental cognitive neuroscience, air pollution and social inequality. In: Paper Presented for the American Society for Investigative Pathology at the Aannual Meeting of Experimental Biology, March 29, 2015, Boston, MA, USA.
De Felice, A., Ricceri, L., Venerosi, A., Chiarotti, F., Calamandrei, G., 2015. Multifactorial origin of neurodevelopmental disorders: approaches to understanding complex etiologies. Toxics 3 (1), 89–129.
Debette, S., Markus, H.S., 2010. The clinical importance of white matter hyperintensities on brain magnetic resonance imaging: systematic review and meta-analysis. BMJ 341, c3666.
Everett-Jones, S., Smith, A.M., Wheeler, L.S., McManus, T., 2010. School policies and practices that improve indoor air quality. J. Sch. Health 80, 280–286.
Farina, M., Avila, D.S., Da Rocha, J.B.T., Aschner, M., 2013. Metals, oxidative stress and neurodegeneration: a focus on iron, manganese and mercury. Neurochem. Int. 62 (5), 575–594.
Fonken, L.K., Xu, X., Weil, Z.M., Chen, G., Sun, Q., Rajagopalan, S., Nelson, R.J., 2011. Air pollution impairs cognition, provokes depressive-like behaviors and alters hippocampal cytokine expression and morphology. Mol. Psychiatry 16, 987–995.
Froehlich, T.E., Anixt, J.S., Loe, I.M., Chirdkiatgumchai, V., Kuan, L., Gilman, R.C., 2011. Update on environmental risk factors for attention-deficit/hyperactivity disorder. Curr. Psychiatry Rep. 13, 333–344.
Gläscher, J., Tranel, D., Paul, L.K., Rudrauf, D., Rorden, C., Hornaday, A., Grabowski, T., Damasio, H., Adolphs, R., 2009. Lesion mapping of cognitive abilities linked to intelligence. Neuron 61, 681–691.
Gläscher, J., Rudrauf, D., Colom, R., Paul, L.K., Tranel, D., Damasio, H., Adolphs, R., 2010. Distributed neural system for general intelligence revealed by lesion mapping. PNAS 107, 4705–4709.
Gouw, A.A., Seewann, A., van der Flier, W.M., Barkhof, F., Rozemuller, A.M., Scheltens, P., Geurts, J.J., 2011. Heterogeneity of small vessel disease: a systematic review of MRI and histopathology correlations. J. Neurol. Neurosurg. Psychiatry 82 (2), 126–135.
Grandjean, P., Landrigan, P.J., 2014. Neurobehavioural effects of developmental toxicity. Lancet Neurol. 13, 330–338.
Guo, L., Zhu, N., Guo, Z., Li, G.K., Chen, C., Sang, N., Yao, Q.C., 2012. Particulate matter (PM10) exposure induces endothelial dysfunction and inflammation in rat brain. J. Hazard. Mater. 213–214, 28–37.
Guxens, M., Sunyer, J., 2012. A review of epidemiological studies on neuropsychological effects of air pollution. Swiss Med. Wkly. 141, w13322. https://doi.org/10.4414/smw.2011.13322.
Haier, R.J., Jung, R.E., Yeo, R.A., Head, K., Alkire, M.T., 2004. Structural brain variation and general intelligence. Neuroimage 23, 425–433.
Haynes, E.N., Chen, A., Ryan, P., Succop, P., Wright, J., Dietrich, K.N., 2011. Exposure to airborne metals and particulate matter and risk for youth adjudicated for criminal activity. Environ. Res. 111, 1243–1248.
Heilman, K.M., Watson, R.T., Valenstein, E., Goldberg, M.E., 1987. Attention: behaviour and neural mechanisms. In: Plum, F., Mountcastle, V.B., Geiger, S.T. (Eds.), The

Handbook of Physiology. Section 1: The Nervous System. Volume V. Higher Functions of the Brain Part 2. American Physiological Society, Bethesda, MD, pp. 461–481.
Ho, Y.S., Yang, X., Yeung, S.C., Chiu, K., Lau, C.F., Tsang, A.W., Mak, J.C., Chang, R.C., 2012. Cigarette smoking accelerated brain aging and induced pre-Alzheimer-like neuropathology in rats. PLoS One 7 (5), e36752.
Huang, Y., Zhang, W., Lin, L., Feng, L., Chen, F., Wei, W., Zhao, X., Guo, W., Li, J., Yin, W., Li, L., 2010. Is endothelial dysfunction of cerebral small vessel responsible for white matter lesions after chronic cerebral hypoperfusion in rats? J. Neurol. Sci. 299, 72–80.
Hukuhara, T., 1988. Organization of the brainstem neural mechanisms for generation of respiratory rhythm—current problems. Jpn. J. Physiol. 38, 753–776.
Jefferson, A.L., Massaro, J.M., Wolf, P.A., Seshadri, S., Au, R., Vasan, R.S., Larson, M.G., Meigs, J.B., Keaney, J.F., Lipinska, I., Kathiresan, S., Benjamin, E.J., Decarli, C., 2007. Inflammatory markers are associated with total brain volume: the Framingham Heart Study. Neurology 68, 1032–1038.
Jian, H., Yi-Fang, W., Qi, L., Xiai-Song, H., Gui-Yun, Z., 2012. Cerebral blood flow and metabolic changes in hippocampal regions of a modified rat model with chronic cerebral hypoperfusion. Acta Neurol. Belg. 113, 313–317.
Jung, R.E., Haier, R.J., 2007. The parieto-frontal integration theory (P-FIT) of intelligence: converging neuroimaging evidence. Behav. Brain Sci. 30, 135–154.
King, M.D., Bearman, P.S., 2011. Socioeconomic status and the increased prevalence of autism in California. Am. Sociol. Rev. 76, 320–346.
Kingsley, S.L., Eliot, M.N., Carlson, L., Finn, J., Macintosh, D.L., Suh, H.H., Wellenius, G.A., 2014. Proximity of US schools to major roadways: a nationwide assessment. J. Expo. Sci. Environ. Epidemiol. 24, 53–259.
Komada, M., Hara, N., Kawachi, S., Kawachi, K., Kagawa, N., Nagao, T., Ikeda, Y., 2017. Mechanisms underlying neuro-inflammation and neurodevelopmental toxicity in the mouse neocortex following prenatal exposure to ethanol. Sci. Rep. 7 (1), 4934.
Kulesza Jr., R.J., Lukose, R., Stevens, L.V., 2011. Malformation of the human superior olive in autistic spectrum disorders. Brain Res. 1367, 360–371.
Kulesza, R.J., Mangunay, K., 2008. Morphological features of the medial superior olive in autism. Brain Res. 1200, 132–137.
Larsson, M., Weiss, B., Janson, S., Sundell, J., Bornehag, C.G., 2009. Associations between indoor environmental factors and parental-reported autistic spectrum disorders in children 6–8 years of age. Neurotoxicology 30, 822–831.
Lein, P.J., Supasai, S., Guignet, M., 2018. Apoptosis as a mechanism of developmental neurotoxicity. In: Slikker Jr., W., Paule, M.G., Wang, C. (Eds.), Handbook of Developmental Neurotoxicology, second ed. Academic Press Elsevier Inc., pp. 91–112
Lelieveld, J., Evans, J.S., Fnais, M., Giannadaki, D., Pozzer, A., 2015. The contribution of outdoor air pollution sources to premature mortality on a global scale. Nature 525 (7569), 367–371.
Levesque, S., Surace, M.J., McDonald, J., Block, M.L., 2011. Air pollution and the brain: subchronic diesel exhaust exposure causes neuroinflammation and elevates early markers of neurodegenerative disease. J. Neuroinflammation 8, 105. https://doi.org/10.1186/1742-2094-8-105.
Levesque, S., Taetzsch, T., Lull, M.E., Johnson, J.A., McGraw, C., Block, M.L., 2013. The role of MAC1 in diesel exhaust particle-induced microglial activation and loss of dopaminergic neuron function. J. Neurochem. 125, 756–765.
Linsinger, T.P., Chaudhry, Q., Dehalu, V., Delahaut, P., Dudkiewicz, A., Grombe, R., von der Kammer, F., Larsen, E.H., Legros, S., Loeschner, K., Peters, R., Ramsch, R., Roebbe, G., Tiede, K., Weigel, S., 2013. Validation of methods for the detection and quantification of engineered nanoparticles in food. Food Chem. 138, 1959–1966.

Litscher, G., Schwarz, G., Reimann, R., 1996. Abnormal brainstem auditory evoked potentials in a girl with the central alveolar hypoventilation syndrome. Int. J. Neurosci. 87, 113–117.

Liu, J., 2011. Early risk factors for violence: conceptualization, review of the evidence and implications. Agress. Violent Behav. 16, 63–73.

Lopez Gonzalez, I., Garcia-Esparcia, P., Llorens, F., Ferrer, I., 2016. Genetic and transcriptomic profiles of inflammation in neurodegenerative diseases: Alzheimer, Parkinson, Creutzfeldt-Jakob and Tauopathies. Int. J. Mol. Sci. 17 (2).

Lukose, R., Kevin, B., Barber, C.M., Kulesza Jr., R.J., 2013. Quantification of the stapedial reflex reveals delayed responses in autism. Autism Res. 6 (5), 344–353.

Lyall, K., Schmidt, R.J., Hertz-Picciotto, I., 2014. Maternal lifestyle and environmental risk factors for autism spectrum disorders. Int. J. Epidemiol. 43, 443–464.

Maffini, M.V., Neltner, T.G., 2015. Brain drain: the cost of neglected responsibilities in evaluating cumulative effects of environmental chemicals. J. Epidemiol. Community Health 69, 496–499.

Martens, D.S., Cox, B., Janssen, B.G., Clemente, D.B.P., Gasparrini, A., Vanpoucke, C., Lefebvre, W., Roels, H.A., Plusquin, M., Nawrot, T.S., 2017. Prenatal air pollution and newborns' predisposition to accelerated biological aging. JAMA Pediatr. 171 (12), 1160–1167.

Mesulam, M., 2000. Brain, mind, and the evolution of connectivity. Brain Cogn. 42 (1), 4–6.

Miller, M.W., 1986. Effects of alcohol on the generation and migration of cerebral cortical neurons. Science 233, 1308–1311.

Mohai, P., Kweon, B.S., Lee, S., Ard, K., 2011. Air pollution around schools is linked to poorer student health and academic performance. Health Aff. 30, 852–862.

Moldes, O., Sobrino, T., Millan, M., Castellanos, M., Perez de la Ossa, N., Leira, R., Serena, J., Vivancos, J., Davalos, A., Castillo, J., 2008. High serum levels of endothelin-1 predict severe cerebral edema in patients with acute ischemic stroke treated with t-PA. Stroke 39, 2006–2010.

Moore, C.F., 2009. Children and Pollution: Why Scientists Disagree. Oxford University Press, New York, NY.

Morris, Z., Whiteley, W.N., Longstreth Jr., W.T., Weber, F., Lee, Y.C., Alphs, H., Ladd, S.C., Warlow, C., Wardlow, J.M., Salman, A., 2009. Incidental findings on brain magnetic resonance imaging: systematic review and meta-analysis. BMJ 339, b3016.

Mostafa, G.A., Bjørklund, G., Urbina, M.A., Al-ayadhi, L.Y., 2016. The levels of blood mercury and inflammatory-related neuropeptides in the serum are correlated in children with autism spectrum disorder. Metab. Brain Dis. 31 (3), 593–599.

Murata, K., Weihe, P., Budtz-Jorgensen, E., Jorgensen, P.J., Grandjean, P., 2004. Delayed brainstem auditory evoked potential latencies in 14-year-old children exposed to methylmercury. J. Pediatr. 144 (2), 177–183.

Murray, M.E., Senjem, M.L., Petersen, R.C., Hollman, J.H., Preboske, G.M., Weigand, S.D., Knopman, D.S., Ferman, D.S., Dickson, D.W., Jack Jr., C.R., 2010. Functional impact of white matter hyperintensities in cognitively normal elderly subjects. Arch. Neurol. 67, 1379–1385.

NAS, 2009. Science and Decisions: Advancing Risk Assessment. The National Academies Press, Washington, DC, pp. 213–239.

Neal, A.P., Worley, P.F., Guilarte, T.R., 2011. Lead exposure during synaptogenesis alters NMDA receptor targeting via NMDA receptor inhibition. Neurotoxicology 32, 281–289.

O'Leary, J.D., Warner, D.O., 2017. What do recent human studies tell us about the association between anaesthesia in young children and neurodevelopmental outcomes? Br. J. Anaesth. 119 (3), 458–464.

Oppenheim, H.A., Lucero, J., Guyot, A.C., Herbert, L.M., McDonald, J.D., Mabondzo, A., Lund, A.K., 2013. Exposure to vehicle emissions results in altered blood brain barrier permeability and expression of matrix metalloproteinases and tight junction proteins in mice. Part. Fibre Toxicol. 10, 62. https://doi.org/10.1186/1743-8977-10-62.

Organisation for Economic Co-operation and Development, 2016. The Economic Consequences of Outdoor Air Pollution. OECD Publishing, Paris. https://doi.org/10.1787/9789264257474-en.

Oudin, A., Bråbäck, L., Åström, D.O., Strömgren, M., Forsberg, B., 2016. Association between neighbourhood air pollution concentrations and dispensed medication for psychiatric disorders in a large longitudinal cohort of Swedish children and adolescents. BMJ Open 6 (6), e010004.

Paul, D., Cowan, A.E., Ge, S., Pachter, J.S., 2013. Novel 3D analysis of Claudin-5 revelas significant endothelial heterogeneity among CNS microvessels. Microvasc. Res. 86, 1–10.

Perera, F.P., Li, Z., Whyatt, R., Hoepner, L., Wang, S., Camann, D., Rauh, V., 2009. Prenatal airborne polycyclic aromatic hydrocarbon exposure and child IQ at age 5 years. Pediatrics 124, e195–e202.

Pujol, J., Martínez-Vilavella, G., Macià, D., Fenoll, R., Alvarez-Pedrerol, M., Rivas, I., Forns, J., Blanco-Hinojo, L., Capellades, J., Querol, X., Deus, J., 2016. Traffic pollution exposure is associated with altered brain connectivity in school children. Neuroimage 129, 175–184.

Rai, D., Lewis, G., Lundberg, M., Araya, R., Svensson, A., Dalman, C., Carpenter, P., Magnusson, C., 2012. Parental socioeconomic status and risk of offspring autism spectrum disorders in a Swedish population-based study. J. Am. Acad. Child. Adolesc. Psychiatry 51, 467–476.e6.

Rivas-Arancibia, S., Guevara-Guzman, R., Lopez-Vidal, Y., Rodriguez-Martinez, E., Zanardo-Gomes, M., Angoa-Perez, M., Raisman-Vozari, R., 2010. Oxidative stress caused by ozone exposure induces loss of brain repair in the hippocampus of adult rats. Toxicol. Sci. 113, 187–197.

Roher, A.E., Debbins, J.P., Malek-Ahmadi, M., Chen, K., Pipe, J.G., Maze, S., Belden, C., Maarouf, C.L., Thiyyagura, P., Mo, H., Hunter, J.M., Kokjohn, T.A., Walker, T.A., Walker, D.G., Kruchowsky, J.C., Belohlavek, M., Sabbagh, M.N., Beach, T.G., 2012. Cerebral blood flow in Alzheimer's disease. Vasc. Health Risk Manag. 8, 599–611.

Roy, T.S., Seidler, F.J., Slotkin, T.A., 2004. Morphologic effects of subtoxic neonatal chlorpyrifos exposure in developing rat brain: regionally selective alterations in neurons and glia. Brain Res. Dev. Brain Res. 148, 197–206.

Rubin, B.S., 2011. Bisphenol A: an endocrine disruptor with widespread exposure and multiple effects. J. Steroid Biochem. Mol. Biol. 127, 27–34.

Salgado, J.V., Costa-Silva, M., Malloy-Diniz, L.F., Siqueira, J.M., Teixeira, A.L., 2007. Prefrontal cognitive dysfunction following brainstem lesion. Clin. Neurol. Neurosurg. 109 (4), 379–382.

Salonia, R., Empey, P.E., Poloyac, S.M., Wisniewski, S.R., Klamerus, M., Ozawa, H., Wagner, A.K., Ruppel, R., Bell, M.J., Feldman, K., Adelson, P.D., Clark, R.S., Kochanek, P.M., 2010. Endothelin-1 is increased in cerebrospinal fluid and associated with unfavorable outcomes in children after severe traumatic brain injury. J. Neurotrauma 27, 1819–1825.

Sampson, N., 2012. Environmental justice at school: understanding research, policy and practice to improve our children's health. J. Sch. Health 82, 246–252.

Schibli, K., Wong, K., Hedayati, N., D'Angiulli, A., 2017. Attending and learning under socioeconomic disadvantage: developmental cognitive and social neuroscience of resilience and vulnerability. Ann. N. Y. Acad. Sci. 1396, 19–38.

Schröder, N., Silva-Figueiredo, L., Martins de Lima, M.N., 2013. Role of brain iron accumulation in cognitive dysfunction: evidence from animal models and human studies. J. Alzheimers Dis. 34, 797–812.

Schwarz, G., Litscher, G., Pfurtscheller, G., List, W., 1990. Multimodal evozierte Potentiale und Herzratenvariabiliat bei komatosen Patienten-Teil 3: Elektrophysiologische Befunde bei entziindlichen Erhankungen des Gehirns. Zeirschrijr EEG-EMG 21, 35–41.

Schwarz, G., Litscher, G., Pfurtscheller, G., Rumpl, E., Lechner, A., Offenbacher, J., List, W.F., 1994. "Akustisch evozierte Hirnstammpotentiale bei schwerer Enzephalitis", Beirruge zurAnuesthesiologie. Inrensiv- und-fullmedizin 43 (17), 1–180.

Schwarz, G., Litscher, G., Rumpl, E., Pfurtscheller, G., Reimann, R., 1996. Brainstem auditory evoked potentials in respiratory insufficiency following encephalitis. Int. J. Neurosci. 84, 3544.

Siddique, S., Banerjee, M., Ray, M.R., Lahiri, T., 2011. Attention-deficit hyperactivity disorder in children chronically exposed to high level of vehicular pollution. Eur. J. Pediatr. 170, 923–929.

Silbert, L.C., Howieson, D.B., Dodge, H., Kaye, J.A., 2009. Cognitive impairment risk: white matter hyperintensity progression matters. Neurology 73, 120–125.

Singhal, G., Jaehne, E.J., Corrigan, F., Toben, C., Baune, B.T., 2014. Inflammasomes in neuroinflammation and changes in brain function: a focused review. Front. Neurosci. 8, 315. https://doi.org/10.3389/fnins.2014.00315.

Slotkin, T.A., Seidler, F.J., 2009. Oxidative and excitatory mechanisms of developmental neurotoxicity: transcriptional profiles for chlorpyrifos, diazinon, dieldrin, and divalent nickel in PC12 cells. Environ. Health Perspect. 117, 587–596.

Suglia, S.F., Gryparis, A., Wright, R.O., Schwartz, J., Wright, R.J., 2008. Association of black carbon with cognition among children in a prospective birth cohort study. Am. J. Epidemiol. 167, 280–286.

Takahashi, Y., Mizuo, K., Shinkai, Y., Oshio, S., Takeda, K., 2010. Prenatal exposure to titanium dioxide nanoparticles increases dopamine levels in the prefrontal cortex and neostriatum of mice. J. Toxicol. Sci. 35, 749–756.

Tansey, M.G., Goldberg, M.S., 2010. Neuroinflammation in Parkinson's disease: its role in neuronal death and implications for therapeutic intervention. Neurobiol. Dis. 37, 510–518.

Umezawa, M., Tainaka, H., Kawashima, N., Shimizu, M., Takeda, K., 2012. Effect of fetal exposure to titanium dioxide nanoparticle on brain development-brain region information. J. Toxicol. Sci. 37, 1247–1252.

United Nations Children's Funds (UNICEF), 2016. Clean Air for Children. The Impact of Air Pollution on Children. UNICEF, New York, NY.

van Donkelaar, A., Martin, R.V., Brauer, M., Hsu, N.C., Kahn, R.A., Levy, R.C., Lyapustin, A., Sayer, A.M., Winker, D.M., 2016. Global estimates of fine particulate matter using a combined geophysical-statistical method with information from satellites, models, and monitors. Environ. Sci. Technol. 50 (7), 3762–3772. https://doi.org/10.1021/acs.est.5b05833.

Vanos, J.K., 2015. Children's health and vulnerability in outdoor microclimates: a comprehensive review. Environ. Int. 76, 1–15. https://doi.org/10.1016/j.envint.2014.11.016.

Villarreal-Calderón, A., Acuña, H., Villarreal-Calderón, J., Garduño, M., Henríquez-Roldán, C.F., Calderón-Garcidueñas, L., Valencia-Salazar, G., 2002. Assessment of physical education time and after-school outdoor time in elementary and middle school students in South Mexico City: the dilemma between physical fitness and the adverse health effects of outdoor pollutant exposure. Arch. Environ. Health 57, 450–460.

Volk, H.E., Lurmann, F., Penfold, B., Hertz-Picciotto, I., McConnell, R., 2013. Traffic-related air pollution, particulate matter, and autism. JAMA Psychiat. 70, 71–77.

Voordecker, P., Brunko, E., de Beyl, Z., 1988. Selective unilateral absence or attenuation of wave V of brainstem auditory evoked potentials with intrinsic brainstem lesions. Arch. Neurol. 45, 1272–1276.
Vrijheid, M., 2014. The exposome: a new paradigm to study the impact of environment on health. Thorax 69, 876–878.
Wallin, A., Fladby, T., 2010. Do white matter hyperintensities on MRI matter clinically? BMJ 341, c3400. https://doi.org/10.1136/bmj.c3400.
Wild, C.P., 2012. The exposome: from concept to utility. Int. J. Epidemiol. 41, 24–32.
Woolgar, A., Parr, A., Cusack, R., Thompson, R., Mimmo-Smith, I., Torralva, T., Roca, M., Antoun, N., Manes, F., Duncan, J., 2010. Fluid intelligence loss linked to restricted regions of damage within frontal and parietal cortex. PNAS 107 (33), 14899–14902.
World Health Organization, 2016. Concentration of Particulate Matter with an Aerodynamic Diameter of 10 μm or Less (PM10) in Nearly 3000 Urban Areas, 2008–2015. Accessed online at: http://gamapserver.who.int/mapLibrary/Files/Maps/Global_pm10_cities_2008_2015.png, on June 22 2017.
Xu, H., Stamova, B., Jickling, G., Tian, Y., Zhan, X., Ander, B.P., Liu, D., Turner, R., Rosand, J., Goldstein, L.B., Furie, K.L., 2010. Distinctive RNA expression profiles in blood associated with white matter hyperintensities in brain. Stroke 41 (12), 2744–2749.
Yadav, A., Saini, V., Arora, S., 2010. MCP-1: chemoattractant with a role beyond immunity: a review. Clin. Chim. Acta 411, 1570–1579.
Zhang, X., Lv, C.C., Tian, J., Miao, R.J., Hertz-Picciotto, I., Qi, L., 2010. Prenatal and perinatal risk factors for autism in China. J. Autism Dev. Disord. 40, 1311–1321.
Zhang, X.-L., Guariglia, S.R., McGlothan, J.L., Stansfield, K.H., Stanton, P.K., Guilarte, T.R., 2015. Presynaptic mechanisms of lead neurotoxicity: effects on vesicular release, vesicle clustering and mitochondria number. PLoS One 10 (5), e0127461.
Zlokovic, B.V., 2008. The blood-brain-barrier in health and chronic neurodegenerative diseases. Neuron 57, 178–201.

Further reading

Caldera-Alvarado, G., Khan, D.A., Defina, L.F., Pieper, A., Brown, E.S., 2013. Relationship between asthma and cognition: the Cooper Center Longitudinal Study. Allergy 68, 545–548.
Charles, M.A., 2013. Developmental origins of adult health and disease: an important concept for social inequalities in health. Rev. Epidemiol. Sante Publique 61, S133–S138. https://doi.org/10.1016/j.respe.2013.05.013.
Clausen, G., Bekö, G., Corsi, R.L., Gunnarsen, L., Nazaroff, W.W., Olesen, B.W., Sigsgaard, T., Sundell, J., Toftum, J., Weschler, C.J., 2011. Reflections on the state of research: indoor environmental quality. Indoor Air 21, 219–230.
Wright, R.J., 2011. Psychological stress: a social pollutant that may enhance environmental risk. Am. J. Respir. Crit. Care Med. 184, 752–754.

CHAPTER NINE

Role of neuroinflammation in ethanol neurotoxicity

Consuelo Guerri[a,*], María Pascual[a,b]
[a]Department of Molecular and Cellular Pathology of Alcohol, Príncipe Felipe Research Center, Valencia, Spain
[b]Department of Physiology, School of Medicine and Dentistry, University of Valencia, Valencia, Spain
*Corresponding author: e-mail address: cguerri@cipf.es

Contents

1. Introduction

Excessive alcohol consumption is the third leading preventable cause of death in many countries that is associated with multiple adverse health consequences, including liver cirrhosis, various cancers, unintentional injuries, addiction, brain damage and violence. The World Health Organization (WHO) estimates that alcohol causes approximately 3.3 million

Advances in Neurotoxicology, Volume 3
ISSN 2468-7480
https://doi.org/10.1016/bs.ant.2018.10.009

deaths every year (or 5.9% of all deaths), and that 5.1% of the global burden of disease is attributable to alcohol consumption (WHO, 2018). Yet despite the health consequences of alcohol abuse, the brain is an important target of alcohol effects, and alcohol is one of the commonest drugs and neurotoxic compounds that can lead to brain damage and behavioral dysfunctions. Even uncomplicated alcoholics, with no specific neurological or hepatic problems, show signs of regional brain damage and cognitive dysfunction. Experimental studies have further demonstrated that alcohol is a neurotoxic compound, and heavy alcohol consumption can induce structural and functional brain alterations, as well as cognitive dysfunction (Harper and Matsumoto, 2005). Neuroimaging and postmortem studies have confirmed reduced brain weight in alcoholics, along with effects associated with loss of gray and white matter (Agartz et al., 2003; Sutherland et al., 2014).

Nevertheless, one of the most detrimental effects of the neurotoxic effects of alcohol is its action on the developing brain. Ethanol is considered one of the commonest substances that impacts brain ontogeny, and drinking alcohol during pregnancy can produce a wide range of cognitive, behavioral and physical anomalies, and is one of the leading preventable causes of birth defects and neurodevelopmental disorders (Guerri et al., 2009; Moore et al., 2014; Warren et al., 2011). However, prenatal alcohol exposure effects lie in a continuum of physical anomalies and behavioral and cognitive deficits, and the term fetal alcohol spectrum disorders (FASDs) have been adopted (Sokol et al., 2003).

Studies conducted in the last decade have also demonstrated that the brain is not completely mature until the age of 21–22 years, and that adolescence is an important brain maturation stage in which plastic and dynamic processes drive adolescent brain development by creating flexibility that allows the brain to refine itself, and to sharpen its functions for specific demands (Giedd, 2004). These dynamic processes take place in some immature regions, such as the limbic system involved in self-control, decision making, emotions and risk-taking behaviors, and in the frontal lobe implicated in cognitive processes during adolescence (Spear, 2007). A high prevalence of binge ethanol drinking in the younger generation, 14–21 years, is a particular concern in many countries given the neurotoxic effects that alcohol has on the developing adolescent brain. This alcohol consumption pattern triggers high ethanol levels in the brain by disrupting brain plasticity and causing important structural and functional changes in immature brain areas (prefrontal cortex, limbic system), which result in long-term cognitive and behavioral deficits (Crews et al., 2007; Montesinos et al., 2016b).

The mechanisms involved in the neurotoxic effects of ethanol in the brain are not completely understood and the following have been classically implicated: excitotoxicity, free radical formation and oxidative stress, nutritional deficiencies, growth factor dysfunctions, increased cell death and DNA damage (Alfonso-Loeches and Guerri, 2011; Harper and Matsumoto, 2005; Kruman et al., 2012).

Studies conducted in the last decade evidence the new role of the immune system in the brain's function in health and disease. These studies have demonstrated that the immune response takes place in the central nervous system (CNS), and inflammation can occur in response to bacteria or viral infection or brain disorders, including not only in autoimmune diseases, traumatic brain injuries, and neurodegenerative diseases (Chen et al., 2016; Steinman, 2010), but also in some neurodevelopmental defects, such as schizophrenia and autism (Meyer et al., 2011).

The importance of the neuroimmune system in the neuropathological consequences associated with alcohol abuse has also been supported in both human and experimental animals, as well as in neural cells in culture, and evidences that alcohol abuse can activate the innate response in glial cells by triggering inflammatory mediators and causing neuroinflammation and neural damage, effects that contribute to alcohol-induced neuropathology (Alfonso-Loeches et al., 2010, 2012; Crews and Vetreno, 2014; Montesinos et al., 2016a). Different studies have also evidenced the participation of the immune response in many actions of alcohol in the CNS, such as behavioral dysfunctions (Crews et al., 2017; Montesinos et al., 2015, 2016b), addiction (Coleman et al., 2017; Mayfield et al., 2013) and FASD neuropathology (Pascual et al., 2017c).

In this chapter, we review the functions of immune cells in the CNS, such as glial cells, and the molecular mechanisms by which ethanol is capable of activating different signaling pathways associated with innate immune signaling, including receptors Toll-like receptors (TLRs) and (NOD)-like receptors (inflammasome NLRs) (see rev. Montesinos et al., 2016a). The activation of these receptors allows the induction of cytokines and chemokines, which promotes neuroinflammation, brain damage, behavioral and cognitive dysfunction, and alcohol consumption (Coleman and Crews, 2018). The potential therapeutic interventions that may target neuroinflammation, and the neuropathological and behavioral consequences of alcohol abuse, are also discussed.

2. The neuroimmune system

Until recently, the CNS has been considered an immune-privileged site. However, current studies evidence that immune surveillance occurs under physiological and pathological conditions (Shrestha et al., 2013). The immune responses generated in the nervous system are, however, distinct from those that occur in the periphery because the blood brain barrier (BBB) and tissue-resident glial cells exert some protection of nervous cells from potentially harmful immune-cell mediated damage. Recent studies have also shown that the peripheral immune system and the nervous system can communicate using common molecular signaling cues (Prinz and Priller, 2017). Accordingly, stress, environmental and psychological stimuli can activate the neuroimmune system by modulating the communication pathways from the peripheral immune system to the brain to promote neuronal plasticity and neurogenesis (Yirmiya and Goshen, 2011). However, sustained inflammation (neuroinflammation) and the persistent stimulus that result from either environmental factors, or the formation of endogenous factors (e.g., pathogens, protein aggregates, toxins, psychological stress), can lead to memory impairment and neural plasticity dysfunctions (Yirmiya and Goshen, 2011), as observed in neurodegenerative diseases (Chen et al., 2016) and neuropsychiatric disorders (Rosenblat and McIntyre, 2016).

2.1 Glial cells: A hallmark of neuroinflammation

Glial cells, astroglia and microglia, are the principal cells that participate in the immune response in the CNS, which not only serve as supportive and nutritive roles for neurons by contributing in remodeling circuit connectivity and plasticity, but also defend the CNS from stress and pathogenic insults by transiently up-regulating inflammatory processes (Tian et al., 2012). Both types of glial cells play different and common roles in the development, homeostasis and diseases of the CNS.

2.1.1 Microglia

Microglia cells are non parenchymal macrophages in the brain or mononuclear phagocytes that are increasingly recognized as essential players in the development, homeostasis and diseases of the CNS. During development and neurogenesis, microglia influence the survival of newborn neurons and progenitors, the clearance of apoptotic cells and the formation of dendritic spines by influencing synaptic development and connectivity of neural circuits by facilitating circuit maturation and the remodeling of postsynaptic sites (Salter and Stevens, 2017; Tay et al., 2017; Weinhard et al., 2018). Mature microglia utilize several surface molecules to quickly respond to their extracellular environment and are capable of responding to cytokines, chemokines, purines, hormones and neurotransmitters, among others (see rev. Tay et al., 2017). Similarly to other tissue resident macrophages, microglia express common markers, such as fractalkine receptor CX_3CR_1, Csf-1R, integrin CD11b, and glycoproteins F4/80 and CD68, ionized calcium-binding adapter molecule 1 (Iba-1), and pan-hematopoietic CD45. Nonetheless, the levels of these markers are lower in perivascular macrophages and blood monocytes in the steady state (Greter and Merad, 2013) (see rev. Tay et al., 2017). Microglia also show some regional heterogeneity and sex-related differences (Lenz et al., 2013). Microglia like macrophages express some immune receptors, such as TLRs and NLRs, and are capable of responding to pathogens and other insults.

Under physiological conditions, microglia exhibit a deactivated phenotype and appear as an oval cell body with long branched processes. Under stress stimuli or pathological events however, these cells are activated and their morphology changes and transforms into an ameboid phenotype, which is functionally similar to macrophages (Davis et al., 1994; Streit, 2000). Activated microglia produce and release a variety of factors, including cytokines, prostanoids, free radical and pro-inflammatory mediators, which make these cells a good marker of neuroinflammation (Kettenmann et al., 2011). Activation of microglia, along with the production of inflammatory mediators, causes the dysfunction of the neural network by participating in many neurological and neurodegenerative diseases (see rev. Cunningham, 2013; Du et al., 2017).

Ethanol triggers microglia activation, as demonstrated in in vitro studies of cells in culture (Fernandez-Lizarbe et al., 2009), in in vivo ones in alcoholics' brains (He and Crews, 2008), and in experimental animals with chronic or acute alcohol intake (Alfonso-Loeches et al., 2010; Walter et al., 2017). These studies have clearly demonstrated that alcohol induces microgliosis by contributing to neuroinflammation, and it participates in the neuropathological changes associated with alcohol abuse (Alfonso-Loeches et al., 2010; Fernandez-Lizarbe et al., 2009; Marshall et al., 2016).

2.1.2 Astroglial cells

Astroglial cells are one of the most abundant cell types in the brain that perform many important physiological functions. They are produced largely in final neurogenesis stages, and they also perform important functions in many processes while the brain develops, such as neuronal survival, synaptogenesis and axonal guidance (Reemst et al., 2016). These glial cells also participate in the control of extracellular ion and neurotransmitter concentrations by providing neurotrophic support to neurons and controlling synapse formation, functioning, and pruning (Allen, 2014; Liddelow and Barres, 2015). Neurotransmitters, such as glutamate and GABA, are taken up by astrocytes to thereby terminate their action (Attwell et al., 2010). Astrocytes also contribute to establish and maintain the BBB (Allen, 2014; Liddelow and Barres, 2015).

Morphologically, astrocytes are multiprocess-bearing stellate cells and two types of astrocyte exist: protoplasmic (gray matter) and fibrous (white matter). The expression of GFAP, a protein associated with intermediary filaments, is commonly used as a specific marker for these cells (Liddelow and Barres, 2015). These cells respond vigorously to brain injury and are postulated to play an important role in the fine tuning of brain inflammation (Dietrich et al., 2003). Indeed injury to CNS damage is accompanied by astrocytic hypertrophy, proliferation and altered gene expression, a process commonly referred to as reactive astrogliosis (Maragakis and Rothstein, 2006; Sofroniew and Vinters, 2010). This process is associated with inflammation (Little and O'Callagha, 2001). However depending on the disease context, astrogliosis can be seen as a positive event that promotes neuronal and glial survival via the production of neurotrophins and growth factors, or as a negative influence on regeneration via the inhibition of neuronal and glial growth and migration.

Activation of astrocytes, in response to a neuropathologic process, triggers the production of inflammatory cytokines, increases the expression of major histocompatibility complex-II and augments the production of free radicals (Sofroniew and Vinters, 2010). The cytokines, for which evidence is most compelling in the initiation and modulation of reactive astrogliosis, include IL-1β, TNF-α, IFN-γ and TGF-β (Pekny and Pekna, 2014). Glial cells express receptors for all these cytokines, and each cytokine appears to perform a different functional role in the astrocytic response (Pekny and Pekna, 2014). Among cytokines, IL-1β has been considered an important mediator of inflammatory responses in the CNS, which is involved in Alzheimer's diseases (Mrak and Griffin, 2005) and is produced mainly under brain damage, disease or stress conditions. Current data demonstrate that the activation of glial cells, along with the up-regulation of pro-inflammatory cytokines, contributes to the neuroinflammation associated with many neurological and neurodegenerative diseases, such as Huntington's, Parkinson's, and Alzheimer's diseases and also in amyotrophic lateral sclerosis (McManus and Heneka, 2017; Ransohoff, 2016).

Substantial evidence also demonstrates the impact of alcohol on astroglial cells to cause astrogliosis (Alfonso-Loeches et al., 2010; Hayes et al., 2013) and loss of astrocytes in some brain areas, such as the prefrontal cortex and the hippocampus of human alcoholics' brains (Adermark and Bowers, 2016). Changes in myelination can also occur during chronic alcoholism as the gene expression of several myelin-associated genes is down-regulated in alcoholics' brains (Lewohl et al., 2000).

Finally, although glial cells are the main innate immune cells found in the CNS, neurons also appear to play some roles in the innate immune response as they express some immune receptors, inflammasome AIM-2 and IL-1β production by regulating neuronal morphology (Wu et al., 2016).

2.2 Innate immune receptors in the CNS

Evidence from the last decade demonstrates that the CNS is immunologically active and is able to recognize and respond to pathogens, toxins and danger molecules by activating the innate immune response (Lampron et al., 2013). Inflammatory responses to infectious agents are initiated by pattern recognition receptors (PRRs), which detect *pathogen-associated molecular patterns* (PAMPs, stranger signals), molecular "signatures" of whole microorganism classes, including bacteria, fungi, viruses, or *endogenous damage-associated molecular patterns* (DAMPs) endogenous molecules or damaged tissue (Akira et al., 2001). This immune response does not provide long-lasting immunity or memory, unlike what happens in the adaptive immune system.

Among PRRs, *Toll-like receptors* (TLRs) and NOD-like receptors (NLRs) (nucleotide-binding domain leucine-rich repeat-containing), have emerged as important regulators of the immune response in the CNS and contribute to brain damage in both neurodegenerative diseases (Amor et al., 2010) and alcohol-related neuropathologies (Montesinos et al., 2016a).

2.2.1 TLRs

These receptors were the first to be identified and are the best characterized by being an evolutionarily conserved family of receptors discovered in the *Drosophila melanogaster* as a major defense against microbial infection (Medzhitov et al., 1997). TLRs, type I integral membrane glycoproteins, are members of a large superfamily that includes interleukin-1 receptors

(IL-1Rs) (Akira and Takeda, 2004; O'Neill and Bowie, 2007). TLRs and IL-1Rs have a conserved region of ~200 amino acids in their cytoplasmic tails known as the Toll/IL-1R (TIR) domain (Takeda et al., 2003). In contrast, the extracellular regions of TLRs and IL-1Rs markedly differ: the extracellular region of TLRs contains leucine-rich repeats motifs, but the extracellular region of IL-1Rs comprises three immunoglobulin-like domains (see rev. Akira and Takeda, 2004). The activation of both IL-1R and TLRs shares downstream signaling molecules, which culminate in the activation of nuclear factors NF-κB and AP-1, which are transcription factors that regulate the expression of a wide array of genes involved in immune responses (see Fig. 1).

The TLRs, a family of transmembrane glycoproteins, comprise 10 members in humans (TLR1–TLR10) and 13 in mice. Among these receptors, TLR4 is the most widely studied and recognizes the endotoxin *lipopolysaccharide* (LPS), a part of the external membrane of *gram*-negative bacteria. Among TLRs, TLR1, TLR2, TLR4, TLR5 and TLR6 are located in the cell plasma membrane, while TLR3, TLR7, TLR8 and TLR9 recognize viral nucleic acids and are located in endosome/lysosome membranes (Nishiya and DeFranco, 2004).

Stimulation of TLRs by PAMPs or DAMs evokes the activation of different signaling pathways, including nuclear factor-κB (NF-κB) and mitogen-activated protein kinases (MAPKs), which results in the production of cytokines, chemokines and reactive oxygen species (ROS) that serve to destroy pathogens (Fig. 1). Some TLRs, like TLR4, TLR3 and TLR9 and TLR7, also induce interferon (IFN)-α or IFN-β responses by activating distinct interferon regulator factors (IRF) to allow the immune system to tailor its responses to viral pathogens (Colonna, 2007).

TLRs are expressed mainly in immune system-related cell types, including B cells, mast cells, natural killer cells, regulatory T-cells, macrophages, monocytes, dendritic cells, neutrophils and basophils (Kokkinopoulos et al., 2005). However, the expression of different TLRs has also been demonstrated in the CNS (Blanco et al., 2005; Fernandez-Lizarbe et al., 2009; Kielian, 2006), and several studies support the impact of TLRs response on infectious and non-infectious CNS diseases/injury (Hanke and Kielian, 2011) by influencing inflammatory responses in tissue injury and autoimmunity (Farrugia and Baron, 2017).

Both microglia and astroglia express TLR2, TLR4, TLR5 and TLR9 (Bowman et al., 2003) respond functionally with cytokine and chemokine production, and are capable of contributing to an inflammatory

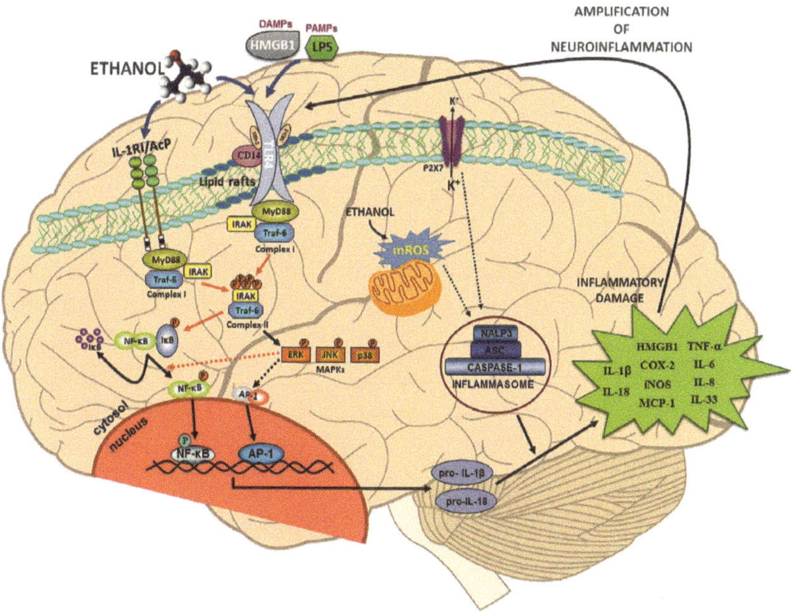

Fig. 1 Schematic of alcohol-induced neuroinflammation through the activation of immune receptors TLR4, IL-1RI/AcP and inflammasome NLRP3 receptor. Ethanol or PAMPs (e.g., LPS) or DAMPs (e.g., HMGB1) activate/s the TLR4 by its recruitment with adaptor proteins CD14 and MD2 in lipid rafts. This activation, along with IL-1RI/AcP activation, leads to the recruitment of complex I (MyD88, IRAK 1–4, TRIF) to trigger rapid downstream signaling pathways (e.g., IRAK/Traf-6 and MAPKs) and transcription nuclear factors (NF-κB and AP-1), which terminate in the generation of cytokines (e.g., pro-IL-1β, IL-6, IL-8, TNF-α, etc.), chemokines (e.g., MCP-1) and inflammatory mediators (e.g., iNOS, COX-2), which lead to neuroinflammation. The expression of pro-IL-1β induced by TLR4 activation, along with other stimuli such as the mitochondrial ROS (mROS) generation or the K^+ efflux, is able to trigger NLRP3 inflammasome multiprotein complex recruitment by activating cysteine-protease caspase-1, which cleaves pro-IL-1β to active IL-1β, as well as other cytokines such as IL-18 and alarmins (IL-33, HMGB1) which, in turn, can amplify the neuroinflammatory response. *IL-1RI/AcP*, interleukin-1 receptor accessory protein; *MyD88*, myeloid differentiation primary response gene 88; *IRAK*, interleukin-1 receptor-associated kinase-1; *ROS*, reactive oxygen species; *TRIF*, TIR-domain-containing adaptor-inducing IFN-β; *MAPK*, mitogen-activated protein kinase; *IRF3*, Interferon regulatory factor 3; *ASC*, apoptosis-associated speck-like protein containing a CARD; *AP-1*, activator protein 1.

environment in the CNS following a variety of infectious or inflammatory insults (Arroyo et al., 2011; Hanke and Kielian, 2011).

Brain endothelial cells, an important component of the BBB, also express TLR4 (Nagyoszi et al., 2010). These cells can receive immune signals from either the brain or peripheral blood by releasing cytokines in the brain which, in turn, can activate glial cells or neurons. However, it is uncertain whether CNS inflammation results from a neural or peripheral TLR4 response because some evidence suggests a cross-talk between both systems since peripheral TLR4 signaling can mediate the CNS immune response (Roach et al., 2013).

2.2.2 NLRs

Another type of important innate immune PRRs are NOD-like receptors (NLRs) (nucleotide-binding domain leucine-rich repeat-containing), which are cytoplasmic receptors that sense and detect invading pathogens, and can initiate an innate immune response. NLR family members NLRP1, NLRP3, NLRC4 and AIM-2 can also be activated by PAMPs, non microbial danger signals or damaged-associated molecular patterns (DAMPs) (Jo et al., 2016).

The link between TLRs and NLRs occurs through the activation of NF-κB by TLRs, which induces pro-IL-1β until a second stimulus (e.g., ions or membrane perturbations, ROS, or extracellular ATP) promotes the recruitment of the large multiprotein complex, called inflammasome, by activating caspase-1. This, in turn, cleaves pro-ILβ into its bioactive molecule IL-1β, along with other cytokines of this family (IL-18, IL-33) (Guo et al., 2015). The discovery of inflammasome (Martinon, 2012) has advanced our knowledge of the role played by the innate immune system in several immune pathologies, including neurodegeneration (Guo et al., 2015) (Fig. 1), since IL-1β is a potent pro-inflammatory cytokine that can promote sickness and depression (Konsman et al., 2008; Raison et al., 2006). Among the different NLRs inflammasomes, NLRP3 (or cryopyrin) is the best characterized, whose activation is evidenced in cryopyrin-associated periodic syndromes (CAPS) (Willingham et al., 2007). NLRP3 is expressed in microglia (Hanamsagar et al., 2012), astrocytes (Lu et al., 2014) and neurons (Zou and Crews, 2012). Experimental evidence demonstrates that ethanol activates different inflammasomes, including NLRP3, in neural cells by triggering the production of IL-1β, IL-18 and IL-33 (Alfonso-Loeches et al., 2014, 2016) (Fig. 1).

Both membrane TLRs and cytoplasmic NLRs are not only activated by pathogens or PAMPs but are also highly effective at sensing and responding to the released components of DAMPs, or "alarmins" (Bianchi, 2007; Hanamsagar et al., 2012). DAMP-triggered "sterile inflammation" is most important in the neuroinflammatory disorders that occur in the absence of any foreign pathogens, as in neurodegenerative diseases (Glass et al., 2010). One relevant alarmin is the high-mobility group box 1 (HMGB1) protein, a nuclear protein which, given a certain stimulus, is hyperacetylated in lysine residues and translocates into the cytosol. HMGB1 is an endogenous ligand for different TLRs (TLR2, TLR4, TLR9), RAGE (receptor for advanced glycation products) and TREM-1 (triggering receptor expressed in myeloid cells 1) receptors, and leads to the activation of NF-κB along with cytokine and chemokine production (El Mezayen et al., 2007; Ibrahim et al., 2013).

Alarmins or DAMPs participate in many neurological and neurodegenerative disorders (Amor et al., 2010; Hanamsagar et al., 2012), and in alcohol consumption-associated neuroinflammation. Some studies have demonstrated that alcohol directly or indirectly releases alarmin HMGB1 (Vetreno and Crews, 2012) by activating TLRs and NLRs response and contributing to not only ethanol-induced neuroinflammation, behavior and neural damage (Alfonso-Loeches et al., 2010, 2012, 2014; Montesinos et al., 2016a), but also to alcoholic liver injury (Szabo and Lippai, 2014).

3. Mechanisms that participate in ethanol-induced neuroinflammation, neurotoxicity and brain damage

3.1 The ethanol-induced innate immune system response

It has been well established that alcohol abuse impairs immune and inflammatory systems in humans (Szabo and Mandrekar, 2009), which results in specific defects in the cellular components of the innate immune responses to bacterial and viral pathogens (Messingham et al., 2002). Acute ethanol impairs immune functions by interfering with inflammatory response, with an increasing risk for infectious diseases, and bacterial pneumonia is observed in particular in alcoholics (Ruiz et al., 1999). Conversely, chronic ethanol treatment increases the innate immune response in the liver by participating in the development of alcoholic liver disease. The gastrointestinal tract seems to be the initial target of ethanol since alcohol consumption impairs the intestinal mucosa's barrier function (Parlesak et al., 2000) by increasing the plasma levels of bacteria endotoxin, lipopolysaccharide or LPS which, in turn, activate TLR4 in Kupffer cells (resident macrophages in the liver) to thus contribute to inflammation in alcohol-liver disease (Miller et al., 2011; Nagy, 2003). Recent studies have also shown that alcohol consumption is associated with gut microbiome alterations (Lowe et al., 2017), and the imbalance of pathogenic and commensal organisms in the intestinal microbiome may contribute to the abnormal gut-liver axis in alcohol liver disease (Szabo and Saha, 2015). Some data suggest that ethanol-induced peripheral inflammation, along with monocytes activation and production of pro-inflammatory cytokines, might cross the BBB via diffusion or active transport to contribute to neuroinflammation (see below).

3.2 Contribution of TLRs and NLRs response in the neuroinflammation and neuropathological changes associated with alcohol abuse

3.2.1 Involvement of TLRs in alcohol neurotoxicity

Early studies have demonstrated that ethanol is capable of promoting the activation of signaling pathways, which triggers the production of cytokines in the brain to cause neural death (Valles et al., 2004). More recently, in vitro and in vivo studies have further showed that ethanol can activate innate immune receptors, TLRs, by triggering signaling pathways, which causes neuroinflammation. For instance, studies in vitro using cortical astroglia and microglia cells in culture have demonstrated that ethanol, in the absence of any toxin, can recruit TLR4 receptors into membrane lipid rafts by activating their signaling response (MAPK, NF-κB, AP-1), along with the release of cytokines and inflammatory mediators in the medium of microglia and astrocytes in culture (Blanco et al., 2005, 2008; Fernandez-Lizarbe et al., 2009) (Fig. 1). These studies have also shown that ethanol, like LPS (the TLR4 natural ligand), is capable of promoting TLR4 receptor endocytosis, which is dependent on clathrin and rafts/caveolae, by inducing the TLR4 signaling associated with receptor internalization and trafficking (Pascual-Lucas et al., 2014). Blocking TLR4 in glial cells with either siRNA or cells from TLR4-deficient mice (TLR4-knockout [KO]) abolishes ethanol effects on the inflammatory response in cultured astrocytes and microglia (Alfonso-Loeches et al., 2010; Fernandez-Lizarbe et al., 2009).

The studies conducted in in vitro glial cells have been confirmed using mice with chronic ethanol intake and demonstrate that this treatment induces gliosis (activation of astrocytes and microglia), MAPK and NF-κB signaling activation, along with the up-regulation of inflammatory cytokines and mediators (e.g., COX-2, iNOS). These neuroinflammatory effects have been associated with myelin disruptions and apoptosis in the cerebral cortex of WT mice (Alfonso-Loeches et al., 2010, 2012).

As mentioned before, TLRs are also activated by tissue damage or DAMPS. Therefore, ethanol-induced neural damage can further activate TLR4 or other TLRs by triggering additional neural damage, which occurs in many neurodegenerative diseases (Arroyo et al., 2011). Abnormal protein accumulation and the aggregation of specific proteins can result for deficient clearance systems by the dysregulation of proteolytic processes to contribute to the pathology of several neurodegenerative diseases (Todde et al., 2009). Chronic ethanol intake can also impair degradation processes by contributing to alcohol abuse-associated neuropathology (Pla et al., 2014) (see Section 3.2.4.2).

It is noteworthy that most of the above effects of ethanol are dependent of TLR4 response since TLR4-KO mice are protected against neuroinflammation, neuropathological alterations and autophagy dysfunctions induced by chronic ethanol abuse (Alfonso-Loeches et al., 2010, 2012; Pla et al., 2014). As to whether the neuroimmune effects of alcohol are reversible when alcohol is eliminated, experimental data indicate that the neuroimmune effects of alcohol remain for long periods after alcohol exposure, as demonstrated by Qin et al. (2008). Their study showed that ethanol potentiates the LPS-induced increase in TNF-α, monocyte chemoattractant protein-1 (MCP-1 or CCL2) and IL-1β in the liver, brain and serum. However, the up-regulation of cytokines observed in the mouse brain remains high, whereas the peripheral levels return to normal (Qin et al., 2008). Long-term effects of ethanol on neuroimmune system activation have also been observed in adult animals exposed to alcohol in adolescence (Montesinos et al., 2016b; Vetreno et al., 2013) (see Section 5.2). It is noteworthy that the role of neuroinflammation and TLRs has also been observed in humans as changes in the expression of neuroimmune genes, such as microglial transcripts MCP-1 (He and Crews, 2008), the specific microRNAs involved in immune function regulation (Lewohl et al., 2011) and the up-regulation of several TLRs (TLR2, TLR3, TLR4), have been found in postmortem alcoholic brains (Coleman et al., 2017; Vetreno et al., 2013), including a recent study showing increased TLR7 expression and microglial activation in hippocampal human tissue (Coleman et al., 2017). These findings suggest the participation of TLRs in ethanol-induced neuroinflammation and neurotoxicity.

3.2.2 Participation of NLRP3 in alcohol-induced neuroinflammation

Ethanol also activates other immune receptors, such as NLRP3 inflammasome, in the astroglia and microglia of the cerebral cortex of chronic ethanol-treated mice, and in primary cultured astroglial and microglial cells (Alfonso-Loeches et al., 2014, 2016). NLRP3 activation triggers the production of IL-1β, IL-18 and IL-33, which leads to cell death by pyroptosis, a type of necrotic cell death. The NLRP3 inflammasome response is TLR4-dependent as it is initiated by pro-IL-1β being generated by TLR4 signaling (priming signal) which, along with ethanol-induced mitochondria (m)ROS production (second signal), triggers the recruitment of the inflammasome complex, caspase-1 activation and pro-inflammatory cytokine IL-1β release (Alfonso-Loeches et al., 2014) (Fig. 1). Notably, the activation of TLR4/NLRP3-inflammasome signaling in microglial cells promotes leukocyte infiltration, which compromises BBB permeability (Alfonso-Loeches et al., 2016). The activation of NLRP1 and NLRP3 inflammasomes also contributes to the inhibitory effects of ethanol during hippocampal neurogenesis (Zou and Crews, 2012). Studies in vivo have further demonstrated that alcohol up-regulates and activates NLRP3/ASC inflammasome, which leads to caspase-1 activation and IL-1β production in the cerebellum. IL-1β amplifies neuroinflammation, while the disruption of IL-1/IL-1R signaling prevents alcohol-induced inflammasome activation and neuroinflammation (Lippai et al., 2013). These results support the key role of NLRP3/ASC and IL-1β production in ethanol-induced neuroinflammation.

The question that rises from the above findings is whether the alcohol-induced activation of TLR4 and NLRs response is caused by direct ethanol interactions with these receptors or by DAMPs such as HGMB1. Despite ethanol rapidly crossing the BBB and being able to initiate inflammatory events by activating TLR4/NLRs in glial cells, endogenous TLR4 ligands, such as HGMB1, can also activate these receptors and thus contribute to brain damage (Crews and Vetreno, 2014). Ethanol also induces the release of LPS from the gut into the bloodstream (Mandrekar and Szabo, 2009) and can activate TLR4 in liver Kupffer cells and peripheral monocytes, which may trigger the production of cytokines and, in turn, can cross the BBB via diffusion or active transport, or even by BBB alterations. Disruption of BBB integrity has been recently reported in the postmortem alcoholic brain (Rubio-Araiz et al., 2017), but TLR4-KO mice were protected against ethanol-induced BBB disruption (Alfonso-Loeches et al., 2016; Rubio-Araiz et al., 2017). It is presently uncertain whether one mechanism or more can be involved in ethanol-induced brain cytokine production, and if this depends on an alcohol drinking pattern (acute vs. chronic), or not. A recent study has found increased levels of peripheral pro-inflammatory cytokines in non cirrhotic alcohol-dependent subjects compared with control subjects, which correlated with alcohol craving (Leclercq et al., 2012). This suggests a role for the gut-brain axis in alcohol dependence. Further research is needed to understand the contribution of not only the peripheral vs. central immune response in ethanol-induced neuroimmune activation, but also of BBB integrity.

Finally, although chronic ethanol consumption can cause neuroinflammation and brain damage, transient exposure to low/moderate levels of ethanol increases the levels of anti-inflammatory cytokines, such as IL-10 and TGF-β in monocytes (Mandrekar et al., 2006) and peritoneal macrophages (Pascual et al., 2011b), which suggests that low/moderate ethanol intake may offer some beneficial actions for the immune system response. Indeed some studies (Doremus-Fitzwater et al., 2014; Gano et al., 2016) support the anti-inflammatory role of acute ethanol intoxication as the gene expression of IL-1β and TNF-α was suppressed in the hippocampus, amygdala and bed nucleus of the striatum terminalis after 3h, while the IL-1β mRNA levels increased during withdrawal (18h post-injection).

3.2.3 TLRs response is involved in the myelin dysfunctions associated with alcohol abuse

One type of glial cells, which do not participate in the immune response, is the glial-subtype responsible for myelin production called oligodendrocytes. These cells constitute the main brain white matter components (White and Kramer-Albers, 2014) by contributing in the myelination process, and synthesizing a lipid rich membrane around the axon to form multiple myelin internodes. In addition to lipid composition, the most abundant proteins of the myelin sheaths in the CNS are the myelin basic protein (MBP), and the proteolipid protein (PLP) that participates in the maintenance of its structural integrity. During this process, the structure and saltatory localization of myelin in nerve fibers is fundamental in the CNS to provide correct impulse conduction along axons to rapidly process information (Simons and Trotter, 2007). Consequently, white matter dysfunction or demyelination, or alterations during the myelin development process, can impair interneuronal communication to cause cognitive and behavioral dysfunctions, which occurs in many neurodegenerative and autoimmune disorders (Bells et al., 2017).

Clinical evidence indicates that chronic alcohol consumption alters the structure and functions of myelin. For instance, loss of frontal, temporal and cerebellar white matter (de la Monte and Kril, 2014) along with the down-regulation of several myelin-related genes (Lewohl et al., 2000) have been reported in human alcoholics' postmortem brains. Neuroimaging studies that have used diffusion tensor imaging (DTI) and matrix-assisted laser desorption ionization imaging mass spectrometry (MALDI-IMS) confirm that chronic alcohol abuse disrupts white matter microstructural integrity (Pfefferbaum et al., 2006; Schulte et al., 2005), and affects connecting brain areas, such as the corpus callosum, cortico-limbic frontal, superior fronto-occipital fasciculus. These effects may contribute to deficits in executive and cognitive functioning, motor impairments and information processing in alcoholics (Bagga et al., 2014; Durkee et al., 2013; Pfefferbaum et al., 2010).

Studies in experimental animals have also demonstrated that chronic alcohol abuse induces (Alfonso-Loeches et al., 2012; Samantaray et al., 2015) important ultrastructural and biochemical alterations in myelin fibers by lowering the expression of the proteins associated with myelination (e.g., MBP, PLP, CNPase or NG2), inducing axon and myelin degeneration, as well as disrupting the compact disposition of total myelin sheaths in 47.3% of the axons in the corpus callosum and cortex (Alfonso-Loeches et al., 2012). Notably the effects of ethanol on myelin dysfunctions were dependent mostly on the TLR4 response as the same ethanol treatment did not cause significant changes in TLR4-KO mice. These results suggest that the ethanol-induced TLR4 response triggers an inflammatory environment (e.g., production of proinflammatory mediators and cytokines) which can, in turn, induce myelin sheaths damage and dysfunctions (Alfonso-Loeches et al., 2012) (Fig. 2).

Ethanol-induced myelin disruptions have also been observed during brain ontogeny, including fetal and neonatal and adolescence stages (see Sections 5.1 and 5.2).

3.2.4 Neuroimmune response participates in ethanol-induced autophagy dysfunction by contributing to neuropathological alteration in alcoholism

Autophagy is a highly regulated process involved in the turnover of long-lived proteins and damaged organelles that involves the sequestration of cytosol regions within double-membrane-bound compartments and the delivery of contents to the lysosome for degradation (Chen and Klionsky, 2011; Yorimitsu and Klionsky, 2005). This process plays an important physiological role in several events, including cell survival, differentiation, development or immunity. The autophagy pathway is regulated by the

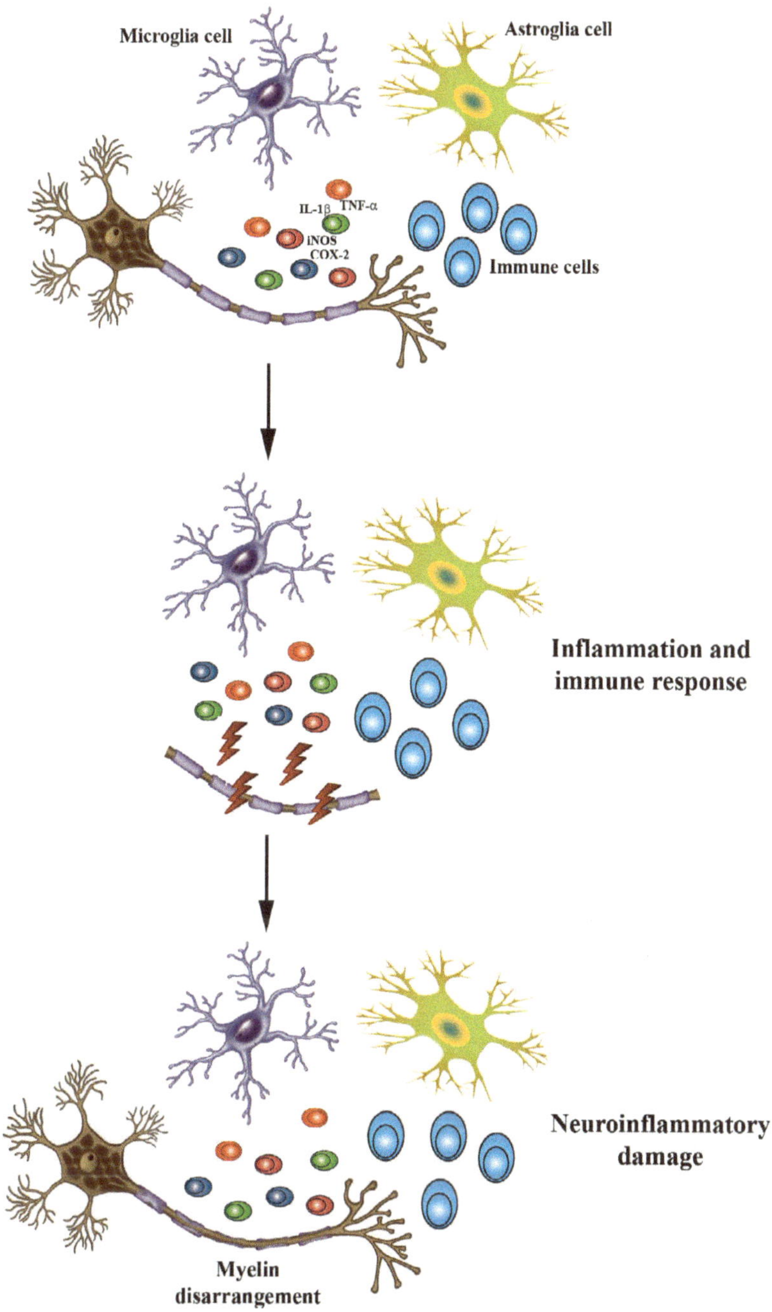

Fig. 2 The role of the immune response in myelin disruptions. Activation of astroglial and microglial cells induces the release of inflammatory mediators and cytokines, and activated immune cells, which can cause neuroinflammatory damage that leads to myelin disarrangements.

phosphorylation of several proteins, although the main regulator of this proteolytic pathway is mTOR (mechanistic target of rapamycin) (see rev. Pattingre et al., 2008).

Recent studies have demonstrated that the actions of ethanol in the autophagy function depend on both dose and treatment. Thus whereas an acutely high ethanol dose can increase autophagy by acting as a protective mechanism against ethanol toxicity in astrocytes and neurons, chronic ethanol consumption impairs the autophagy process that leads to protein aggregation and accumulation in the cytoplasm, and aggravates ethanol neurotoxicity. We will shortly describe both processes.

3.2.4.1 Autophagy protects from ethanol neurotoxicity

Autophagy is an important mechanism that controls cell protein homeostasis and can act as cell protection against the toxicity caused by abnormal proteins or protein accumulation by helping their degradation and, thereby, allowing to adapt immune and cellular responses to inflammation stress (Giordano et al., 2014; Wang and Mandelkow, 2012). Accordingly, some studies have demonstrated that ethanol at high doses (50 mM) increases the autophagy process in astrocytes and microglia in primary culture (Pla et al., 2016), as well as SH-SY5Y neuroblastoma cells (Chen et al., 2012). During this process, ethanol increases the levels of several proteins (e.g., LC3-II/Atg8, Atg12 and the cathepsin) for short times (30 min), which then decrease after 2 h of incubation (Pla et al., 2016). The alteration of autophagic proteins is modulated by mTOR phosphorylation as the down-regulation of the phosphorylation of this protein is associated with the activation of autophagic catabolism. Additionally, an ethanol-induced increase in autophagy processes correlates with an increase in electrodense autolysosomes, and they all contribute to activate the autophagy pathway (Pla et al., 2016). The protective effect of autophagy has also been shown in gastric mucosal epithelial cells as the suppression of autophagy with pharmacological agents results in a significant increase in cell apoptosis and gastric mucosa injury, which suggests that autophagy can protect cells from ethanol toxicity (Chang et al., 2017).

3.2.4.2 Ethanol treatment and autophagy dysfunction

Although it is generally assumed that the induction of autophagy is beneficial for clearing toxic compounds and protein aggregates (Giordano et al., 2014; Wang and Mandelkow, 2012), autophagy machinery deficiency generates a toxic environment, named "proteotoxicity," and characterized different defective proteins accumulating, which can contributes to brain damage in several neurodegenerative diseases (e.g., Alzheimer's disease, Parkinson's disease, Huntington's disease or multiple sclerosis) (Yamamoto and Simonsen, 2011).

Some studies have demonstrated that chronic ethanol consumption inhibits the autophagy process by lowering the expression of the proteins involved in this pathway (Atg proteins, such as Atg5, Atg12 and LC3-II/Atg8), and it also reduces the activity of lysosomal hydrolase and cathepsin B in the cerebral cortex of WT mice. These biochemical changes have been associated with an increase in size of autophagic vacuoles, which suggest lack of ability to fuse vacuoles with lysosomes and to, hence, trigger the effective degradation of their cargoes (Pla et al., 2014).

Disruptions in the autophagy process also occur in brain microvessels of the human fetuses prenatally exposed to alcohol (Girault et al., 2017). Similarly, autophagy impairment has also been shown in ethanol-treated embryonic-derived primary cortical neuroepithelial cells in which ethanol induces a reduction in autophagic lysosomes that leads to a significant destruction of cellular organelles to result in cellular material accumulation and programmed cell death (Prock and Miranda, 2007).

4. Neuroimmune activation in alcohol drinking

Evidence from animal studies also indicates the impact of TLRs on alcohol drinking and behaviors. For instance, some studies have demonstrated that null mutations in either TLR4 or MyD88 block the motor impairment recovery time induced by a single ethanol injection to BALB/c wild-type (WT) mice (Wu et al., 2012). C3H/HeJ mice, a strain that is naturally TLR4-deficient, display reduced operant ethanol self-administration compared to controls (Mayfield et al., 2016). However, naïve C57BL/6-TLR4-KO mice present a similar alcohol intake and preference to their WT counterparts (Harris et al., 2017; Pascual et al., 2011a), although TLR4-KO mice are protected against the escalation of alcohol preference caused by intermittent-ethanol treatment (Montesinos et al., 2016b).

Activation of TLR4 signaling triggers cytokine and chemokine production, which can also influence alcohol drinking and behavioral patterns as the mice lacking chemokine MCP-1 or CCL2, or its receptor CCR2, display a less marked preference for alcohol (Blednov et al., 2005). The targeted elimination of MCP-1/CCL2 in the central amygdala (CeA) or the ventral tegmental area (VTA) blunts binge drinking (June et al., 2015), while a chronic intracerebroventricular infusion of MCP-1/CCL2 increases alcohol operant self-administration (Valenta and Gonzales, 2016). In summary, experimental evidence mainly from knockout mice indicates that TLR4, along with some cytokines and chemokines, can participate in alcohol intake modulation, but not in excessive alcohol consumption. Recent studies have suggested the participation of the innate immune response in alcohol addiction (Coleman and Crews, 2018).

5. Alcohol induces neuroinflammation in the developing brain

As mentioned in the Introduction, the developing brain is most vulnerable to the toxic effects of ethanol during fetal and neonatal development, as well as in adolescence, which is a brain maturation stage. In these stages, the damaging effects of ethanol are usually irreversible and cause long-term cognitive and behavioral dysfunctions. Although different mechanisms have been proposed, recent data suggest the role of the innate immune response in some of the effects of ethanol during brain development.

5.1 Ethanol effects during fetal and neonatal development

One of the most disturbing and lasting effects of prenatal alcohol exposure occurs in the developing brain, in which structural and functional defects, along with a range of neuropsychological and behavioral consequences, are observed in children with FASD (see rev. Alfonso-Loeches and Guerri, 2011; Guerri et al., 2009; Moore et al., 2014; Warren et al., 2011). Several mechanisms have been proposed as possible candidates to cause the brain anomalies observed in FASD, including alterations to neurotrophic factors and cell adhesion molecules, higher oxidative stress and free radical formation damage, epigenetic factors and dysfunctions in chromatic remodeling, etc. (see revs. Gavin et al., 2017; Pascual et al., 2017a). Although ethanol possibly induces neurotoxicity through several mechanisms in different brain development stages (Alfonso-Loeches and Guerri, 2011), current evidence suggests that the role of the immune system during both physiological and pathological processes occurs during brain development. For instance, the immune system participates in important brain functions, such as synaptogenesis, neuronal migration, the activity-dependent refinement of circuits or synaptic plasticity (Marc, 2013). However, the overactivation of the inflammatory response can affect normal brain processes by increasing the risk of neurological and neuropsychiatric diseases (Hagberg et al., 2015).

With different animal models of FASD, several studies (Drew and Kane, 2014; Goodfellow et al., 2018; Kane et al., 2011; Tiwari and Chopra, 2012) have demonstrated that ethanol exposure in the neonatal stage, a human third trimester-equivalent period, causes microglial activation and releases pro-inflammatory compounds, which lead to neuronal damage (Boschen et al., 2016; Topper et al., 2015). These results suggest that exposure to ethanol in some developing brain stages may trigger immune system responses and potentially contribute to long-term alcohol-related changes to cognition, behavior and the immune function. Accordingly, studies in humans have indicated that maternal plasma levels of IL-10 lower in the third trimester of gestation and change the balance between pro- and anti-inflammatory cytokines which increases the risk of having an FASD child (Sowell et al., 2018).

Alcohol-induced alterations in the myelin structure and function have also been demonstrated during brain development since exposure in the neonatal stage, in which myelin develops, reduces the number of mature oligodendrocytes and the proliferating oligodendrocyte progenitor cells in the corpus callosum. Some of these alterations, such as myelin microstructure changes in the corpus callosum, persist for long periods (Newville et al., 2017).

The mechanisms involved in ethanol-induced pro-inflammatory cytokines during fetal/neonatal brain development are presently uncertain, but some studies suggest the role of the innate immune response in the action of ethanol in brain development. A recent study has demonstrated that alcohol consumption during gestation and lactation triggers the release of cytokines and chemokines in maternal blood, amniotic fluid and the brain of fetuses and postnatal pups (Pascual et al., 2017c). The up-regulation of brain cytokines/chemokines is associated with the activation of microglia, along with alterations in the myelin structure and synaptic dysfunctions that lead to long-lasting cognitive and behavioral dysfunctions. These effects have been associated with the TLR4 response since TLR4 deficiency, by using TLR4-KO mice, protects mostly against not only ethanol-induced neuroinflammation, but also the neuropathological and behavioral changes associated with FASD (Pascual et al., 2017c).

In summary, current studies suggest the participation of the TLR4 innate immune response in the action of ethanol on the developing brain. Some results also hint at ethanol-induced neurodevelopmental alterations not only being mediated by the direct neurotoxic effects of ethanol in the developing brain, but that the fetal pro-inflammatory environment can also participate by being induced by maternal peripheral cytokines and chemokines, and by the release of fetal or placental tissues (Pascual et al., 2017c). Indeed inflammation is a critical contributor to both normal development and injury outcome in the immature brain. Moreover, perinatal immune activation in the mother and/or fetus adversely affects neurodevelopment and increases the risk of developing behavioral psychiatric disorders (see revs. Bauman et al., 2014; Hagberg et al., 2015).

5.2 Ethanol-induced neurotoxic effect and behavioral dysfunction in the adolescent brain

Adolescence is a brain maturation period during which plastic and dynamic processes take place in some regions, such as the prefrontal cortex and hippocampus (Toga et al., 2006). Experimental evidence indicates the physiological role of the immune response in brain development and maturation by participating in synaptic pruning, myelination, neurite outgrowth or synaptic neurotransmission (Brenhouse and Schwarz, 2016). However, the overexpression of the innate immune response can cause neuroinflammation and brain injury.

A large body of evidence in human and animal studies demonstrates the role of the immune system response in the neurotoxicity, neuropathological processes and behavioral dysfunction associated with binge alcohol drinking in adolescence (see revs. Brenhouse and Schwarz, 2016; Crews et al., 2016; Montesinos et al., 2016a; Pascual et al., 2017a). For instance, several studies demonstrate that repeated ethanol binge drinking in adolescence triggers the release of proinflammatory molecules, such as cytokines, chemokines (TNF-α, IL-1β, MCP-1, MIP-1α) or alarmins (HMGB1), by activating immune receptors TLRs (e.g., TLR2, TLR3 and TLR4) (Pascual et al., 2014; Vetreno and Crews, 2012). Some of these studies show that the ethanol-induced up-regulation of the expression of inflammatory mediators (iNOS, COX-2), cytokines and chemokines (TNF-α, IL-1β, MCP-1, MIP-1α, IL-17A) is mediated by the TLR4 response since mice deficient in the functional TLR4 protein can protect against ethanol-induced cytokine production (Montesinos et al., 2015). Gender differences in peripheral and brain pro-inflammatory cytokine levels have been observed in human and mice adolescents (Pascual et al., 2017b).

Innate immune response activation by either ethanol or other immune molecules, such as HGB1, also causes a persistent loss of neurogenesis in the hippocampi (Vetreno and Crews, 2015) of adolescent animals, and has been associated with epigenetic changes by long-term cognitive and behavioral impairments (Montesinos et al., 2015, 2016b; Pascual et al., 2007; Vetreno and Crews, 2015), and might predispose to alcohol drinking (Montesinos et al., 2016b). Indeed a single inflammatory insult increases voluntary alcohol intake (Blednov et al., 2011). What this suggests is that inflammation can facilitate reward/addictive behavior. In addition, most neuropathological and long-term cognitive and behavioral dysfunctions are associated with the innate immune response and TLR4 signaling (Coleman et al., 2017; Crews et al., 2016; Montesinos et al., 2016b).

Some neuropathological consequences of binge ethanol drinking are myelin dysfunctions and long-term cognitive and behavioral dysfunctions. Indeed human evidence has demonstrated marked myelin alterations in adolescents who participate in binge ethanol drinking. For instance, neuroimaging studies have indicated frontal myelin fiber tract alterations (Bava et al., 2013), as well as reduced white matter integrity in the prefrontal cortex and superior longitudinal fasciculus (De Bellis et al., 2005; Elofson et al., 2013; Medina et al., 2008). These events are related with long-term memory, learning and executive dysfunctions (Jacobus and Tapert, 2013). Experimental studies have also demonstrated that the inflammatory environment (e.g., production of proinflammatory mediators and cytokines) induced by binge ethanol administration in the adolescence down-regulates the expression of several proteins associated with myelination (e.g., MBP, PLP, CNPase or NG2), along with ultrastructural myelin sheath disarrangements, observed by electron microscopy in the prefrontal cortices of adolescent WT mice treated with binge ethanol drinking. These effects have been associated with the TLR4 immune response as TLR4-deficient mice are protected mostly against both myelin alterations (Fig. 3) and the long-term cognitive dysfunctions induced by binge ethanol treatment in adolescence (Montesinos et al., 2015).

Another important neuropathological consequence of binge-ethanol drinking in adolescence is an alteration to synaptic pruning. In adolescence synapses and neural circuits are remodeled to acquire a normal cognitive function (Glantz et al., 2007). The pruning of synaptic connections in the adolescent prefrontal cortex is associated with a reduction in dendritic spines, and recent findings have demonstrated that cytokines and the TLR4 response can modulate synaptic transmission and plasticity (see the review Vezzani and Viviani, 2015). Recent evidence reveals that binge

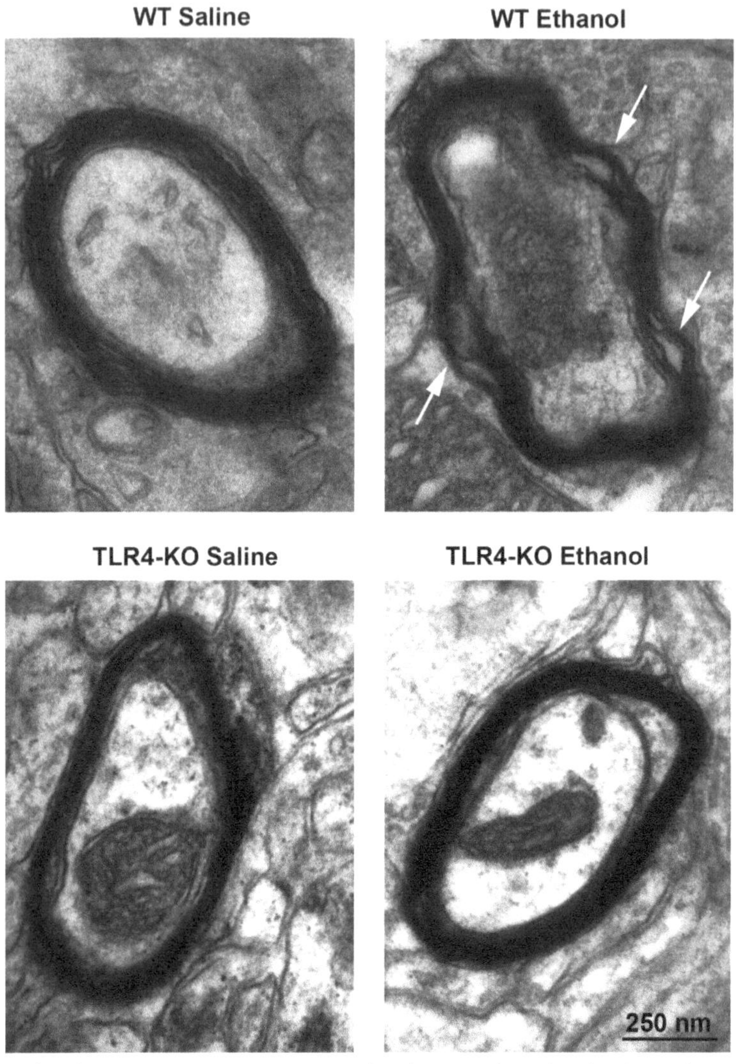

Fig. 3 The electron microscopy micrographs showing the role of TLR4 in the myelin sheath disarrangements induced by intermittent ethanol treatment in adolescent mice. The micrographs were obtained from the prefrontal cortices of the WT and TLR4-KO adolescent mice (at postnatal day 44) treated intermittently with saline or ethanol (3 g/kg) in adolescence. Arrows indicate the inter-laminar splitting of the myelin sheaths in the ethanol-treated WT adolescent mice.

ethanol treatment in adolescence affects structural synaptic proteins (PSD-95 and SHANK3) by impairing synaptic pruning maturation and that the TLR4 signaling response is involved in these effects (Montesinos et al., 2018).

6. Potential treatments and biomarkers of neuroinflammation

As described in this chapter, neuroimmune signaling is involved in many actions of ethanol on the brain, including neural damage, behavior and addiction. Moreover, the drugs used to reduce microglia activation or anti-inflammatory compounds would be good candidates to treat the neuroinflammation associated with alcohol abuse. In fact early studies have demonstrated that indomethacin, a non steroidal anti-inflammatory drug that inhibits cyclooxygenase enzyme (COX-2), reduces neuroinflammation and blocks behavioral deficits in alcohol-exposed adolescent rats (Pascual et al., 2007), and protects against loss of neurogenesis in the hippocampi of adolescents who participate in binge ethanol drinking (Vetreno et al., 2018). Blocking HMGB1 by means of antagonists such as glycyrrhizin and ethyl pyruvate prevents chronic ethanol consumption from causing brain damage (Whitman et al., 2013). Pioglitazone, a peroxisome proliferator-activated type gamma receptor (PPAR) agonist, is able to block neuroinflammation and prevent ethanol-induced microglia activation and cytokine production in several neonatal brain regions (Drew et al., 2015), and also protects against the neuronal and cognitive behavioral dysfunction elicited by binge ethanol exposure in adult rats (Cippitelli et al., 2017). Another anti-inflammatory compound used is oleoylethanolamide, whose administration prevents neuroimmune HMGB1/TLR4/NF-κB danger signaling in the rat frontal cortex and the depressive-like behavior induced by ethanol binge administration (Anton et al., 2017).

Another experimental approach was to reduce microglial activation by using minocycline, a tetracycline antibiotic. For instance, administering minocycline reduces neuroinflammation and ameliorates ethanol neurotoxicity in the developing brain (Wang et al., 2018). Minocycline also mitigates motor impairments and the cortical neuronal loss induced by focal ischemia in adolescent rats chronically exposed to ethanol (Oliveira et al., 2014).

Regarding biomarkers for neuroinflammation, some studies have shown that during chronic alcohol intake, some cytokines (IL-1β, IL-17, TNF-α) and chemokines (MCP-1, MIP-1α, CX_3CL_1) are up-regulated in the striatum and serum in mice (Pascual et al., 2015). Similarly, IL-17A, MCP-1 and MIP-1α have also been observed in the prefrontal cortex and serum of adolescent mice after binge ethanol treatment, which suggests they are potential biomarkers of neuroinflammation (Pascual et al., 2017b).

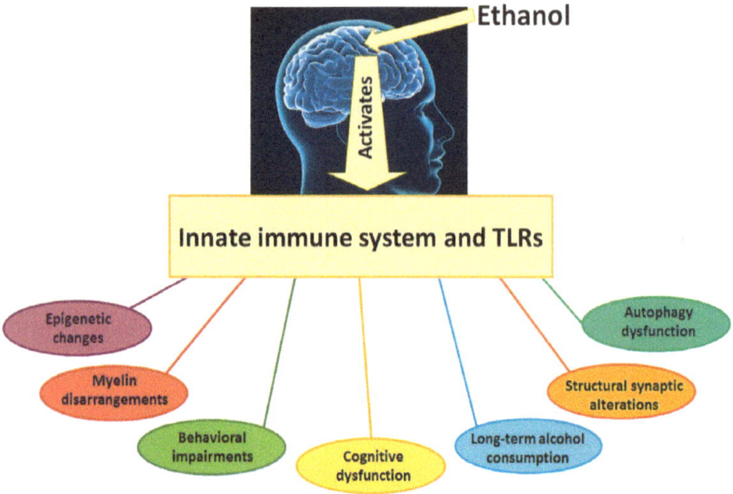

Fig. 4 Neurochemical, pathological and behavioral consequences of ethanol-induced innate immune activation and TLRs response in the adult and developing brain.

7. Concluding remarks

Human and animal studies have demonstrated that alcohol is a neurotoxic compound whose abuse induces neuropathology and cognitive/behavioral dysfunctions. Current evidence demonstrates the role of the immune response in many actions of ethanol on the brain. Ethanol activates the innate immune response in glial cells through immune receptor TLRs by leading to the release of cytokines and inflammatory mediators and causing neural damage, degeneration and long-term behavioral impairments. The ethanol-induced innate immune response occurs with either chronic consumption or binge drinking and affects the adult and developing brain. There is also evidence to demonstrate that the inflammatory environment induced by ethanol in the brain is associated with alterations to myelin, the autophagy pathway, epigenetic changes, behavioral impairments, and cognitive and synaptic dysfunctions (Fig. 4). New insights into the molecular actions of ethanol on the brain may contribute to develop effective therapeutic strategies to treat disorders associated with alcohol use.

Acknowledgments

The author's experimental work has been supported by grants from Plan Nacional sobre Drogas (20141010), Institute Carlos III and FEDER funds (RTA-Network, RD16 0017 0004) and the Spanish Ministry of Science and Innovation (SAF2015-69187R).

References

Adermark, L., Bowers, M.S., 2016. Disentangling the role of astrocytes in alcohol use disorder. Alcohol. Clin. Exp. Res. 40 (9), 1802–1816.

Agartz, I., Brag, S., Franck, J., Hammarberg, A., Okugawa, G., Svinhufvud, K., Bergman, H., 2003. MR volumetry during acute alcohol withdrawal and abstinence: a descriptive study. Alcohol Alcohol. 38 (1), 71–78.

Akira, S., Takeda, K., 2004. Toll-like receptor signalling. Nat. Rev. Immunol. 4 (7), 499–511.

Akira, S., Takeda, K., Kaisho, T., 2001. Toll-like receptors: critical proteins linking innate and acquired immunity. Nat. Immunol. 2 (8), 675–680.

Alfonso-Loeches, S., Guerri, C., 2011. Molecular and behavioral aspects of the actions of alcohol on the adult and developing brain. Crit. Rev. Clin. Lab. Sci. 48 (1), 19–47.

Alfonso-Loeches, S., Pascual-Lucas, M., Blanco, A.M., Sanchez-Vera, I., Guerri, C., 2010. Pivotal role of TLR4 receptors in alcohol-induced neuroinflammation and brain damage. J. Neurosci. 30 (24), 8285–8295.

Alfonso-Loeches, S., Pascual, M., Gomez-Pinedo, U., Pascual-Lucas, M., Renau-Piqueras, J., Guerri, C., 2012. Toll-like receptor 4 participates in the myelin disruptions associated with chronic alcohol abuse. Glia 60 (6), 948–964.

Alfonso-Loeches, S., Urena-Peralta, J.R., Morillo-Bargues, M.J., Oliver-De La Cruz, J., Guerri, C., 2014. Role of mitochondria ROS generation in ethanol-induced NLRP3 inflammasome activation and cell death in astroglial cells. Front. Cell. Neurosci. 8, 216.

Alfonso-Loeches, S., Urena-Peralta, J., Morillo-Bargues, M.J., Gomez-Pinedo, U., Guerri, C., 2016. Ethanol-induced TLR4/NLRP3 neuroinflammatory response in microglial cells promotes leukocyte infiltration across the BBB. Neurochem. Res. 41 (1–2), 193–209.

Allen, N.J., 2014. Astrocyte regulation of synaptic behavior. Annu. Rev. Cell Dev. Biol. 30, 439–463.

Amor, S., Puentes, F., Baker, D., van der Valk, P., 2010. Inflammation in neurodegenerative diseases. Immunology 129 (2), 154–169.

Anton, M., Alen, F., Gomez de Heras, R., Serrano, A., Pavon, F.J., Leza, J.C., Garcia-Bueno, B., Rodriguez de Fonseca, F., Orio, L., 2017. Oleoylethanolamide prevents neuroimmune HMGB1/TLR4/NF-kB danger signaling in rat frontal cortex and depressive-like behavior induced by ethanol binge administration. Addict. Biol. 22 (3), 724–741.

Arroyo, D.S., Soria, J.A., Gaviglio, E.A., Rodriguez-Galan, M.C., Iribarren, P., 2011. Toll-like receptors are key players in neurodegeneration. Int. Immunopharmacol. 11 (10), 1415–1421.

Attwell, D., Buchan, A.M., Charpak, S., Lauritzen, M., Macvicar, B.A., Newman, E.A., 2010. Glial and neuronal control of brain blood flow. Nature 468 (7321), 232–243.

Bagga, D., Sharma, A., Kumari, A., Kaur, P., Bhattacharya, D., Garg, M.L., Khushu, S., Singh, N., 2014. Decreased white matter integrity in fronto-occipital fasciculus bundles: relation to visual information processing in alcohol-dependent subjects. Alcohol 48 (1), 43–53.

Bauman, M.D., Iosif, A.M., Smith, S.E., Bregere, C., Amaral, D.G., Patterson, P.H., 2014. Activation of the maternal immune system during pregnancy alters behavioral development of rhesus monkey offspring. Biol. Psychiatry 75 (4), 332–341.

Bava, S., Jacobus, J., Thayer, R.E., Tapert, S.F., 2013. Longitudinal changes in white matter integrity among adolescent substance users. Alcohol. Clin. Exp. Res. 37 (Suppl. 1), E181–E189.

Bells, S., Lefebvre, J., Prescott, S.A., Dockstader, C., Bouffet, E., Skocic, J., Laughlin, S., Mabbott, D.J., 2017. Changes in white matter microstructure impact cognition by disrupting the ability of neural assemblies to synchronize. J. Neurosci. 37 (34), 8227–8238.

Bianchi, M.E., 2007. DAMPs, PAMPs and alarmins: all we need to know about danger. J. Leukoc. Biol. 81 (1), 1–5.
Blanco, A.M., Valles, S.L., Pascual, M., Guerri, C., 2005. Involvement of TLR4/type I IL-1 receptor signaling in the induction of inflammatory mediators and cell death induced by ethanol in cultured astrocytes. J. Immunol. 175 (10), 6893–6899.
Blanco, A.M., Perez-Arago, A., Fernandez-Lizarbe, S., Guerri, C., 2008. Ethanol mimics ligand-mediated activation and endocytosis of IL-1RI/TLR4 receptors via lipid rafts caveolae in astroglial cells. J. Neurochem. 106 (2), 625–639.
Blednov, Y.A., Bergeson, S.E., Walker, D., Ferreira, V.M., Kuziel, W.A., Harris, R.A., 2005. Perturbation of chemokine networks by gene deletion alters the reinforcing actions of ethanol. Behav. Brain Res. 165 (1), 110–125.
Blednov, Y.A., Benavidez, J.M., Geil, C., Perra, S., Morikawa, H., Harris, R.A., 2011. Activation of inflammatory signaling by lipopolysaccharide produces a prolonged increase of voluntary alcohol intake in mice. Brain Behav. Immun. 25 (Suppl. 1), S92–S105.
Boschen, K.E., Ruggiero, M.J., Klintsova, A.Y., 2016. Neonatal binge alcohol exposure increases microglial activation in the developing rat hippocampus. Neuroscience 324, 355–366.
Bowman, C.C., Rasley, A., Tranguch, S.L., Marriott, I., 2003. Cultured astrocytes express toll-like receptors for bacterial products. Glia 43 (3), 281–291.
Brenhouse, H.C., Schwarz, J.M., 2016. Immunoadolescence: neuroimmune development and adolescent behavior. Neurosci. Biobehav. Rev. 70, 288–299.
Chang, W., Bai, J., Tian, S., Ma, M., Li, W., Yin, Y., Deng, R., Cui, J., Li, J., Wang, G., Zhang, P., Tao, K., 2017. Autophagy protects gastric mucosal epithelial cells from ethanol-induced oxidative damage via mTOR signaling pathway. Exp. Biol. Med. (Maywood) 242 (10), 1025–1033.
Chen, Y., Klionsky, D.J., 2011. The regulation of autophagy—unanswered questions. J. Cell Sci. 124 (Pt. 2), 161–170.
Chen, G., Ke, Z., Xu, M., Liao, M., Wang, X., Qi, Y., Zhang, T., Frank, J.A., Bower, K.A., Shi, X., Luo, J., 2012. Autophagy is a protective response to ethanol neurotoxicity. Autophagy 8 (11), 1577–1589.
Chen, W.W., Zhang, X., Huang, W.J., 2016. Role of neuroinflammation in neurodegenerative diseases (review). Mol. Med. Rep. 13 (4), 3391–3396.
Cippitelli, A., Domi, E., Ubaldi, M., Douglas, J.C., Li, H.W., Demopulos, G., Gaitanaris, G., Roberto, M., Drew, P.D., Kane, C.J.M., Ciccocioppo, R., 2017. Protection against alcohol-induced neuronal and cognitive damage by the PPARgamma receptor agonist pioglitazone. Brain Behav. Immun. 64, 320–329.
Coleman Jr., L.G., Crews, F.T., 2018. Innate immune signaling and alcohol use disorders. Handb. Exp. Pharmacol. 248, 369–396.
Coleman Jr., L.G., Zou, J., Qin, L., Crews, F.T., 2017. HMGB1/IL-1beta complexes regulate neuroimmune responses in alcoholism. Brain Behav. Immun. 72, 61–77.
Colonna, M., 2007. TLR pathways and IFN-regulatory factors: to each its own. Eur. J. Immunol. 37 (2), 306–309.
Crews, F.T., Vetreno, R.P., 2014. Neuroimmune basis of alcoholic brain damage. Int. Rev. Neurobiol. 118, 315–357.
Crews, F., He, J., Hodge, C., 2007. Adolescent cortical development: a critical period of vulnerability for addiction. Pharmacol. Biochem. Behav. 86 (2), 189–199.
Crews, F.T., Vetreno, R.P., Broadwater, M.A., Robinson, D.L., 2016. Adolescent alcohol exposure persistently impacts adult neurobiology and behavior. Pharmacol. Rev. 68 (4), 1074–1109.
Crews, F.T., Walter, T.J., Coleman Jr., L.G., Vetreno, R.P., 2017. Toll-like receptor signaling and stages of addiction. Psychopharmacology (Berl) 234 (9–10), 1483–1498.

Cunningham, C., 2013. Microglia and neurodegeneration: the role of systemic inflammation. Glia 61 (1), 71–90.
Davis, E.J., Foster, T.D., Thomas, W.E., 1994. Cellular forms and functions of brain microglia. Brain Res. Bull. 34 (1), 73–78.
De Bellis, M.D., Narasimhan, A., Thatcher, D.L., Keshavan, M.S., Soloff, P., Clark, D.B., 2005. Prefrontal cortex, thalamus, and cerebellar volumes in adolescents and young adults with adolescent-onset alcohol use disorders and comorbid mental disorders. Alcohol. Clin. Exp. Res. 29 (9), 1590–1600.
de la Monte, S.M., Kril, J.J., 2014. Human alcohol-related neuropathology. Acta Neuropathol. 127 (1), 71–90.
Dietrich, P.Y., Walker, P.R., Saas, P., 2003. Death receptors on reactive astrocytes: a key role in the fine tuning of brain inflammation? Neurology 60 (4), 548–554.
Doremus-Fitzwater, T.L., Buck, H.M., Bordner, K., Richey, L., Jones, M.E., Deak, T., 2014. Intoxication- and withdrawal-dependent expression of central and peripheral cytokines following initial ethanol exposure. Alcohol. Clin. Exp. Res. 38 (8), 2186–2198.
Drew, P.D., Kane, C.J., 2014. Fetal alcohol spectrum disorders and neuroimmune changes. Int. Rev. Neurobiol. 118, 41–80.
Drew, P.D., Johnson, J.W., Douglas, J.C., Phelan, K.D., Kane, C.J., 2015. Pioglitazone blocks ethanol induction of microglial activation and immune responses in the hippocampus, cerebellum, and cerebral cortex in a mouse model of fetal alcohol spectrum disorders. Alcohol. Clin. Exp. Res. 39 (3), 445–454.
Du, L., Zhang, Y., Chen, Y., Zhu, J., Yang, Y., Zhang, H.L., 2017. Role of microglia in neurological disorders and their potentials as a therapeutic target. Mol. Neurobiol. 54 (10), 7567–7584.
Durkee, C.A., Sarlls, J.E., Hommer, D.W., Momenan, R., 2013. White matter microstructure alterations: a study of alcoholics with and without post-traumatic stress disorder. PLoS One 8 (11), e80952.
El Mezayen, R., El Gazzar, M., Seeds, M.C., McCall, C.E., Dreskin, S.C., Nicolls, M.R., 2007. Endogenous signals released from necrotic cells augment inflammatory responses to bacterial endotoxin. Immunol. Lett. 111 (1), 36–44.
Elofson, J., Gongvatana, W., Carey, K.B., 2013. Alcohol use and cerebral white matter compromise in adolescence. Addict. Behav. 38 (7), 2295–2305.
Farrugia, M., Baron, B., 2017. The role of toll-like receptors in autoimmune diseases through failure of the self-recognition mechanism. Int. J. Inflamm. 2017, 8391230.
Fernandez-Lizarbe, S., Pascual, M., Guerri, C., 2009. Critical role of TLR4 response in the activation of microglia induced by ethanol. J. Immunol. 183 (7), 4733–4744.
Gano, A., Doremus-Fitzwater, T.L., Deak, T., 2016. Sustained alterations in neuroimmune gene expression after daily, but not intermittent, alcohol exposure. Brain Res. 1646, 62–72.
Gavin, D.P., Grayson, D.R., Varghese, S.P., Guizzetti, M., 2017. Chromatin switches during neural cell differentiation and their dysregulation by prenatal alcohol exposure. Genes (Basel) 8 (5).
Giedd, J.N., 2004. Structural magnetic resonance imaging of the adolescent brain. Ann. N. Y. Acad. Sci. 1021, 77–85.
Giordano, S., Darley-Usmar, V., Zhang, J., 2014. Autophagy as an essential cellular antioxidant pathway in neurodegenerative disease. Redox Biol. 2, 82–90.
Girault, V., Gilard, V., Marguet, F., Lesueur, C., Hauchecorne, M., Ramdani, Y., Laquerriere, A., Marret, S., Jegou, S., Gonzalez, B.J., Brasse-Lagnel, C., Bekri, S., 2017. Prenatal alcohol exposure impairs autophagy in neonatal brain cortical microvessels. Cell Death Dis. 8 (2), e2610.

Glantz, L.A., Gilmore, J.H., Hamer, R.M., Lieberman, J.A., Jarskog, L.F., 2007. Synaptophysin and postsynaptic density protein 95 in the human prefrontal cortex from mid-gestation into early adulthood. Neuroscience 149 (3), 582–591.

Glass, C.K., Saijo, K., Winner, B., Marchetto, M.C., Gage, F.H., 2010. Mechanisms underlying inflammation in neurodegeneration. Cell 140 (6), 918–934.

Goodfellow, M.J., Shin, Y.J., Lindquist, D.H., 2018. Mitigation of postnatal ethanol-induced neuroinflammation ameliorates trace fear memory deficits in juvenile rats. Behav. Brain Res. 338, 28–31.

Greter, M., Merad, M., 2013. Regulation of microglia development and homeostasis. Glia 61 (1), 121–127.

Guerri, C., Bazinet, A., Riley, E.P., 2009. Foetal alcohol spectrum disorders and alterations in brain and behaviour. Alcohol Alcohol. 44 (2), 108–114.

Guo, H., Callaway, J.B., Ting, J.P., 2015. Inflammasomes: mechanism of action, role in disease, and therapeutics. Nat. Med. 21 (7), 677–687.

Hagberg, H., Mallard, C., Ferriero, D.M., Vannucci, S.J., Levison, S.W., Vexler, Z.S., Gressens, P., 2015. The role of inflammation in perinatal brain injury. Nat. Rev. Neurol. 11 (4), 192–208.

Hanamsagar, R., Hanke, M.L., Kielian, T., 2012. Toll-like receptor (TLR) and inflammasome actions in the central nervous system. Trends Immunol. 33 (7), 333–342.

Hanke, M.L., Kielian, T., 2011. Toll-like receptors in health and disease in the brain: mechanisms and therapeutic potential. Clin. Sci. (Lond.) 121 (9), 367–387.

Harper, C., Matsumoto, I., 2005. Ethanol and brain damage. Curr. Opin. Pharmacol. 5 (1), 73–78.

Harris, R.A., Bajo, M., Bell, R.L., Blednov, Y.A., Varodayan, F.P., Truitt, J.M., de Guglielmo, G., Lasek, A.W., Logrip, M.L., Vendruscolo, L.F., Roberts, A.J., Roberts, E., George, O., Mayfield, J., Billiar, T.R., Hackam, D.J., Mayfield, R.D., Koob, G.F., Roberto, M., Homanics, G.E., 2017. Genetic and pharmacologic manipulation of TLR4 has minimal impact on ethanol consumption in rodents. J. Neurosci. 37 (5), 1139–1155.

Hayes, D.M., Deeny, M.A., Shaner, C.A., Nixon, K., 2013. Determining the threshold for alcohol-induced brain damage: new evidence with gliosis markers. Alcohol. Clin. Exp. Res. 37 (3), 425–434.

He, J., Crews, F.T., 2008. Increased MCP-1 and microglia in various regions of the human alcoholic brain. Exp. Neurol. 210 (2), 349–358.

Ibrahim, Z.A., Armour, C.L., Phipps, S., Sukkar, M.B., 2013. RAGE and TLRs: relatives, friends or neighbours? Mol. Immunol. 56 (4), 739–744.

Jacobus, J., Tapert, S.F., 2013. Neurotoxic effects of alcohol in adolescence. Annu. Rev. Clin. Psychol. 9, 703–721.

Jo, E.K., Kim, J.K., Shin, D.M., Sasakawa, C., 2016. Molecular mechanisms regulating NLRP3 inflammasome activation. Cell. Mol. Immunol. 13 (2), 148–159.

June, H.L., Liu, J., Warnock, K.T., Bell, K.A., Balan, I., Bollino, D., Puche, A., Aurelian, L., 2015. CRF-amplified neuronal TLR4/MCP-1 signaling regulates alcohol self-administration. Neuropsychopharmacology 40 (6), 1549–1559.

Kane, C.J., Phelan, K.D., Han, L., Smith, R.R., Xie, J., Douglas, J.C., Drew, P.D., 2011. Protection of neurons and microglia against ethanol in a mouse model of fetal alcohol spectrum disorders by peroxisome proliferator-activated receptor-gamma agonists. Brain Behav. Immun. 25 (Suppl. 1), S137–S145.

Kettenmann, H., Hanisch, U.K., Noda, M., Verkhratsky, A., 2011. Physiology of microglia. Physiol. Rev. 91 (2), 461–553.

Kielian, T., 2006. Toll-like receptors in central nervous system glial inflammation and homeostasis. J. Neurosci. Res. 83 (5), 711–730.

Kokkinopoulos, I., Jordan, W.J., Ritter, M.A., 2005. Toll-like receptor mRNA expression patterns in human dendritic cells and monocytes. Mol. Immunol. 42 (8), 957–968.
Konsman, J.P., Veeneman, J., Combe, C., Poole, S., Luheshi, G.N., Dantzer, R., 2008. Central nervous action of interleukin-1 mediates activation of limbic structures and behavioural depression in response to peripheral administration of bacterial lipopolysaccharide. Eur. J. Neurosci. 28 (12), 2499–2510.
Kruman, I.I., Henderson, G.I., Bergeson, S.E., 2012. DNA damage and neurotoxicity of chronic alcohol abuse. Exp. Biol. Med. (Maywood) 237 (7), 740–747.
Lampron, A., Elali, A., Rivest, S., 2013. Innate immunity in the CNS: redefining the relationship between the CNS and its environment. Neuron 78 (2), 214–232.
Leclercq, S., Cani, P.D., Neyrinck, A.M., Starkel, P., Jamar, F., Mikolajczak, M., Delzenne, N.M., de Timary, P., 2012. Role of intestinal permeability and inflammation in the biological and behavioral control of alcohol-dependent subjects. Brain Behav. Immun. 26 (6), 911–918.
Lenz, K.M., Nugent, B.M., Haliyur, R., McCarthy, M.M., 2013. Microglia are essential to masculinization of brain and behavior. J. Neurosci. 33 (7), 2761–2772.
Lewohl, J.M., Wang, L., Miles, M.F., Zhang, L., Dodd, P.R., Harris, R.A., 2000. Gene expression in human alcoholism: microarray analysis of frontal cortex. Alcohol. Clin. Exp. Res. 24 (12), 1873–1882.
Lewohl, J.M., Nunez, Y.O., Dodd, P.R., Tiwari, G.R., Harris, R.A., Mayfield, R.D., 2011. Up-regulation of microRNAs in brain of human alcoholics. Alcohol. Clin. Exp. Res. 35 (11), 1928–1937.
Liddelow, S., Barres, B., 2015. SnapShot: astrocytes in health and disease. Cell 162 (5), 1170–1170.e1.
Lippai, D., Bala, S., Petrasek, J., Csak, T., Levin, I., Kurt-Jones, E.A., Szabo, G., 2013. Alcohol-induced IL-1beta in the brain is mediated by NLRP3/ASC inflammasome activation that amplifies neuroinflammation. J. Leukoc. Biol. 94 (1), 171–182.
Little, A.R., O'Callagha, J.P., 2001. Astrogliosis in the adult and developing CNS: is there a role for proinflammatory cytokines? Neurotoxicology 22 (5), 607–618.
Lowe, P.P., Gyongyosi, B., Satishchandran, A., Iracheta-Vellve, A., Ambade, A., Kodys, K., Catalano, D., Ward, D.V., Szabo, G., 2017. Alcohol-related changes in the intestinal microbiome influence neutrophil infiltration, inflammation and steatosis in early alcoholic hepatitis in mice. PLoS One 12 (3), e0174544.
Lu, M., Sun, X.L., Qiao, C., Liu, Y., Ding, J.H., Hu, G., 2014. Uncoupling protein 2 deficiency aggravates astrocytic endoplasmic reticulum stress and nod-like receptor protein 3 inflammasome activation. Neurobiol. Aging 35 (2), 421–430.
Mandrekar, P., Szabo, G., 2009. Signalling pathways in alcohol-induced liver inflammation. J. Hepatol. 50 (6), 1258–1266.
Mandrekar, P., Catalano, D., White, B., Szabo, G., 2006. Moderate alcohol intake in humans attenuates monocyte inflammatory responses: inhibition of nuclear regulatory factor kappa B and induction of interleukin 10. Alcohol. Clin. Exp. Res. 30 (1), 135–139.
Maragakis, N.J., Rothstein, J.D., 2006. Mechanisms of disease: astrocytes in neurodegenerative disease. Nat. Clin. Pract. Neurol. 2 (12), 679–689.
Marc, T., 2013. Brain development and the immune system: an introduction to inflammatory and infectious diseases of the child's brain. Handb. Clin. Neurol. 112, 1087–1089.
Marshall, S.A., Geil, C.R., Nixon, K., 2016. Prior binge ethanol exposure potentiates the microglial response in a model of alcohol-induced neurodegeneration. Brain Sci. 6 (2).
Martinon, F., 2012. Dangerous liaisons: mitochondrial DNA meets the NLRP3 inflammasome. Immunity 36 (3), 313–315.
Mayfield, J., Ferguson, L., Harris, R.A., 2013. Neuroimmune signaling: a key component of alcohol abuse. Curr. Opin. Neurobiol. 23 (4), 513–520.

Mayfield, J., Arends, M.A., Harris, R.A., Blednov, Y.A., 2016. Genes and alcohol consumption: studies with mutant mice. Int. Rev. Neurobiol. 126, 293–355.
McManus, R.M., Heneka, M.T., 2017. Role of neuroinflammation in neurodegeneration: new insights. Alzheimers Res. Ther. 9 (1), 14.
Medina, K.L., McQueeny, T., Nagel, B.J., Hanson, K.L., Schweinsburg, A.D., Tapert, S.F., 2008. Prefrontal cortex volumes in adolescents with alcohol use disorders: unique gender effects. Alcohol. Clin. Exp. Res. 32 (3), 386–394.
Medzhitov, R., Preston-Hurlburt, P., Janeway Jr., C.A., 1997. A human homologue of the Drosophila Toll protein signals activation of adaptive immunity. Nature 388 (6640), 394–397.
Messingham, K.A., Faunce, D.E., Kovacs, E.J., 2002. Alcohol, injury, and cellular immunity. Alcohol 28 (3), 137–149.
Meyer, U., Feldon, J., Dammann, O., 2011. Schizophrenia and autism: both shared and disorder-specific pathogenesis via perinatal inflammation? Pediatr. Res. 69 (5 Pt. 2), 26R–33R.
Miller, A.M., Horiguchi, N., Jeong, W.I., Radaeva, S., Gao, B., 2011. Molecular mechanisms of alcoholic liver disease: innate immunity and cytokines. Alcohol. Clin. Exp. Res. 35 (5), 787–793.
Montesinos, J., Pascual, M., Pla, A., Maldonado, C., Rodriguez-Arias, M., Minarro, J., Guerri, C., 2015. TLR4 elimination prevents synaptic and myelin alterations and long-term cognitive dysfunctions in adolescent mice with intermittent ethanol treatment. Brain Behav. Immun. 45, 233–244.
Montesinos, J., Alfonso-Loeches, S., Guerri, C., 2016a. Impact of the innate immune response in the actions of ethanol on the central nervous system. Alcohol. Clin. Exp. Res. 40 (11), 2260–2270.
Montesinos, J., Pascual, M., Rodriguez-Arias, M., Minarro, J., Guerri, C., 2016b. Involvement of TLR4 in the long-term epigenetic changes, rewarding and anxiety effects induced by intermittent ethanol treatment in adolescence. Brain Behav. Immun. 53, 159–171.
Montesinos, J., Pascual, M., Millan-Esteban, D., Guerri, C., 2018. Binge-like ethanol treatment in adolescence impairs autophagy and hinders synaptic maturation: role of TLR4. Neurosci. Lett. 682, 85–91.
Moore, E.M., Migliorini, R., Infante, M.A., Riley, E.P., 2014. Fetal alcohol spectrum disorders: recent neuroimaging findings. Curr. Dev. Disord. Rep. 1 (3), 161–172.
Mrak, R.E., Griffin, W.S., 2005. Potential inflammatory biomarkers in Alzheimer's disease. J. Alzheimers Dis. 8 (4), 369–375.
Nagy, L.E., 2003. Recent insights into the role of the innate immune system in the development of alcoholic liver disease. Exp. Biol. Med. (Maywood) 228 (8), 882–890.
Nagyoszi, P., Wilhelm, I., Farkas, A.E., Fazakas, C., Dung, N.T., Hasko, J., Krizbai, I.A., 2010. Expression and regulation of toll-like receptors in cerebral endothelial cells. Neurochem. Int. 57 (5), 556–564.
Newville, J., Valenzuela, C.F., Li, L., Jantzie, L.L., Cunningham, L.A., 2017. Acute oligodendrocyte loss with persistent white matter injury in a third trimester equivalent mouse model of fetal alcohol spectrum disorder. Glia 65 (8), 1317–1332.
Nishiya, T., DeFranco, A.L., 2004. Ligand-regulated chimeric receptor approach reveals distinctive subcellular localization and signaling properties of the Toll-like receptors. J. Biol. Chem. 279 (18), 19008–19017.
Oliveira, G.B., Fontes Ede Jr., A., de Carvalho, S., da Silva, J.B., Fernandes, L.M., Oliveira, M.C., Prediger, R.D., Gomes-Leal, W., Lima, R.R., Maia, C.S., 2014. Minocycline mitigates motor impairments and cortical neuronal loss induced by focal ischemia in rats chronically exposed to ethanol during adolescence. Brain Res. 1561, 23–34.

O'Neill, L.A., Bowie, A.G., 2007. The family of five: TIR-domain-containing adaptors in Toll-like receptor signalling. Nat. Rev. Immunol. 7 (5), 353–364.
Parlesak, A., Schafer, C., Schutz, T., Bode, J.C., Bode, C., 2000. Increased intestinal permeability to macromolecules and endotoxemia in patients with chronic alcohol abuse in different stages of alcohol-induced liver disease. J. Hepatol. 32 (5), 742–747.
Pascual, M., Blanco, A.M., Cauli, O., Minarro, J., Guerri, C., 2007. Intermittent ethanol exposure induces inflammatory brain damage and causes long-term behavioural alterations in adolescent rats. Eur. J. Neurosci. 25 (2), 541–550.
Pascual, M., Balino, P., Alfonso-Loeches, S., Aragon, C.M., Guerri, C., 2011a. Impact of TLR4 on behavioral and cognitive dysfunctions associated with alcohol-induced neuroinflammatory damage. Brain Behav. Immun. 25 (Suppl. 1), S80–S91.
Pascual, M., Fernandez-Lizarbe, S., Guerri, C., 2011b. Role of TLR4 in ethanol effects on innate and adaptive immune responses in peritoneal macrophages. Immunol. Cell Biol. 89 (6), 716–727.
Pascual, M., Pla, A., Minarro, J., Guerri, C., 2014. Neuroimmune activation and myelin changes in adolescent rats exposed to high-dose alcohol and associated cognitive dysfunction: a review with reference to human adolescent drinking. Alcohol Alcohol. 49 (2), 187–192.
Pascual, M., Balino, P., Aragon, C.M., Guerri, C., 2015. Cytokines and chemokines as biomarkers of ethanol-induced neuroinflammation and anxiety-related behavior: role of TLR4 and TLR2. Neuropharmacology 89, 352–359.
Pascual, M., Montesinos, J., Guerri, C., 2017a. Role of the innate immune system in the neuropathological consequences induced by adolescent binge drinking. J. Neurosci. Res. 96 (5), 765–780.
Pascual, M., Montesinos, J., Marcos, M., Torres, J.L., Costa-Alba, P., Garcia-Garcia, F., Laso, F.J., Guerri, C., 2017b. Gender differences in the inflammatory cytokine and chemokine profiles induced by binge ethanol drinking in adolescence. Addict. Biol. 22 (6), 1829–1841.
Pascual, M., Montesinos, J., Montagud-Romero, S., Forteza, J., Rodriguez-Arias, M., Minarro, J., Guerri, C., 2017c. TLR4 response mediates ethanol-induced neurodevelopment alterations in a model of fetal alcohol spectrum disorders. J. Neuroinflammation 14 (1), 145.
Pascual-Lucas, M., Fernandez-Lizarbe, S., Montesinos, J., Guerri, C., 2014. LPS or ethanol triggers clathrin- and rafts/caveolae-dependent endocytosis of TLR4 in cortical astrocytes. J. Neurochem. 129 (3), 448–462.
Pattingre, S., Espert, L., Biard-Piechaczyk, M., Codogno, P., 2008. Regulation of macroautophagy by mTOR and Beclin 1 complexes. Biochimie 90 (2), 313–323.
Pekny, M., Pekna, M., 2014. Astrocyte reactivity and reactive astrogliosis: costs and benefits. Physiol. Rev. 94 (4), 1077–1098.
Pfefferbaum, A., Adalsteinsson, E., Sullivan, E.V., 2006. Supratentorial profile of white matter microstructural integrity in recovering alcoholic men and women. Biol. Psychiatry 59 (4), 364–372.
Pfefferbaum, A., Rosenbloom, M.J., Fama, R., Sassoon, S.A., Sullivan, E.V., 2010. Transcallosal white matter degradation detected with quantitative fiber tracking in alcoholic men and women: selective relations to dissociable functions. Alcohol. Clin. Exp. Res. 34 (7), 1201–1211.
Pla, A., Pascual, M., Renau-Piqueras, J., Guerri, C., 2014. TLR4 mediates the impairment of ubiquitin-proteasome and autophagy-lysosome pathways induced by ethanol treatment in brain. Cell Death Dis. 5, e1066.
Pla, A., Pascual, M., Guerri, C., 2016. Autophagy constitutes a protective mechanism against ethanol toxicity in mouse astrocytes and neurons. PLoS One 11 (4), e0153097.
Prinz, M., Priller, J., 2017. The role of peripheral immune cells in the CNS in steady state and disease. Nat. Neurosci. 20 (2), 136–144.

Prock, T.L., Miranda, R.C., 2007. Embryonic cerebral cortical progenitors are resistant to apoptosis, but increase expression of suicide receptor DISC-complex genes and suppress autophagy following ethanol exposure. Alcohol. Clin. Exp. Res. 31 (4), 694–703.
Qin, L., He, J., Hanes, R.N., Pluzarev, O., Hong, J.S., Crews, F.T., 2008. Increased systemic and brain cytokine production and neuroinflammation by endotoxin following ethanol treatment. J. Neuroinflammation 5, 10.
Raison, C.L., Capuron, L., Miller, A.H., 2006. Cytokines sing the blues: inflammation and the pathogenesis of depression. Trends Immunol. 27 (1), 24–31.
Ransohoff, R.M., 2016. How neuroinflammation contributes to neurodegeneration. Science 353 (6301), 777–783.
Reemst, K., Noctor, S.C., Lucassen, P.J., Hol, E.M., 2016. The indispensable roles of microglia and astrocytes during brain development. Front. Hum. Neurosci. 10, 566.
Roach, J.M., Racioppi, L., Jones, C.D., Masci, A.M., 2013. Phylogeny of Toll-like receptor signaling: adapting the innate response. PLoS One 8 (1), e54156.
Rosenblat, J.D., McIntyre, R.S., 2016. Bipolar disorder and inflammation. Psychiatr. Clin. North Am. 39 (1), 125–137.
Rubio-Araiz, A., Porcu, F., Perez-Hernandez, M., Garcia-Gutierrez, M.S., Aracil-Fernandez, M.A., Gutierrez-Lopez, M.D., Guerri, C., Manzanares, J., O'Shea, E., Colado, M.I., 2017. Disruption of blood-brain barrier integrity in postmortem alcoholic brain: preclinical evidence of TLR4 involvement from a binge-like drinking model. Addict. Biol. 22 (4), 1103–1116.
Ruiz, M., Ewig, S., Torres, A., Arancibia, F., Marco, F., Mensa, J., Sanchez, M., Martinez, J.A., 1999. Severe community-acquired pneumonia. Risk factors and follow-up epidemiology. Am. J. Respir. Crit. Care Med. 160 (3), 923–929.
Salter, M.W., Stevens, B., 2017. Microglia emerge as central players in brain disease. Nat. Med. 23 (9), 1018–1027.
Samantaray, S., Knaryan, V.H., Patel, K.S., Mulholland, P.J., Becker, H.C., Banik, N.L., 2015. Chronic intermittent ethanol induced axon and myelin degeneration is attenuated by calpain inhibition. Brain Res. 1622, 7–21.
Schulte, T., Sullivan, E.V., Muller-Oehring, E.M., Adalsteinsson, E., Pfefferbaum, A., 2005. Corpus callosal microstructural integrity influences interhemispheric processing: a diffusion tensor imaging study. Cereb. Cortex 15 (9), 1384–1392.
Shrestha, R., Millington, O., Brewer, J., Bushell, T., 2013. Is central nervous system an immune-privileged site? Kathmandu Univ. Med. J. (KUMJ) 11 (41), 102–107.
Simons, M., Trotter, J., 2007. Wrapping it up: the cell biology of myelination. Curr. Opin. Neurobiol. 17 (5), 533–540.
Sofroniew, M.V., Vinters, H.V., 2010. Astrocytes: biology and pathology. Acta Neuropathol. 119 (1), 7–35.
Sokol, R.J., Delaney-Black, V., Nordstrom, B., 2003. Fetal alcohol spectrum disorder. JAMA 290 (22), 2996–2999.
Sowell, K.D., Uriu-Adams, J.Y., Van de Water, J., Chambers, C.D., Coles, C.D., Kable, J.A., Yevtushok, L., Zymak-Zakutnya, N., Wertelecki, W., Keen, C.L., 2018. Implications of altered maternal cytokine concentrations on infant outcomes in children with prenatal alcohol exposure. Alcohol 68, 49–58.
Spear, L.P., 2007. Assessment of adolescent neurotoxicity: rationale and methodological considerations. Neurotoxicol. Teratol. 29 (1), 1–9.
Steinman, L., 2010. Mixed results with modulation of TH-17 cells in human autoimmune diseases. Nat. Immunol. 11 (1), 41–44.
Streit, W.J., 2000. Microglial response to brain injury: a brief synopsis. Toxicol. Pathol. 28 (1), 28–30.

Sutherland, G.T., Sheedy, D., Sheahan, P.J., Kaplan, W., Kril, J.J., 2014. Comorbidities, confounders, and the white matter transcriptome in chronic alcoholism. Alcohol. Clin. Exp. Res. 38 (4), 994–1001.
Szabo, G., Lippai, D., 2014. Converging actions of alcohol on liver and brain immune signaling. Int. Rev. Neurobiol. 118, 359–380.
Szabo, G., Mandrekar, P., 2009. A recent perspective on alcohol, immunity, and host defense. Alcohol. Clin. Exp. Res. 33 (2), 220–232.
Szabo, G., Saha, B., 2015. Alcohol's effect on host defense. Alcohol Res. 37 (2), 159–170.
Takeda, K., Kaisho, T., Akira, S., 2003. Toll-like receptors. Annu. Rev. Immunol. 21, 335–376.
Tay, T.L., Savage, J.C., Hui, C.W., Bisht, K., Tremblay, M.E., 2017. Microglia across the lifespan: from origin to function in brain development, plasticity and cognition. J. Physiol. 595 (6), 1929–1945.
Tian, L., Ma, L., Kaarela, T., Li, Z., 2012. Neuroimmune crosstalk in the central nervous system and its significance for neurological diseases. J. Neuroinflammation 9, 155.
Tiwari, V., Chopra, K., 2012. Attenuation of oxidative stress, neuroinflammation, and apoptosis by curcumin prevents cognitive deficits in rats postnatally exposed to ethanol. Psychopharmacology (Berl) 224 (4), 519–535.
Todde, V., Veenhuis, M., van der Klei, I.J., 2009. Autophagy: principles and significance in health and disease. Biochim. Biophys. Acta 1792 (1), 3–13.
Toga, A.W., Thompson, P.M., Sowell, E.R., 2006. Mapping brain maturation. Trends Neurosci. 29 (3), 148–159.
Topper, L.A., Baculis, B.C., Valenzuela, C.F., 2015. Exposure of neonatal rats to alcohol has differential effects on neuroinflammation and neuronal survival in the cerebellum and hippocampus. J. Neuroinflammation 12, 160.
Valenta, J.P., Gonzales, R.A., 2016. Chronic intracerebroventricular infusion of monocyte chemoattractant protein-1 leads to a persistent increase in sweetened ethanol consumption during operant self-administration but does not influence sucrose consumption in Long-Evans rats. Alcohol. Clin. Exp. Res. 40 (1), 187–195.
Valles, S.L., Blanco, A.M., Pascual, M., Guerri, C., 2004. Chronic ethanol treatment enhances inflammatory mediators and cell death in the brain and in astrocytes. Brain Pathol. 14 (4), 365–371.
Vetreno, R.P., Crews, F.T., 2012. Adolescent binge drinking increases expression of the danger signal receptor agonist HMGB1 and Toll-like receptors in the adult prefrontal cortex. Neuroscience 226, 475–488.
Vetreno, R.P., Crews, F.T., 2015. Binge ethanol exposure during adolescence leads to a persistent loss of neurogenesis in the dorsal and ventral hippocampus that is associated with impaired adult cognitive functioning. Front. Neurosci. 9, 35.
Vetreno, R.P., Qin, L., Crews, F.T., 2013. Increased receptor for advanced glycation end product expression in the human alcoholic prefrontal cortex is linked to adolescent drinking. Neurobiol. Dis. 59, 52–62.
Vetreno, R.P., Lawrimore, C.J., Rowsey, P.J., Crews, F.T., 2018. Persistent adult neuroimmune activation and loss of hippocampal neurogenesis following adolescent ethanol exposure: blockade by exercise and the anti-inflammatory drug indomethacin. Front. Neurosci. 12, 200.
Vezzani, A., Viviani, B., 2015. Neuromodulatory properties of inflammatory cytokines and their impact on neuronal excitability. Neuropharmacology 96 (Pt. A), 70–82.
Walter, T.J., Vetreno, R.P., Crews, F.T., 2017. Alcohol and stress activation of microglia and neurons: brain regional effects. Alcohol. Clin. Exp. Res. 41 (12), 2066–2081.
Wang, Y., Mandelkow, E., 2012. Degradation of tau protein by autophagy and proteasomal pathways. Biochem. Soc. Trans. 40 (4), 644–652.

Wang, X., Zhang, K., Yang, F., Ren, Z., Xu, M., Frank, J.A., Ke, Z.J., Luo, J., 2018. Minocycline protects developing brain against ethanol-induced damage. Neuropharmacology 129, 84–99.
Warren, K.R., Hewitt, B.G., Thomas, J.D., 2011. Fetal alcohol spectrum disorders: research challenges and opportunities. Alcohol Res. Health 34 (1), 4–14.
Weinhard, L., di Bartolomei, G., Bolasco, G., Machado, P., Schieber, N.L., Neniskyte, U., Exiga, M., Vadisiute, A., Raggioli, A., Schertel, A., Schwab, Y., Gross, C.T., 2018. Microglia remodel synapses by presynaptic trogocytosis and spine head filopodia induction. Nat. Commun. 9 (1), 1228.
White, R., Kramer-Albers, E.M., 2014. Axon-glia interaction and membrane traffic in myelin formation. Front. Cell. Neurosci. 7, 284.
Whitman, B.A., Knapp, D.J., Werner, D.F., Crews, F.T., Breese, G.R., 2013. The cytokine mRNA increase induced by withdrawal from chronic ethanol in the sterile environment of brain is mediated by CRF and HMGB1 release. Alcohol. Clin. Exp. Res. 37 (12), 2086–2097.
WHO, 2018. Global Status Report on Alcohol and Health. WHO Press, Geneva.
Willingham, S.B., Bergstralh, D.T., O'Connor, W., Morrison, A.C., Taxman, D.J., Duncan, J.A., Barnoy, S., Venkatesan, M.M., Flavell, R.A., Deshmukh, M., Hoffman, H.M., Ting, J.P., 2007. Microbial pathogen-induced necrotic cell death mediated by the inflammasome components CIAS1/cryopyrin/NLRP3 and ASC. Cell Host Microbe 2 (3), 147–159.
Wu, Y., Lousberg, E.L., Moldenhauer, L.M., Hayball, J.D., Coller, J.K., Rice, K.C., Watkins, L.R., Somogyi, A.A., Hutchinson, M.R., 2012. Inhibiting the TLR4-MyD88 signalling cascade by genetic or pharmacological strategies reduces acute alcohol-induced sedation and motor impairment in mice. Br. J. Pharmacol. 165 (5), 1319–1329.
Wu, P.J., Liu, H.Y., Huang, T.N., Hsueh, Y.P., 2016. AIM 2 inflammasomes regulate neuronal morphology and influence anxiety and memory in mice. Sci. Rep. 6, 32405.
Yamamoto, A., Simonsen, A., 2011. The elimination of accumulated and aggregated proteins: a role for aggrephagy in neurodegeneration. Neurobiol. Dis. 43 (1), 17–28.
Yirmiya, R., Goshen, I., 2011. Immune modulation of learning, memory, neural plasticity and neurogenesis. Brain Behav. Immun. 25 (2), 181–213.
Yorimitsu, T., Klionsky, D.J., 2005. Autophagy: molecular machinery for self-eating. Cell Death Differ. 12 (Suppl. 2), 1542–1552.
Zou, J., Crews, F.T., 2012. Inflammasome-IL-1beta signaling mediates ethanol inhibition of hippocampal neurogenesis. Front. Neurosci. 6, 77.

CHAPTER TEN

Blood-brain barrier at the interface of air pollution-associated neurotoxicity and neuroinflammation

Tamara L. Young, Katherine E. Zychowski, Jesse L. Denson, Matthew J. Campen*
Department of Pharmaceutical Sciences, College of Pharmacy, University of New Mexico, Albuquerque, NM, United States
*Corresponding author: e-mail address: mcampen@salud.unm.edu

Contents

Advances in Neurotoxicology, Volume 3
ISSN 2468-7480
https://doi.org/10.1016/bs.ant.2018.10.010

1. Introduction

The brain is the complex command center of the nervous system. Making up just 2% of the total body weight, it utilizes 20% of the body's energy resources and requires 15% of the total cardiac output to facilitate the delivery of oxygen and nutrients and removal of waste products (Fantini et al., 2016; Joris et al., 2018; Raichle and Gusnard, 2002). The central nervous system (CNS) microenvironment is very strictly regulated to maintain homeostasis. One of the main lines of defense from external insult is the blood-brain barrier (BBB), which separates the brain circulation from the systemic. Despite this safeguard, neurological dysfunction and neurodegenerative diseases are on the rise, and BBB dysfunction seems to be a key feature in numerous studied pathologies. Environmental exposures have been implicated in observed increases in CNS-related illnesses. Efforts are underway to determine both the environmental agents with the greatest impact on CNS dysregulation along with the mechanisms underlying environmental influences on neurodegeneration and neurocognitive/neurodevelopmental deficits. In this chapter, we will explore the role of the BBB in the relationship between environmental exposures and CNS dysfunction.

2. Anatomy and physiology of the BBB

In 1885, Professor Dr. Paul Ehrlich made the discovery that dyes injected into the blood of experimental animals stained various organs, excluding brain tissue (Ehrlich, 1885). Later, Edwin Goldmann, his student, showed that dyes injected directly into the cerebrospinal fluid (CSF) permeated into brain tissue (Bentivoglio and Kristensson, 2014; Goldmann, 1913). These finding led to the understanding that a physical barrier exists between the brain and the circulating blood and this became known as the BBB. The BBB is a specialized, highly selective, and tightly regulated barrier at the brain microvessel level that performs various homeostatic functions, including regulation of the passage of cells, ions, macromolecules, neurotransmitters, and nutrients between the systemic circulation and the central nervous system, as well as limiting neural exposure to circulating neurotoxicants (Abbott et al., 2010; Ballabh et al., 2004). The BBB is comprised of endothelial cells with specialized barrier properties (Almutairi et al., 2016; Coronado-Velazquez et al., 2018; Serlin et al., 2015) and a capillary basement membrane with embedded neighboring pericytes and astrocyte end-feet surrounding the neurovascular unit (NVU) (Ballabh et al., 2004; Giannoni et al., 2018) (Fig. 1) that operates in a complex interplay to maintain proper neurophysiologic functioning. These physical barriers ensure that transport across the vasculature into the brain parenchyma occurs via transcellular mechanisms, rather than paracellularly as is the case in other endothelial beds (Wolburg and Lippoldt, 2002).

2.1 Endothelial cells

Endothelial cells of the BBB have specialized features that facilitate the separation of systemic and CNS compartments. They lack the typical fenestrations of other endothelial cells that facilitate paracellular transport between adjacent cells (Petty and Lo, 2002), a feature established by the presence of specialized tight junction proteins, comprised primarily of claudin

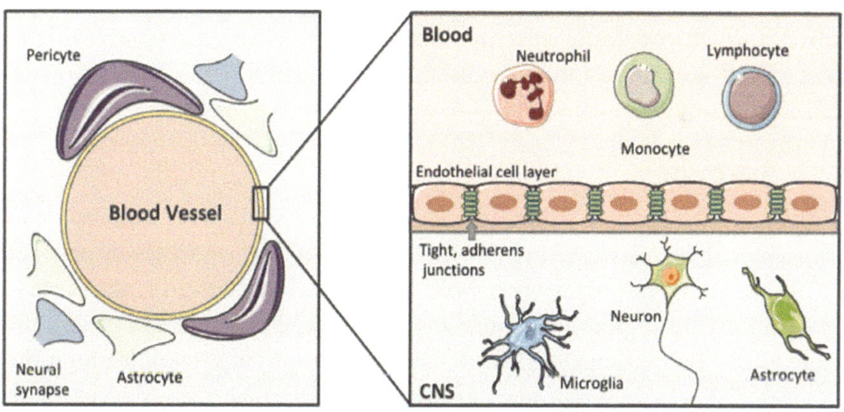

Fig. 1 Neurovascular unit and blood-brain barrier components. The blood-brain barrier is composed of endothelial cells connected by tight junctions. Pericytes embedded in the basement membrane and astrocytic end-feet surround the microvessel walls and are in close proximity to axonal projections from neurons. Microglia reside in the surrounding area and are CNS-specific immune cells. Neutrophils, monocytes, and lymphocytes can migrate from the systemic circulation across the BBB to the CNS.

and occludin proteins at the cell-cell interface (Greene and Campbell, 2016; Kniesel and Wolburg, 2000; Luissint et al., 2012; Petty and Lo, 2002; Wolburg et al., 2009). BBB passage is facilitated by membrane-bound transporters and receptors that allow nutrients and waste to move between the brain parenchyma and the systemic circulation (Greene and Campbell, 2016; Obermeier et al., 2016). Cerebrovascular endothelial cells also possess a large number of mitochondria to support elevated metabolic activity of the NVU. The overall interrelation between the NVU cell types is complex and can vary across vessel size and region (Abbott et al., 2006; Zlokovic, 2008). In addition to being a first layer of barrier protection, the endothelial cells are also key regulators of vessel diameter, releasing dilating, and constricting substances to maintain adequate regional perfusion. Endothelial nitric oxide synthase (eNOS) generates the vasodilator molecule nitric oxide and is a key enzyme in the regulation of vascular tone through the modulation of both vasoconstriction and angiogenesis (Faraci, 2006). Chen et al. demonstrated that eNOS-deficient mice display altered angiogenesis and neurogenesis, which led to loss of recovery of neuronal function following experimentally-induced stroke (Chen et al., 2005). Under oxidative conditions, however, eNOS functionality can be impaired and potentially have a negative role in BBB maintenance. Transient focal cerebral ischemia models have shown that application of NOS inhibitor L-NAME reduced extravasation of Evans blue dye from the cerebral vasculature into the cerebral cortex and that neuronal NOS-specific blockade significantly reduced BBB disruption without affecting vasodilation and thus may play a role in early BBB disruption following insult (Jiang et al., 2014).

2.2 Astrocytes

Astrocytes are the most abundant glial cells in the CNS with a multitude of functions. They play important roles in CNS homeostasis by modulation of CNS development, repair, metabolism, neuronal transmission, and participation in CNS response to injury and disease (Quintana, 2017). Additionally, astrocytes participate in BBB formation by physically ensheathing the endothelial cell/pericyte unit. Astrocytes may facilitate the development of the microenvironment required for cerebrovascular endothelial cells to develop the barrier properties unique to the BBB (Blanchette and Daneman, 2015; Dehouck et al., 1990; Hamm et al., 2004; Luissint et al., 2012). Janzer and Raff showed that astrocytes isolated from neonatal Sprague-Dawley rat cerebral cortex induced vascularization in the anterior eye chamber of syngeneic rats and, further, that these vessels formed tight connections that prevented the passage of a dye tracer (Janzer and Raff, 1987). Nakagawa et al. elegantly demonstrated, in their triple cell culture model of the BBB, that astrocytes in coculture with endothelial cells could increase endothelial cell tightness more than twofold (Nakagawa et al., 2009). In addition to barrier properties, astrocytes, due to their unique connection with both the NVU and neuronal synapses, link the vasculature to neural activity, contributing to immune surveillance at the BBB.

2.3 Pericytes

Pericytes are the cells in the most intimate contact with endothelial cells of the BBB, covering approximately 60–70% of the abluminal endothelial cell surface (Giannoni et al., 2018). They share a basement membrane in common with cerebrovascular endothelial cells and encircle them to provide structural support (Ballabh et al., 2004). Pericytes and cerebrovascular endothelial cells communicate via platelet-derived growth factor β (PDGF-β)-mediated signaling to control endothelial cell numbers and organization in the NVU, demonstrating the essential nature of pericytes for BBB function (Hellstrom et al., 1999). Pericytes, like astrocytes, play a crucial role in modulating CNS immune responses and pericytes may themselves act as immune cells (Balabanov et al., 1996; Giannoni et al., 2018). Several studies have demonstrated that pericyte modification and reactivity changes are associated with inflammation in many neurological conditions, including Alzheimer's disease, stroke, and traumatic brain injury (Fernandez-Klett et al., 2013; Zehendner et al., 2015; Zlokovic, 2011). Proebstl et al. have shown that pericytes are necessary for neutrophil trafficking across the cerebrovascular endothelium (Proebstl et al., 2012).

2.4 Microglia

Microglia are the resident macrophages of the CNS and the key innate defense mechanism in the brain (Kettenmann et al., 2013; Kettenmann and Verkhratsky, 2011; Rothhammer and Quintana, 2015). Via their surveillance, they can detect the first signs of infection or damage to neurological tissue (Domingues et al., 2016; Petrelli et al., 2016; Reemst et al., 2016). Once activated, they convert from their surveillance (resting) phenotype to facilitate the removal of damaged tissue and recruitment of other immune cells to mediate return to homeostasis (Kettenmann et al., 2011). Microglial dysfunction is reported in several neurological conditions.

Microglia have been shown to play both a protective role in Alzheimer's disease via the clearance of β-amyloid, and in neuropathology via β-amyloid-dependent activation of the NLRP3 inflammasome (Gold and El Khoury, 2015). Additionally, a key role for "surveillance" microglia may be to assist in synaptic pruning and plasticity (Wake et al., 2009), and thus toxicant-induced changes in this phenotype may have important implications for cognition and neurodevelopment.

2.5 Glymphatic system

The "glymphatic system" is another important, and recently characterized, component of the brain that facilitates proper functioning by limiting accumulation of neurotoxic substances and facilitating nutrient, growth factor, and neuromodulator distribution (Jessen et al., 2015). Macroscopic waste clearance is mediated via the perivascular tunnels formed by astroglial cells and glial aquaporin 4 channels that enable fluid exchanges between the CSF and the brain parenchyma interstitial fluid (Albargothy et al., 2017; Iliff et al., 2012; Jessen et al., 2015; Plog and Nedergaard, 2018). The glymphatic system is most active during sleep (Bakker et al., 2016), and several recent studies have implicated glymphatic dysfunction in neurodegenerative diseases (Iliff et al., 2013; Kress et al., 2014; Weller et al., 2008; Zuroff et al., 2017). Sleep disturbance accompanied by reduced clearance of waste is common in patients suffering from traumatic brain injury (Sullan et al., 2018). The association of sleep deprivation with reduced glymphatic clearance may lead to the accumulation of neurodegenerative proteins such as tau, β-amyloid, GFAP, and S100-B. The BBB and lymphatic systems are likely highly integrated and even overlapping in certain functions.

3. Transport across the BBB

Due to the unique configuration of cerebrovascular endothelial cells to limit or regulate the passage of molecules and cells from the periphery into the brain parenchyma it is necessary to develop specialized transport mechanisms for the delivery of nutrients, ions, neurotransmitters, and other molecules into the CNS, as well as for the removal of waste products. The BBB has a number of membrane-bound influx and efflux transporters and channels to accommodate these requirements. The ATP-binding cassette (ABC) transporters and the solute carrier (SLC) transposers are the main transporters found in the BBB in addition to the aquaporin water channels (Al Feteisi et al., 2018).

3.1 Aquaporins

Aquaporins are membrane proteins that form the channels by which water travels into and between cells across all organ systems (Badaut et al., 2002, 2014; Hill et al., 2004). In the BBB, aquaporin-4 is the channel protein that facilitates movement of water into the CNS. Highly expressed in astrocyte end-feet surrounding the endothelial cells in the NVU, aquaporin-4 may be involved in maintaining potassium homeostasis (Binder et al., 2006) and facilitating water-mediated clearance of waste, accumulated peptides, and toxic substances from the CNS (Iliff et al., 2012). Aberrant expression of aquaporin-4 has been reported in neurodegenerative disorders such as amyotrophic lateral sclerosis, Alzheimer's disease, post juvenile-traumatic brain injury, and Parkinson's disease (Bataveljic et al., 2014). A lack of aquaporin-4 in astrocyte end-feet is shown to decrease migration toward sites of neuronal injury, as well as reduce clearance of toxins such as the Alzheimer's associated β-amyloid peptide (Badaut et al., 2014; Iliff et al., 2012; Wilcock et al., 2009). In models of multiple sclerosis, the correlation between aquaporin-4 upregulation with BBB disruption and tight junction protein loss was observed in several studies.

3.2 ABC transporters

The ATP-binding cassette (ABC) transporters, unlike the aquaporins, are responsible for efflux from the CNS. They protect the brain from toxicity of xenobiotics by facilitating the transport of drugs into the systemic circulation (Brzica et al., 2017). ABC transporters have been implicated in epilepsy, brain tumors, and neurodegenerative disorders such as Alzheimer's disease (Hartz and Bauer, 2010).

3.3 Solute carrier family transporters

Solute carrier (SLC) transporter family has dual roles in BBB transport with members functioning in the capacity of influx or efflux. They transport a wide variety of solutes, including glucose and amino acids (Bai et al., 2018).

Glucose transporters: Glucose is the main source of energy for the brain and must be obtained from the systemic circulation in a steady uninterrupted supply for optimal brain function. Several glucose transporters can be found in the NVU, namely the facilitative diffusion GLUT transporters and the sodium-dependent SGLT (Navale and Paranjape, 2016). The glucose transporter 1 (GLUT1/SLC2A1), a member of the SLC2 family of solute carrier transporters, is present in both the endothelium and in astrocyte end-feet and cell bodies (Jurcovicova, 2014). GLUT1 is the main transporter responsible for glucose uptake at the BBB (Mueckler and Thorens, 2013; Patching, 2017). Although SGLT1 is also present in the endothelium of the BBB, its function seems to be related to maintaining glucose levels during conditions of stress (Nishizaki and Matsuoka, 1998). Altered glucose metabolism (Harik and LaManna, 1991; Mosconi et al., 2008b) and transport (Friedland et al., 1989) are a well-established finding in Alzheimer's disease and are used as a parameter for diagnostic testing (Samuraki et al., 2007) and will be discussed in more detail further in the chapter.

Amino acid transporters: The CNS utilizes neurotransmitters to mediate the effector functions of neurons. Amino acids, particularly the aromatic and acidic members, serve as precursor molecules for the biosynthesis of these neurotransmitters. To facilitate the uptake of amino acids for biosynthesis and the removal of excess, the BBB has specialized facilitative, and sodium-dependent transporters (Nalecz, 2017). These transporters are distributed across microvascular endothelial cells in either luminal or abluminal membranes (Oldendorf, 1973). Two facilitative transporters of neutral amino acids can be found on both the luminal and abluminal membranes to facilitate the uptake of essential amino acids (Hawkins et al., 2006). Sodium-dependent transporters located on the abluminal membrane set up a concentration gradient by ensuring that amino acid concentrations in the extracellular fluid are maintained at 10% of the plasma concentration. The abluminal membranes also possess sodium-dependent transporters of acidic amino acids and glutamine (Hawkins et al., 2006). The large neutral amino acid transporters (LAT-1) (Geier et al., 2013; Lee et al., 1996) are currently targeted in the treatment of Parkinson's disease and seizure disorders (Alexander et al., 1994; Wang and Welty, 1996). Studies have also shown that certain neurotransmitters, such as serotonin and glutamate, are decreased in Alzheimer's disease (Marcos et al., 2008; Scott et al., 2002; Thai, 2002). Defects in amino acid metabolism have been implicated in dementia and other neurological pathologies but these studies look at the neuronal context and not much is known about the role of BBB transporters in these disease processes.

4. Neuroinflammation and the BBB

Neuroinflammation is a coordinated response in the brain involving microglia, astrocytes, other CNS cells, and the peripheral immune system, which infiltrate into the CNS following injury. Mounting evidence points to a central role of cerebrovascular dysfunction in neuropathology. Neuroinflammation is a key component of both neuroprotective and neuropathologic processes and is described as the activation of reactive CNS elements in response to altered brain homeostasis caused by infections and other initiators of cell death (Kempuraj et al., 2016; Ransohoff et al., 2015). The neuroinflammatory process is mediated by microglia, astrocytes, neutrophils, mast cells, T-cells, neurons themselves, and the mediators released by these cells (Kempuraj et al., 2016; Ransohoff, 2016; Ransohoff et al., 2015; Shabab et al., 2017). Under normal physiological conditions, the BBB prevents the translocation of systemic immune cells into the CNS (Petty and Lo, 2002). There are several distinctions between peripheral and neuroinflammatory responses, mainly the lack of memory T-cells in the brain parenchyma (Tohidpour et al., 2017). It is now well established that neuroinflammation plays a significant, if not primary, role in the process of neurodegeneration and the development of many neurological conditions. Loss of sleep has been shown to produce low-level neuroinflammation. Hurtado-Alvarado et al. demonstrated that sleep restricted C57BL/6 mice had elevated serum TNF-α and IFN-γ, increased BBB permeability as well as increased hippocampal neuroinflammatory markers Iba-1, A_{2A} adenosine receptor and matrix metalloproteinase (MMP)-9 compared to controls (Hurtado-Alvarado et al., 2018).

5. Blood-brain barrier disruption

5.1 Detecting BBB disruption

An increasing number of studies point to BBB breakdown as a key pathogenic feature of several neurological diseases (Sweeney et al., 2018). Therefore, interest in various methods to measure cerebral permeability has risen in the past decade. There are several non-invasive and/or non-destructive methods to assess leakage across the BBB including CSF analysis using enzyme-linked immunosorbent assays (ELISA), magnetic resonance imaging (MRI), computerized tomography (CT), and positron emission tomography (PET). Additionally for preclinical models, there are specific markers and dyes that can be assessed directly in brain tissue by histological or biochemical approaches.

Early CSF studies investigating the role of BBB leakage in neurodegeneration found that subjects with dementia and Alzheimer's had higher albumin ratios of CSF to blood compared to healthy subjects (Blennow et al., 1990; Frolich et al., 1991; Mecocci et al., 1991; Skoog et al., 1998).

Additionally, CSF levels of plasminogen, fibrinogen, immunoglobulin G (IgG), and the oxidative stress marker cytochrome *b*-245 are indicators of BBB permeability (Bell et al., 2012; Ryu and McLarnon, 2009). The pericyte injury marker, soluble platelet-derived growth factor receptor (PDGFRβ), correlates positively with BBB breakdown (Montagne et al., 2015; Sagare et al., 2015), while upregulated MMPs in the CSF have been reported in conjunction with diseases involving BBB leakage including multiple sclerosis, dementia, and Alzheimer's disease (Adair et al., 2004; Bjerke et al., 2011; Fainardi et al., 2006). While CSF analysis can be used as an indicator of BBB disruption in a variety of diseases, it does not give information about the localization of permeability.

Dynamic contrast-enhanced (DCE)-MRI is currently the most widely used imaging method to assess BBB integrity (Heye et al., 2014; Raja et al., 2018). Although spatial resolution is lower compared to other neuroimaging methods, a major advantage to DCE-MRI is rapid acquisition. DCE-MRI is extensively used to assess dramatic changes to BBB functionality induced by intracranial neoplasms, stroke, and multiple sclerosis. However, improvements in imaging techniques and analysis have made it possible to extrapolate its use to additional diseases in which changes to permeability may be subtler, such as diabetes and Alzheimer's disease (Starr et al., 2003, 2009). DCE-MRI measures alterations in brain tissue over time following intravenous injection of a contrasting agent, which is usually gadolinium based (Ewing et al., 2003; Larsson et al., 1990). If the integrity of the BBB is weakened, the gadolinium-based compound, which is otherwise confined to the periphery, is able to pass through the endothelial layer and accumulate, producing T1 shortening and contrast enhancement (Larsson et al., 1990). While less invasive than collection and analysis of CSF, the methodology of these studies is highly heterogeneous and although proposed, there is currently no standard method of data analysis (Heye et al., 2014; Leach et al., 2012; Raja et al., 2018). Methods of analysis range from visual assessment of images to the use of complex pharmacokinetic models (Iannotti et al., 1987; Patlak and Blasberg, 1985; Shames et al., 1993; Tofts et al., 1999; Tofts and Kermode, 1991).

F-2-fluoro-2deoxy-D-glucose (FDG)-PET assesses alterations in central glucose metabolism and has been used to diagnose diseases associated with diminished BBB transport, specifically Alzheimer's disease (Morbelli et al., 2015; Pagani et al., 2015). Similar to glucose, FDG uptake across the BBB is dependent on cerebral blood flow and facilitated by the glucose transporter GLUT1, expressed only in cerebral endothelium (Dick et al., 1984).

Unlike glucose, FDG becomes trapped in metabolically active cells rather than being further metabolized and can thus be visualized using PET (Kuwabara et al., 1990). GLUT1 transporter expression is reduced in Alzheimer's disease and decreased glucose uptake was observed in Alzheimer's disease patients as well as transgenic animal models of Alzheimer's (Hunt et al., 2007; Niwa et al., 2002). A longitudinal follow-up study in humans reported that reductions in central glucose uptake occur prior to the onset of Alzheimer's disease indicating that reduced uptake may be an early biomarker of neurodegeneration (Mosconi et al., 2008a).

In animal studies, analysis of BBB permeability through the injection of Evans blue is still commonly used today due to its relative simplicity. Evans blue dye has a high affinity to non-permeable serum albumin, which enters the CNS when the BBB is compromised (Rossner and Tempel, 1966). While widely used as an indicator of BBB permeability, caution should be taken when using it to quantify magnitude of BBB impairment due to unspecific binding of the dye to connective tissue and possible uptake by astrocytes (Clasen et al., 1970; Saunders et al., 2015). In addition to Evans blue dye, horseradish peroxidase is commonly used as an indicator of alterations to BBB integrity, however its use may be limited depending on the animal model. HRP can induced mast cell degranulation and release of histamine and serotin which can modulate vascular permeability in some rat species (Cotran and Karnovsky, 1967). Taking precautions when choosing the model system and dose can help to avoid this (Clementi, 1970). Sodium fluorescein (376 Da) injection and subsequent antibody detection are often used as an alternative to Evans blue and HRP as it binds only weakly to endogenous proteins, is less toxic, and have been reported to detect more modest alterations in BBB permeability (Kaya and Ahishali, 2011; Wolman et al., 1981). Immunohistochemical methods to label endogenous proteins such as albumin, IgG, and fibrinogen can also be used in the brain parenchyma (Wunder et al., 2012). The advantage of using endogenous proteins is that they do not require injection, thus avoiding a possible non-physiological response and they can be visualized using light microscopy. However the resolution of permeability using this method may not high enough to adequately determine cellular causes of disruption. Biotin- and fluorophore-labeled dextrans, in contrast, are valuable tools for detecting BBB permeability as they can be quantified in the CSF and visualized at high resolution using electron microscopy (Ek et al., 2003, 2006; Johansson et al., 2006; Wunder et al., 2012). Early toxicity concerns have been circumvented by the use of highly purified and small molecular weight dextrans such as biotin-labeled ethylenediamine (60 Da), which is smaller than sucrose (342 Da).

Advanced imaging techniques are available that allow researchers to investigate BBB dysfunction at the microscopic level. Two-photon laser scanning microscopy allows in vivo visualization of neurons and single microvessels up to 500–600 μm in depth (Kleinfeld et al., 1998). Using near-infrared excitation photons that excite fluorescent-based functional dyes or reporters, two-photon microscopy has minimal excitation scattering and thus reducing photobleaching (Denk et al., 1990). Investigations of BBB involvement in cerebral injury using fluorescent labeling of the tight junction protein claudin-5 with enhanced green fluorescent protein (eGFP) enabled the visualization of structural BBB changes during ischemia and more recently, labeling of pericytes revealed activation of MMP-2/9 prior to ischemia-induced BBB capillary leakage (Knowland et al., 2014; Underly et al., 2017).

5.2 Models of BBB disruption

A number of animal models have been developed to study the BBB ranging from transgenic animals with targeted modulation of BBB components to induction models that mimic disease progression. Mouse models with various genetic manipulations of the cellular components of the BBB and transgenic models of diseases involving the BBB are manifold, and have been extensively reviewed elsewhere (Fluri et al., 2015; Hall and Roberson, 2012; Sohet and Daneman, 2013). Here, we briefly discuss common examples of disease induction models in which BBB permeability occurs as a primary or secondary event during pathogenesis.

A major role for systemic and central inflammation in BBB breakdown is becoming increasingly recognized. Infection and inflammation-related BBB breakdown can be modeled in animals by central or peripheral administration of lipopolysaccharide (LPS), the outer membrane of gram-negative bacteria (Banks et al., 2015). Systemic challenge with LPS increases CNS levels of proinflammatory cytokines, nitric oxide, and prostaglandins and disrupts BBB structure and function (Banks et al., 2015; Minami et al., 1998; Sugita et al., 2002). While the exact molecular mechanisms remain unclear, studies indicate that peripherally injected LPS may cross the BBB and directly activate astrocytes, endothelial cells, or microglia (Chakravarty and Herkenham, 2005; Reyes et al., 1999; Verma et al., 2006). Although LPS is commonly used as a more global inflammation model it can also be used to investigate specific elements of disease. For example, LPS injection in mice increases brain influx and decreases brain efflux of amyloid-β, characteristic traits of Alzheimer's progression (Jaeger et al., 2009).

Multiple sclerosis (MS) is a neurodegenerative and inflammatory immune condition that alters the BBB (discussed in Section 6.2). Experimental autoimmune (or allergic) encephalomyelitis (EAE) animal models exhibit the central inflammation and demyelination features of MS and are commonly used in mice and rats (Baxter, 2007; Mannie et al., 2009; Miller and Karpus, 2007). EAE is induced either by administration of myelin or by adoptive transfer of encephalitogenic T-cells and can be combined with transgenic lines depending on the experimental question (Mannie et al., 2009; Miller and Karpus, 2007). EAE models have been used to investigate the mechanisms of immune cell trafficking (Male et al., 1995) and have provided critical evidence identifying BBB disruption as a consequence of immune cell invasion rather than the cause (Wolburg et al., 2005). In addition to MS, EAE can be used as to study BBB transport of immune cells in cerebral injury (Popovich et al., 1996).

Breakdown of the BBB following the rupture or blockage of a cerebral blood vessel (stroke) may lead to central influx of inflammatory molecules and chronic cognitive deficits post-stroke. The majority of stroke cases are caused by ischemic stroke from thromboembolic occlusion of a cerebral artery (Bogousslavsky et al., 1988; Chung and Caplan, 2007). Arterial occlusion induces oxygen deprivation and the production of ROS and inflammation. Studies investigating ischemic stroke often use the middle cerebral artery occlusion model. In this model, ligation of the middle cerebral artery is performed for a predetermined period of time. The ligation can then be released and blood flow to the injured tissue can be restored mimicking physiological events following a stroke (Tamura et al., 1981). Four different forms of middle cerebral artery occlusion models exist, including embolic, endovascular filament, permanent transcranial, and transient transcranial, each with various impacts on the BBB. Additionally, "second event" stroke models exist where a major component of stroke is reproduced. For example, the hypoxia/reoxygenation model employs a time period of global oxygen deprivation. This model does not incorporate decreased blood flow or physical damage to endothelial cells, but instead is focused on the reoxygenation period when damage from oxidative stress is likely to occur after a stroke (Maiti et al., 2006; Ramanathan et al., 2005). A recent study using severe acute hypoxia/reoxygenation in rats reported that while oxidative damage from hypoxia is attenuated after 24 h, initiation of apoptotic pathways involving apoptosis-inducing factor (AIF), cytochrome *c*, and caspase-3 persist and may drive post-stroke detrimental central effects (Coimbra-Costa et al., 2017).

A further disease involving inflammation and BBB disruption is diabetes mellitus (DM). Streptozotocin, a toxic agent that is able to enter pancreatic beta cells using the GLUT2 transporter, is used to induce an insulinopenic/ hyperglycemic state characteristic of Type I DM (Huber et al., 2006). This model allows for downstream investigations into secondary effects of hyperglycemia including alterations to the BBB. Rats treated with 60 mg/kg streptozotocin exhibited permeability to sucrose, decreased occludin and ZO-1 expression, and increased plasma MMP-2/9 activity 14 days following administration (Hawkins et al., 2007). Exposure paradigms can also be used to study the role of external factors on CNS pathologies that may lead to BBB involvement such as high fat diet. Exposure of rodents to a high fat diet is commonly used to model obesity-associated Type II DM. High fat diet-fed animals exhibit prolonged hyperglycemia and chronic, low-grade inflammation, the hallmark features of DM (Hotamisligil et al., 1995; Surwit et al., 1988). Although both streptozotocin and high fat diet induce hyperglycemia, unlike streptozotocin, high fat diet allows for studies into disease etiology and genetic involvement. The molecular mechanisms underlying disease-induced cerebral abnormalities in the above-mentioned models are discussed in the following section.

6. The BBB in disease

Impairment of the BBB has been reported in several diseases of the CNS ranging from the more dramatic BBB breakdown seen in ischemia/ reperfusion and multiple sclerosis to the persistent barrier permeability reported in diabetes mellitus and Alzheimer's disease. Breakdown of the BBB can have damaging effects on the CNS including decreased perfusion, neuronal injury, altered immune cell function, and impaired clearance (Fig. 2). Moreover, BBB impairment can render the CNS vulnerable to attacks by exogenous or systemic particles and neuroinflammation. Below we discuss the role of the BBB in disease pathogenesis.

6.1 Stroke and related vascular disorders

Among the most common cerebrovascular diseases are those caused by vascular occlusion arising from several etiologies, including thromboembolic depositions and cerebral vasospasm (Turner and Sharp, 2016). The cerebrovasculature can be damaged through a single large occlusive

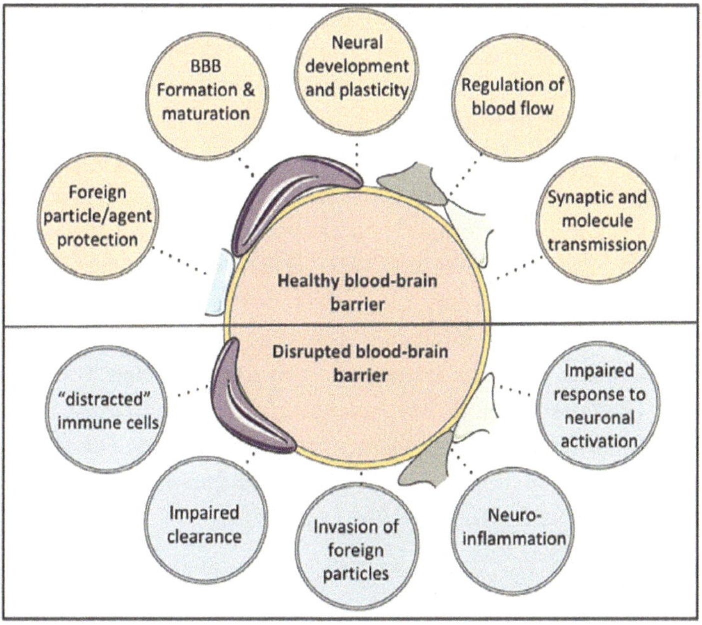

Fig. 2 Comparison of a healthy versus an impaired blood-brain barrier. Top half of the schematic shows the normal functions of an intact BBB including protection of the brain from external or systemically circulating particles and cells, selective transport, and proper neural development and function. The bottom half represents BBB dysfunction including neuroinflammation, altered immune cell function, foreign particle invasion, and impaired debris clearance.

event (stroke) or through the occurrence of numerous smaller transient ischemic events that over time lead to vascular dementia. Regardless of the origin, several common factors converge to reduced BBB functionality both acutely and chronically. In the acute (0–6h) time period following occlusion, the localized ischemia and generation of reactive oxygen species during reperfusion impairs the ability of endothelial cells to maintain tight junctions. Following this period, other pathways such as apoptosis/necrosis and inflammation become the important drivers of BBB dysfunction. With the initial disruption of the BBB during ischemia, circulating factors can access neurons and neurosupportive cells, leading to inflammatory responses. Among the most critical response is the generation and release of chemotactic factors including CCL2 and expression of vascular adhesion molecules that ultimately promote the recruitment of neutrophils and other inflammatory cells (Strecker et al., 2013; Tei et al., 2013). Neutrophils and likely other cells of the NVU release matrix metalloproteinase 9 (MMP-9), which can specifically degrade basement membrane proteins and tight junction proteins to facilitate inflammatory cell invasion (McKittrick et al., 2015; Turner and Sharp, 2016). MMP-9 is likely not the only peptidase with a role in BBB breakdown following stroke or related pathologies, but it is among the most studied and clearly aligns with pathological severity of stroke (Romanic et al., 1998). Interestingly, MMP-9 plays a clear role both in the lungs and systemically in air pollution-related vascular effects (Aragon et al., 2016; Lund et al., 2009, 2011). MMP-9 activation can be augmented by therapeutically administered fibrinolytics such as tissue plasminogen activator; prophylactic treatment with an MMP-9 inhibitor has been shown to limit the exacerbation of stroke models by fibrinolytic treatment (Pfefferkorn and Rosenberg, 2003). Importantly, following an acute ischemia/reperfusion event, the BBB impairment and neuroinflammatory consequences can last weeks beyond resolution of the vascular occlusion (Turner and Sharp, 2016).

6.2 Multiple sclerosis

Multiple sclerosis is a progressive T-lymphocyte-mediated autoimmune disease characterized by chronic inflammation, immune cell infiltration, and reactive changes in glial cells (gliosis) leading to demyelination and degeneration of axons throughout the CNS (Compston and Coles, 2008; Gold et al., 2006; Mammana et al., 2018). Epidemiological and family studies of patients show a significant genetic basis for MS. Several genetic alterations in the HLA class II DRB1 alleles responsible for antigen recognition in CD4 + T lymphocyte and multiple single-nucleotide polymorphisms (SNPs) have been documented in MS (International Multiple Sclerosis Genetics Consortium et al., 2011; Jakkula et al., 2010; Leray et al., 2016; Raj et al., 2014). However, genetics alone does not account for all cases (Dutta et al., 2006, 2007; Grytten et al., 2013; Holmoy and Hestvik, 2008; Simpson et al., 2011). The EAE model, widely used in the study of MS, has improved our understanding of the role of peripheral immune cells in the inflammatory cascade of MS pathogenesis (Bailey et al., 2007; Baxter, 2007; Fletcher et al., 2010; Gold et al., 2006; Hoglund et al., 2011; Kroenke et al., 2008; Langrish et al., 2005; Murphy et al., 2010). Flugel et al. demonstrated that GFP-labeled $CD4^+$ T lymphocytes from the periphery migrate into the CNS upon disease onset in a modified model of EAE (Flugel et al., 2001). In 1986, Poser theorized that several converging factors were necessary for the development of MS: (1) a genetic predisposition,

(2) an environmental, possibly viral-mediated event, (3) an immune-mediated initiating event, (4) BBB alteration, and (5) MS plaque formation (Poser, 1986). In MS, a distinct correlation between disease relapses, inflammation, and BBB disruption has been noted in several studies and has led to a greater understanding of how infiltrating immune cells play a role in MS pathogenesis (Bell et al., 2018; Claudio and Brosnan, 1992; Foster et al., 2009; Hoglund and Maghazachi, 2014; Lucchinetti et al., 2011; Milo and Kahana, 2010; Ortiz et al., 2014; Pozzilli et al., 1988). Since the brain is an immune privileged organ, movement of immune cells into the CNS is a tightly regulated process requiring traversing the endothelial cell layer and its basement membrane into the CSF, where they are presented by antigen-presenting cells in the CSF-drained spaces without themselves entering the CNS parenchyma (Agrawal et al., 2006; Engelhardt and Ransohoff, 2012). In fact, clinical symptomology of MS is shown to present only when leukocytes move out of the perivascular cavities directly into the brain parenchyma (Agrawal et al., 2006). A dynamic contrast-enhanced MRI comparison of the periventricular normal appearing white matter of MS patients and healthy controls showed significantly higher perivascular permeability in normal appearing white matter of MS patients with recent MS relapse compared to healthy controls. The observed permeability changes were attenuated with the administration of either β-interferon or Glatiramer acetate immunomodulatory therapies (Cramer et al., 2014). The precise mechanisms of BBB disruption in MS are currently unknown, however early studies looking at the BBB in MS concluded that there was increased transport of blood-borne compounds across the BBB via pinocytic vesicles without alteration of tight junction structures or capillary morphology (Brown, 1978; Hawkins et al., 1992). More recently, Bennett et al. showed, using the EAE model, that the tight junction protein, zona occludens-1 was relocalized over the course of disease progression. This reorganization preceded clinical disease presentation and correlated with sites of inflammatory activation and increased endothelial layer permeability (Bennett et al., 2010). Uchida et al. demonstrated that tight junction protein, claudin-11, was downregulated in the brain and spinal cord in both human subjects and animal models of disease (Uchida et al., 2018). Current therapeutics for the management of MS focus on the reduction of inflammation and have been shown to also improve BBB integrity, but to date, no therapies are available that directly impact the BBB. It is postulated that such a therapeutic may be more effective in preventing relapses by preventing immune cell infiltration and autoimmune activity and thus, may halt disease progression (Lengfeld et al., 2017).

6.3 Aging

Aging is a natural phenomenon that contributes numerous changes to the NVU in the absence of overt disease but is also a strong risk factor and/or contributor in the development of many neurological disorders (Ransohoff, 2016). For instance, Alzheimer's disease risk increases >30% between the age of 70 and 80 (Kress et al., 2014). The specific mechanisms underlying the deterioration are incompletely understood; however, age-related deposition of protein-aggregates, combined with decreased or impaired clearance, leads to inflammatory activation that may reflect an internal stimulus for BBB disruption (Goodall et al., 2018; Kress et al., 2014). Aging is associated with histological changes to the vasculature, including capillary thinning in white matter regions (Stewart et al., 1987) and relative loss of endothelial cells in humans (Burns et al., 1981). Pericyte numbers have also been shown to decline with aging (Stewart et al., 1987). Expression of tight junction proteins decreases while MMP-9 expression and activity have been observed to increase in models of aging (Lee et al., 2012). In a recent study of ozone-induced neuroinflammatory responses in young and old mice, older (12–18-month-old) mice were observed to have greater numbers of activated microglia, in terms of F4/80 and MHCII expression, than 8–10-week-old mice (Tyler et al., 2018), which may also play a role in BBB maintenance. Thus, independent of disease or environmental exposure, natural aging causes a multifactorial deterioration of BBB integrity.

6.4 Dementia and Alzheimer's disease

Dementia is a broad classification of reversible or irreversible symptoms that manifest in loss of, or decline in, cognitive functioning. According to the World Health Organization, an estimated 50 million people worldwide suffer from various types of dementia with 10 million new cases being diagnosed each year. Alzheimer's disease is the most common form of dementia in the aged population, accounting for approximately 60–80% of cases (Alzheimer's Association, 2015). It is estimated that 30–50 million people worldwide, including 5 million Americans (Karlawish et al., 2017; Selkoe and Hardy, 2016). Alzheimer's, like other forms of dementia is considered to be an aging disease due to the fact that it occurs more frequently in older populations and the risk of disease greatly increases every 5 years past 65 (Batarseh et al., 2016; De-Paula et al., 2012; Folch et al., 2016; Khan and Bloom, 2016; Rissman et al., 2012). Genetics, aging, environmental and other factors contribute to physiological shifts that lead to the development of primary hallmarks, namely, amyloid-β plaques and neurofibrillary tangles observed in Alzheimer's disease (Alkadhi, 2011; Holland et al., 2009; Lim et al., 2015). While the specific convergence of factors and the underlying mechanisms are not well understood, it is generally accepted that aging is the major risk factor which points to a process of accumulated events in early life preceding disease symptomology. Post mortem analyses of the Alzheimer's brain reveal significant alterations in the BBB as well as accumulation of disease-associated peptides (Provias and Jeynes, 2014). Studies have shown that pericyte numbers are significantly lower in patients with Alzheimer's disease (Sengillo et al., 2013) and, in vitro, amyloid-β plaques are especially toxic to pericytes (Verbeek et al., 1997). Additionally, the oxidative and inflammatory environment of Alzheimer's can negatively impact the function of BBB transport proteins, which in turn impairs clearance of amyloid-β plaques (Erickson and Banks, 2013). As mentioned, the brain clears waste products most effectively while we sleep, so unsurprisingly, correlations have been made between sleep disorders and Alzheimer's disease, and accumulation (Liguori et al., 2014; Xia et al., 2018). Although inflammation in Alzheimer's is most closely associated with neuroinflammatory cell activity, Krstic et al. showed that systemic inflammatory challenge can induce Alzheimer's disease symptomology and pathology in mice (Krstic et al., 2012). Thus, alterations in the BBB may grant access to systemic proinflammatory factors that advance disease and the causes of these systemic events may be wide and varied. Later in this review, we will discuss environmental agents that may affect the BBB and thus facilitate Alzheimer's disease progression.

6.5 Diabetes mellitus

Both Type I and Type II DM are characterized by chronic hyperglycemia, reduced or nonexistent insulin action, and vascular complications. A growing body of evidence from clinical and animal studies suggests that chronic hyperglycemia induces cerebral microvasculature abnormalities resulting in cognitive learning and memory deficits and increased risk of neurological disease (Baker et al., 2011; Curb et al., 1999; Leibson et al., 1997; Ott et al., 1999; Segura et al., 2009; Starr et al., 2009). A major contributor to diabetic encephalopathy involves alterations to the structure and function of the BBB (Brownlee, 2001; Janelidze et al., 2017). Increased BBB permeability has been observed using [^{14}C] sucrose accumulation, DCE-MRI, and IgG staining in animal models of DM (Acharya et al., 2013; Hawkins et al., 2007; Huber et al., 2006; Xu et al., 2017). In a study using gadolinium-enhanced MRI, BBB permeability was increased in diabetic patients compared to controls (Starr et al., 2003). The effects of DM on the BBB are multifaceted and appear to manifest through alterations of BBB components including ECs, astrocytes, and pericytes. Chronic DM leads to diminished central expression of the tight junction proteins occludin and claudin-2 (Ahmad et al., 2017; Chehade et al., 2002). Tight junction degradation during hyperglycemic conditions can occur as a secondary effect of chronic inflammation, increased VEGF expression, and/or upregulated matrix metalloproteases (Argaw et al., 2009; Kamada et al., 2007). Under high glucose conditions, VEGF induces the translocation of GLUT-1 to the cell surface at the BBB, inhibiting tight junction protein expression and enhancing permeability (Argaw et al., 2009; Pardridge et al., 1990). Models of ischemia in diabetic rats demonstrate BBB injury through increased oxidative stress and activation of MMP-9 (Dietrich et al., 1993; Hawkins et al., 2007; Kamada et al., 2007). Long-term hyperglycemia induces the formation of advanced glycosylation end products (AGEs) (Paget et al., 1998). Binding of AGEs to specific receptors (RAGE) enhances vascular inflammation and promotes oxidative stress via generation of ROS (Paget et al., 1998; Serban et al., 2016). DM-induced oxidative stress and accumulation of ROS decrease the expression of eNOS and activate VCAM-1, enhancing monocyte adhesion and vascular permeability (Chakravarthy et al., 1998; Lu et al., 1998).

Beyond endothelial cell disruption, exposure of additional components of the BBB to prolonged inflammation and hyperglycemia results in cell-specific effects in DM. The use of immunohistochemistry in a rat streptozotocin model demonstrated structural changes to astrocytic end-feet and membrane attachment along with increased astrocyte cell death (Jing et al., 2013). Cultured astrocytes exposed to high glucose and brain slices from diabetic animals exhibited increased proinflammatory cytokine and ROS expression and reduced communication between astrocytes (Gandhi et al., 2010; Wang et al., 2012). Pericyte disintegration has been shown to play a key role in blood-retinal barrier disruption associated with DM. Along with RAGE, pericyte disruption leads to the activation of NF-κB, which regulates multiple proinflammatory pathways (Addison et al., 1970; Romeo et al., 2002). The blood-retinal barrier is structurally similar to the BBB, with uncontrolled permeability resulting in retinopathy. Retinal vessel pathology studies are well established and thus may be a useful resource for further investigations into mechanisms of BBB disruption in DM. Taken together, chronic hyperglycemia initiates oxidative stress and inflammation, the downstream effects of which significantly contribute to cerebral permeability. As such, DM is a major risk factor for diseases in which altered BBB function plays a central or primary role such as ischemia/stroke, vascular dementia, and Alzheimer's disease (Benito-Leon et al., 2013; Cheng et al., 2012; Hassing et al., 2009; Kissela et al., 2005).

7. BBB impacts of environmental and occupational agents

Although vascular diseases can affect the CNS differently, they often share similar pathologies including damage to cerebral microvessels, diminished blood flow, neuroinflammation, and loss of BBB integrity. Defects in the BBB can be triggered by internal factors beyond overt illness including age, metabolic disturbances, and genetic predisposition and can be accelerated by external stressors such as exposure to pollution. In most cases, studies related to environmental factors and the BBB are limited to the degree that exogenous toxicants can access the brain, and not assessing direct effects of the toxicants on BBB integrity. However, several environmental factors have been specifically linked to BBB disruption, as described below. Agents selected for discussion were largely those for which in vivo data were available to support a BBB disruption.

7.1 Ambient air pollution: Human studies

Over the past 10 years, numerous epidemiological studies have associated air pollution with neurodevelopmental, neurodegenerative, and neurocognitive outcomes (Chen et al., 2017; Kerin et al., 2018; Tonne et al., 2014; Tzivian et al., 2016; Volk et al., 2013). While many studies indicate a link between exposure to environmental pollutants and neurological consequences, there is less evidence highlighting the potential impact of environmental stressors on BBB permeability. Long-term air pollution resulted in downregulated zonula occludens Ab (tight junction protein-1) in Mexico City subjects (Calderon-Garciduenas et al., 2008b, c). Exposed subjects exhibited significantly greater levels of soluble endothelial cell adhesion molecules (e.g., vascular cell adhesion molecule-1 and intracellular adhesion molecule-1) compared to subjects from low pollution regions. Immunohistochemical and transcriptional analyses of brains from young cadavers revealed evidence suggestive of a neurovascular inflammatory outcome related to air pollution exposure (Calderon-Garciduenas et al., 2008b). One study examined cognitive deficits and MRI abnormalities in both children and dogs following air pollution exposure (Calderon-Garciduenas et al., 2008a). In this study, a cohort of children ($n=55$) from

Mexico City were compared to children ($n = 18$) from a low air pollution area (Polotitlan). Comparisons using the Wechsler Intelligence Scale for Children-Revised (WISC-R) indicated that Mexico City children, but not Polotitlan children were significantly "behind" their normative level of cognitive development. This study also indicated that there was a relationship between city of residency and white matter lesions according to MRI. Interestingly, these white matter lesions were also observed in the frontal cortex of a subset of dogs (57%) from Mexico City. In another study, serum and CSF biomarkers in children exposed to high levels of air pollution showed significantly higher serum actin IgG, occludin/zonulin 1 IgA, IgG, myelin oligodendrocyte glycoprotein IgG and IgM ($P < 0.01$), myelin basic protein IgA and IgG, S-100 IgG and IgM, and cerebellar IgG ($P < 0.001$) (Calderon-Garciduenas et al., 2014).

7.2 Ozone (O_3)

Several studies have indicated that O_3 exposure may contribute to CNS dysfunction (Apte et al., 2008; Gonzalez-Guevara et al., 2014; Paz, 1997). Serum amyloid A (SAA) increased following 3ppm O_3 exposure for 6h in BALB/C mice (Erickson et al., 2017). Acute SAA in the blood was correlated with total leukocytes, macrophages, and neutrophils in bronchoalveolar lavage and both isoforms of acute SAA crossed the BBB. The rate of SAA2.1 uptake was significantly faster than SAA1.1. These data suggest that SAA is an important circulating factor that has the potential to cross the BBB following O_3 exposure.

Recently, Tyler et al. linked increased fluorescein staining in the brain to BBB leakage and neuroinflammatory outcomes in aged (18-month) mice exposed to 1ppm O_3 exposure for 4h (Tyler et al., 2018). Reactive microglia were present in the brain and infiltrating immune cells including neutrophils and macrophages were visualized along the BBB. Younger mice were generally less susceptible to the changes in neutrophil influx into the brain and activation of microglia, but still displayed some of the same trends. These studies highlight potential variation in sensitivity to environmental impacts on the BBB in certain susceptible subpopulations.

7.3 Gasoline and diesel engine exhaust

Diesel exhaust particles, a major component of urban air pollution, were specifically implicated in the upregulation of oxidative stress pathways and production of proinflammatory cytokines related to the BBB using ex vivo capillaries and direct exposure to particulates (Hartz et al., 2008). However, it was unclear if the exposure model appropriately considered the low level of transfer of particulates from the lung to the systemic circulation, or potential secondary factors arising in the lung following particle exposure.

Increases in BBB permeability, as indicated by sodium fluorescein uptake, following inhalation exposure to mixed vehicular emissions (MVE, a combination of gasoline and diesel engine exhausts at 100 μg PM/m^3) was evident when compared to filtered air control mice (Oppenheim et al., 2013). This was attributed to degradation of tight junction proteins, more specifically, occludin and claudin-5, and increased levels of inducible nitric oxide synthase and interleukin-1β in the parenchyma. Sucrose permeability, a measure of BBB permeability, also increased following coculture (brain endothelial cells and glial cells) with serum from mixed vehicular emissions (MVE)-exposed mice. Additionally, serum from ApoE$^{-/-}$ mice incubated with this BBB coculture resulted in alteration of transport activity (Oppenheim et al., 2013). In another study, wild-type C57BL/6 mice fed either a high fat or normal chow diet were assessed for BBB disruption following MVE exposure for 7 days/week, 6 h/day, for 30 days (Suwannasual et al., 2018). Fluoroscein stain indicated significantly more BBB leakage in mice fed a high fat diet compared to normal chow-fed animals, independent of the emissions exposure. Moreover, mice in the MVE plus high fat diet group exhibited significantly greater fluorescein staining than MVE-exposed mice fed a normal chow, suggesting an interaction between diet and exposure. Exposure to high fat diet and MVE also led to significant increases in circulating oxidized low-density lipoprotein (oxLDL) and two key scavenger receptors (lectin-like receptor of oxLDL, or LOX-1, and CD-36) in the cerebral vasculature. Significant decreases in claudin-5 and significant increases in oxidative stress markers were evident in both MVE exposure groups, regardless of diet.

In a follow-up study, MVE exposure increased LOX-1 in ApoE$^{-/-}$ mice (Lucero et al., 2017). ApoE$^{-/-}$ mice exposed to a mixture of gasoline and diesel engine exhaust (100 μg PM/m^3 for 6 h/day for 30 days by inhalation) exhibited significantly less tight junction protein expression in the BBB with greater vascular gelatinase (matrix metalloproteinases 2 and 9) activity and ROS. LOX-1 antibody treatment attenuated lipid peroxidation and gelatinase activities, suggesting that LOX-1 may be a key mediator of MVE exposure. In a similar study, diesel exposure alone was also able to drive BBB dysfunction that was associated with neuroinflammatory responses (Heidari Nejad et al., 2015). BALB/c mice exposed to diesel over a 2-week period exhibited greater brain parenchymal IgG and GFAP compared to control (filtered air-exposed) mice 24 h post-exposure.

In a recent related study, $ApoE^{-/-}$ mice exposed to inhaled ultrafine particles + gas (UFP + G) exhibited significant neuroinflammation compared to UFP exposure alone (or to the fine particle fraction of engine emissions), suggesting that the presence of gas-phase pollutants may contribute to greater health effects than UFP alone (Tyler et al., 2016). These findings support epidemiological evidence suggesting an association between roadway proximity, where fresh emissions exist as a mixture with close affinity between physical particles and semivolatile organic species, and adverse neurological outcomes, but did not specifically detail the potential role for BBB impairments in driving such outcomes.

7.4 Nanomaterial exposure

A recent mechanistically-designed study demonstrated an important contribution of blood-borne factors in driving BBB dysfunction following pulmonary exposure to multiwalled carbon nanotubes (MWCNT) (Aragon et al., 2017). Exposure to MWCNT via oropharyngeal aspiration led to elevations in inflammatory transcripts in different brain regions and immunohistochemical staining highlighted glial and astrocytic activation specifically enveloping the neurovascular unit. By using a Rho kinase inhibitor (fasudil) to restore endothelial barrier integrity after MWCNT exposure in a mouse model, Aragon and colleagues were able to inhibit both penetrance of sodium fluorescein into the brain and, subsequently, reduce neuroinflammatory outcomes. Furthermore, the serum obtained from MWCNT-exposed mice exhibited a clear bioactivity that could promote inflammatory outcomes and reduce cell growth and motility in vitro. Further evidence was provided suggesting a role for circulating thrombospondin-1 and related factors acting on the endothelial cell scavenger receptor CD36. These studies bolstered previous findings that serum obtained after MWCNT exposure could alter endothelial function, in terms of vasomotor activity, in a CD36-dependent manner (Aragon et al., 2016). The bioactivity in serum appeared to be, at least partially, a product of MMP-9 activity in the lung.

Aerosolized nanoparticulate TiO_2 exposure results in age-dependent modulation of BBB integrity in aging rats (Disdier et al., 2017). BBB alteration was associated with a significant increase of cytokines and chemokines in the brain including IL-1β, IFN-γ, fractalkine, and synaptophysin a neuronal activity marker. BBB leakage was quantified using an atenolol-based assay and demonstrated significantly greater leakage in 19-month-old rats, compared to 12-week-old controls following nano-aerosol TiO_2 exposure. While TiO_2 particles in the brain were not detected, BBB dysfunction was still associated with neuroinflammation following exposure.

8. Research gaps

Environmental stressors that indirectly or directly target the BBB may thus contribute to numerous other neurological syndromes. We propose that by "distracting" neurosupportive cells from their essential homeostatic roles in neural plasticity, neurodevelopment, and glymphatic waste clearance, environmentally-induced BBB deficits contribute to a wide range of neurological disorders. Experimental evidence for such conclusions is minimal, however, owing to the challenges in translating neurological outcomes in rodents to humans, as well as to limitations in identifying or imaging low-level perturbations of BBB dysfunction.

As we are in the early stages of appreciating a role for environmental impacts on BBB function, many knowledge gaps exist. Among the most crucial is the temporality of effect—how long do acute insults to the BBB last and how does the neurovasculature compensate in the face of chronic repeated stressors? And what is the toxicodynamic effect of changes in microglial phenotype? It may take some time to recover the normal homeostatic role of synaptic pruning and networking after shoring up defenses for an impaired BBB. Furthermore, as age is an essential component of many neurological disorders, are there specific periods of vulnerability, such as was recently demonstrated (Tyler et al., 2018)?

In addition to timing-related factors, questions regarding environmental pollution, mixtures, and dosimetry remain inadequately addressed. While several toxicants have been shown to impact BBB integrity, a head-to-head comparison of potency has not been pursued to date. Moreover, effects have been seen at operationally-defined "low" doses (Aragon et al., 2017) that are still highly relative to occupational and environmental levels, which simply means that we are unclear about no effects levels for most toxicants that have been studied.

The pathways leading to impaired BBB following pulmonary toxicant exposures also remains relatively unexplored. The concepts we have recently proposed (Aragon et al., 2016, 2017) suggest that indirect generation of circulating, bioactive intermediates is a valid central theme but is by no means thoroughly characterized. The direct uptake of particulates or their soluble components into the brain has been demonstrated in a variety of indirect and direct approaches, but the presence of particulates in the brain has never been mechanistically tied to a pathological outcome such as BBB dysfunction or neuroinflammation, much less neurocognitive decline or neurodegenerative syndromes. The BBB may in fact be a more likely intermediary in driving neurological consequences of particulate matter exposure than direct translocation of particles to the brain.

As somewhat of a cautionary note, given that pharmacological interventions targeting brain cells require transport across the BBB, numerous strategies are emerging to adjunctively open the BBB to improve drug delivery. The BBB remains one of the most significant challenges in neuropharmacology and especially to nanomedicine-based approaches to targeting brain delivery. Nanomaterial constructs may provide unique and selective means of bypassing the BBB without reducing the barrier to other circulating factors, but the risks of approaches to non-selectively open the BBB are clear, and may offset benefits of therapeutics. Utilizing ultrasound or related methodologies to reduce BBB integrity to a specific region for a short time may be the ideal scenario to enhance uptake of drugs. Additionally, nasal administration not only permits small molecule and particulate uptake to the brain but also reduces the role of first pass metabolism and/or scavenging by the reticuloendothelial system, although this route may not allow selective regional delivery to desired regions of the brain.

Most of the studies described in this chapter are related to inhaled toxicants, which may relate to a lung-specific interaction to generate secondary intermediates or damage-associated molecular patterns that interact with the endothelial cells of the NVU. However, many other environmental agents may be of concern for neurological disease and the relative impacts of major classes of toxicants, including heavy metals, pesticides, plasticizers, etc., are simply poorly understood. Several studies have been conducted in cell culture models, but because of the relatively early stages of hazard identification for such agents, they were not discussed here. However, more research is needed to understand how commonly encountered environmental pollutants can impact BBB integrity.

References

Abbott, N.J., Rönnbäck, L., Hansson, E., 2006. Astrocyte-endothelial interactions at the blood-brain barrier. Nat. Rev. Neurosci. 7, 41–53.

Abbott, N.J., Patabendige, A.A., Dolman, D.E., Yusof, S.R., Begley, D.J., 2010. Structure and function of the blood-brain barrier. Neurobiol. Dis. 37, 13–25.

Acharya, N.K., Levin, E.C., Clifford, P.M., Han, M., Tourtellotte, R., Chamberlain, D., Pollaro, M., Coretti, N.J., Kosciuk, M.C., Nagele, E.P., Demarshall, C.,

Freeman, T., Shi, Y., Guan, C., Macphee, C.H., Wilensky, R.L., Nagele, R.G., 2013. Diabetes and hypercholesterolemia increase blood-brain barrier permeability and brain amyloid deposition: beneficial effects of the LpPLA2 inhibitor darapladib. J. Alzheimers Dis. 35, 179–198.

Adair, J.C., Charlie, J., Dencoff, J.E., Kaye, J.A., Quinn, J.F., Camicioli, R.M., Stetler-Stevenson, W.G., Rosenberg, G.A., 2004. Measurement of gelatinase B (MMP-9) in the cerebrospinal fluid of patients with vascular dementia and Alzheimer disease. Stroke 35, e159–e162.

Addison, D.J., Garner, A., Ashton, N., 1970. Degeneration of intramural pericytes in diabetic retinopathy. Br. Med. J. 1, 264–266.

Agrawal, S., Anderson, P., Durbeej, M., VAN Rooijen, N., Ivars, F., Opdenakker, G., Sorokin, L.M., 2006. Dystroglycan is selectively cleaved at the parenchymal basement membrane at sites of leukocyte extravasation in experimental autoimmune encephalomyelitis. J. Exp. Med. 203, 1007–1019.

Ahmad, R., Rah, B., Bastola, D., Dhawan, P., Singh, A.B., 2017. Obesity-induces organ and tissue specific tight junction restructuring and barrier deregulation by claudin switching. Sci. Rep. 7, 5125.

Al Feteisi, H., Al-Majdoub, Z.M., Achour, B., Couto, N., Rostami-Hodjegan, A., Barber, J., 2018. Identification and quantification of blood-brain barrier transporters in isolated rat brain microvessels. J. Neurochem. 146, 670–685.

Albargothy, N.J., Sharp, M.M., Gatherer, M., Morris, A., Weller, R.O., Hawkes, C., Carare, R.O., 2017. Investigating the lymphatic drainage of the brain: essential skills and tools. Methods Mol. Biol. 1559, 343–365.

Alexander, G.M., Schwartzman, R.J., Grothusen, J.R., Gordon, S.W., 1994. Effect of plasma levels of large neutral amino acids and degree of parkinsonism on the blood-to-brain transport of levodopa in naive and MPTP parkinsonian monkeys. Neurology 44, 1491–1499.

Alkadhi, K.A., 2011. Chronic stress and Alzheimer's disease-like pathogenesis in a rat model: prevention by nicotine. Curr. Neuropharmacol. 9, 587–597.

Almutairi, M.M., Gong, C., Xu, Y.G., Chang, Y., Shi, H., 2016. Factors controlling permeability of the blood-brain barrier. Cell. Mol. Life Sci. 73, 57–77.

Apte, M.G., Buchanan, I.S., Mendell, M.J., 2008. Outdoor ozone and building-related symptoms in the BASE study. Indoor Air 18, 156–170.

Aragon, M., Erdely, A., Bishop, L., Salmen, R., Weaver, J., Liu, J., Hall, P., Eye, T., Kodali, V., Zeidler-Erdely, P., Stafflinger, J.E., Ottens, A.K., Campen, M.J., 2016. MMP-9-dependent serum-borne bioactivity caused by multiwalled carbon nanotube exposure induces vascular dysfunction via the CD36 scavenger receptor. Toxicol. Sci. 150, 488–498.

Aragon, M.J., Topper, L., Tyler, C.R., Sanchez, B., Zychowski, K., Young, T., Herbert, G., Hall, P., Erdely, A., Eye, T., Bishop, L., Saunders, S.A., Muldoon, P.P., Ottens, A.K., Campen, M.J., 2017. Serum-borne bioactivity caused by pulmonary multiwalled carbon nanotubes induces neuroinflammation via blood-brain barrier impairment. Proc. Natl. Acad. Sci. U. S. A. 114, E1968–E1976.

Argaw, A.T., Gurfein, B.T., Zhang, Y., Zameer, A., John, G.R., 2009. VEGF-mediated disruption of endothelial CLN-5 promotes blood-brain barrier breakdown. Proc. Natl. Acad. Sci. U. S. A. 106, 1977–1982.

Badaut, J., Lasbennes, F., Magistretti, P.J., Regli, L., 2002. Aquaporins in brain: distribution, physiology, and pathophysiology. J. Cereb. Blood Flow Metab. 22, 367–378.

Badaut, J., Fukuda, A.M., Jullienne, A., Petry, K.G., 2014. Aquaporin and brain diseases. Biochim. Biophys. Acta 1840, 1554–1565.

Bai, X., Moraes, T.F., Reithmeier, R.A.F., 2018. Structural biology of solute carrier (SLC) membrane transport proteins. Mol. Membr. Biol. 34, 1–32.

Bailey, S.L., Schreiner, B., McMahon, E.J., Miller, S.D., 2007. CNS myeloid DCs presenting endogenous myelin peptides 'preferentially' polarize CD4+ T(H)-17 cells in relapsing EAE. Nat. Immunol. 8, 172–180.

Baker, L.D., Cross, D.J., Minoshima, S., Belongia, D., Watson, G.S., Craft, S., 2011. Insulin resistance and Alzheimer-like reductions in regional cerebral glucose metabolism for cognitively normal adults with prediabetes or early type 2 diabetes. Arch. Neurol. 68, 51–57.

Bakker, E.N., Bacskai, B.J., Arbel-Ornath, M., Aldea, R., Bedussi, B., Morris, A.W., Weller, R.O., Carare, R.O., 2016. Lymphatic clearance of the brain: perivascular, paravascular and significance for neurodegenerative diseases. Cell. Mol. Neurobiol. 36, 181–194.

Balabanov, R., Washington, R., Wagnerova, J., Dore-Duffy, P., 1996. CNS microvascular pericytes express macrophage-like function, cell surface integrin alpha M, and macrophage marker ED-2. Microvasc. Res. 52, 127–142.

Ballabh, P., Braun, A., Nedergaard, M., 2004. The blood-brain barrier: an overview: structure, regulation, and clinical implications. Neurobiol. Dis. 16, 1–13.

Banks, W.A., Gray, A.M., Erickson, M.A., Salameh, T.S., Damodarasamy, M., Sheibani, N., Meabon, J.S., Wing, E.E., Morofuji, Y., Cook, D.G., Reed, M.J., 2015. Lipopolysaccharide-induced blood-brain barrier disruption: roles of cyclooxygenase, oxidative stress, neuroinflammation, and elements of the neurovascular unit. J. Neuroinflammation 12, 223.

Batarseh, Y.S., Duong, Q.V., Mousa, Y.M., Al Rihani, S.B., Elfakhri, K., Kaddoumi, A., 2016. Amyloid-beta and astrocytes interplay in amyloid-beta related disorders. Int. J. Mol. Sci. 17, 338.

Bataveljic, D., Milosevic, M., Radenovic, L., Andjus, P., 2014. Novel molecular biomarkers at the blood-brain barrier in ALS. Biomed. Res. Int. 2014, 907545.

Baxter, A.G., 2007. The origin and application of experimental autoimmune encephalomyelitis. Nat. Rev. Immunol. 7, 904–912.

Bell, R.D., Winkler, E.A., Singh, I., Sagare, A.P., Deane, R., Wu, Z., Holtzman, D.M., Betsholtz, C., Armulik, A., Sallstrom, J., Berk, B.C., Zlokovic, B.V., 2012. Apolipoprotein E controls cerebrovascular integrity via cyclophilin A. Nature 485, 512–516.

Bell, J.S., Spencer, J.I., Yates, R.L., Deluca, G.C., 2018. The cortical blood-brain barrier in multiple sclerosis: a gateway to progression? J. Neurol. 265, 966–967.

Benito-Leon, J., Mitchell, A.J., Hernandez-Gallego, J., Bermejo-Pareja, F., 2013. Obesity and impaired cognitive functioning in the elderly: a population-based cross-sectional study (NEDICES). Eur. J. Neurol. 20, 899–906. e76–7.

Bennett, J., Basivireddy, J., Kollar, A., Biron, K.E., Reickmann, P., Jefferies, W.A., Mcquaid, S., 2010. Blood-brain barrier disruption and enhanced vascular permeability in the multiple sclerosis model EAE. J. Neuroimmunol. 229, 180–191.

Bentivoglio, M., Kristensson, K., 2014. Tryps and trips: cell trafficking across the 100-year-old blood-brain barrier. Trends Neurosci. 37, 325–333.

Binder, D.K., Yao, X., Zador, Z., Sick, T.J., Verkman, A.S., Manley, G.T., 2006. Increased seizure duration and slowed potassium kinetics in mice lacking aquaporin-4 water channels. Glia 53, 631–636.

Bjerke, M., Zetterberg, H., Edman, A., Blennow, K., Wallin, A., Andreasson, U., 2011. Cerebrospinal fluid matrix metalloproteinases and tissue inhibitor of metalloproteinases in combination with subcortical and cortical biomarkers in vascular dementia and Alzheimer's disease. J. Alzheimers Dis. 27, 665–676.

Blanchette, M., Daneman, R., 2015. Formation and maintenance of the BBB. Mech. Dev. 138 (Pt 1), 8–16.

Blennow, K., Wallin, A., Fredman, P., Karlsson, I., Gottfries, C.G., Svennerholm, L., 1990. Blood-brain barrier disturbance in patients with Alzheimer's disease is related to vascular factors. Acta Neurol. Scand. 81, 323–326.

Bogousslavsky, J., van Melle, G., Regli, F., 1988. The Lausanne stroke registry: analysis of 1,000 consecutive patients with first stroke. Stroke 19, 1083–1092.

Brown, W.J., 1978. The capillaries in acute and subacute multiple sclerosis plaques: a morphometric analysis. Neurology 28, 84–92.

Brownlee, M., 2001. Biochemistry and molecular cell biology of diabetic complications. Nature 414, 813.

Brzica, H., Abdullahi, W., Ibbotson, K., Ronaldson, P.T., 2017. Role of transporters in central nervous system drug delivery and blood-brain barrier protection: relevance to treatment of stroke. J. Cent. Nerv. Syst. Dis. 9, 1179573517693802.

Burns, E.M., Kruckeberg, T.W., Gaetano, P.K., 1981. Changes with age in cerebral capillary morphology. Neurobiol. Aging 2, 283–291.

Calderon-Garciduenas, L., Mora-Tiscareno, A., Ontiveros, E., Gomez-Garza, G., Barragan-Mejia, G., Broadway, J., Chapman, S., Valencia-Salazar, G., Jewells, V., Maronpot, R.R., Henriquez-Roldan, C., Perez-Guille, B., Torres-Jardon, R., Herrit, L., Brooks, D., Osnaya-Brizuela, N., Monroy, M.E., Gonzalez-Maciel, A., Reynoso-Robles, R., Villarreal-Calderon, R., Solt, A.C., Engle, R.W., 2008a. Air pollution, cognitive deficits and brain abnormalities: a pilot study with children and dogs. Brain Cogn. 68, 117–127.

Calderon-Garciduenas, L., Solt, A.C., Henriquez-Roldan, C., Torres-Jardon, R., Nuse, B., Herritt, L., Villarreal-Calderon, R., Osnaya, N., Stone, I., Garcia, R., Brooks, D.M., Gonzalez-Maciel, A., Reynoso-Robles, R., Delgado-Chavez, R., Reed, W., 2008b. Long-term air pollution exposure is associated with neuroinflammation, an altered innate immune response, disruption of the blood-brain barrier, ultrafine particulate deposition, and accumulation of amyloid beta-42 and alpha-synuclein in children and young adults. Toxicol. Pathol. 36, 289–310.

Calderon-Garciduenas, L., Villarreal-Calderon, R., Valencia-Salazar, G., Henriquez-Roldan, C., Gutierrez-Castrellon, P., Torres-Jardon, R., Osnaya-Brizuela, N., Romero, L., Torres-Jardon, R., Solt, A., Reed, W., 2008c. Systemic inflammation, endothelial dysfunction, and activation in clinically healthy children exposed to air pollutants. Inhal. Toxicol. 20, 499–506.

Calderon-Garciduenas, L., Torres-Jardon, R., Kulesza, R.J., Park, S.B., D'angiulli, A., 2014. Air pollution and detrimental effects on children's brain. The need for a multidisciplinary approach to the issue complexity and challenges. Front. Hum. Neurosci. 8, 613.

Chakravarthy, U., Hayes, R.G., Stitt, A.W., Mcauley, E., Archer, D.B., 1998. Constitutive nitric oxide synthase expression in retinal vascular endothelial cells is suppressed by high glucose and advanced glycation end products. Diabetes 47, 945–952.

Chakravarty, S., Herkenham, M., 2005. Toll-like receptor 4 on nonhematopoietic cells sustains CNS inflammation during endotoxemia, independent of systemic cytokines. J. Neurosci. 25, 1788–1796.

Chehade, J.M., Haas, M.J., Mooradian, A.D., 2002. Diabetes-related changes in rat cerebral occludin and zonula occludens-1 (ZO-1) expression. Neurochem. Res. 27, 249–252.

Chen, J., Zacharek, A., Zhang, C., Jiang, H., Li, Y., Roberts, C., LU, M., Kapke, A., Chopp, M., 2005. Endothelial nitric oxide synthase regulates brain-derived neurotrophic factor expression and neurogenesis after stroke in mice. J. Neurosci. 25, 2366–2375.

Chen, H., Kwong, J.C., Copes, R., Hystad, P., van Donkelaar, A., Tu, K., Brook, J.R., Goldberg, M.S., Martin, R.V., Murray, B.J., Wilton, A.S., Kopp, A., Burnett, R.T., 2017. Exposure to ambient air pollution and the incidence of dementia: a population-based cohort study. Environ. Int. 108, 271–277.

Cheng, G., Huang, C., Deng, H., Wang, H., 2012. Diabetes as a risk factor for dementia and mild cognitive impairment: a meta-analysis of longitudinal studies. Intern. Med. J. 42, 484–491.

Chung, C.S., Caplan, L.R., 2007. Stroke and other neurovascular disorders. In: Goetz, C.G. (Ed.), Textbook of Clinical Neurology, third ed. W.B. Saunders, Philadelphia. Chapter 45.
Clasen, R.A., Pandolfi, S., Hass, G.M., 1970. Vital staining, serum albumin and the blood-brain barrier. J. Neuropathol. Exp. Neurol. 29, 266–284.
Claudio, L., Brosnan, C.F., 1992. Effects of prazosin on the blood-brain barrier during experimental autoimmune encephalomyelitis. Brain Res. 594, 233–243.
Clementi, F., 1970. Effect of horseradish peroxidase on mice lung capillaries' permeability. J. Histochem. Cytochem. 18, 887–892.
Coimbra-Costa, D., Alva, N., Duran, M., Carbonell, T., Rama, R., 2017. Oxidative stress and apoptosis after acute respiratory hypoxia and reoxygenation in rat brain. Redox Biol. 12, 216–225.
Compston, A., Coles, A., 2008. Multiple sclerosis. Lancet 372, 1502–1517.
Coronado-Velazquez, D., Betanzos, A., Serrano-Luna, J., Shibayama, M., 2018. An in vitro model of the blood-brain barrier: *Naegleria fowleri* affects the tight junction proteins and activates the microvascular endothelial cells. J. Eukaryot. Microbiol. 65 (6), 804–819.
Cotran, K.S., Karnovsky, M.J., 1967. Vascular leakage induced by horseradish peroxidase in the rat. Proc. Soc. Exp. Biol. Med. 126, 557–561.
Cramer, S.P., Simonsen, H., Frederiksen, J.L., Rostrup, E., Larsson, H.B., 2014. Abnormal blood-brain barrier permeability in normal appearing white matter in multiple sclerosis investigated by MRI. Neuroimage Clin. 4, 182–189.
Curb, J.D., Rodriguez, B.L., Abbott, R.D., Petrovitch, H., Ross, G.W., Masaki, K.H., Foley, D., Blanchette, P.L., Harris, T., Chen, R., White, L.R., 1999. Longitudinal association of vascular and Alzheimer's dementias, diabetes, and glucose tolerance. Neurology 52, 971–975.
Dehouck, M.P., Meresse, S., Delorme, P., Fruchart, J.C., Cecchelli, R., 1990. An easier, reproducible, and mass-production method to study the blood-brain barrier in vitro. J. Neurochem. 54, 1798–1801.
Denk, W., Strickler, J.H., Webb, W.W., 1990. Two-photon laser scanning fluorescence microscopy. Science 248, 73–76.
De-Paula, V.J., Radanovic, M., Diniz, B.S., Forlenza, O.V., 2012. Alzheimer's disease. Subcell. Biochem. 65, 329–352.
Dick, A.P., Harik, S.I., Klip, A., Walker, D.M., 1984. Identification and characterization of the glucose transporter of the blood-brain barrier by cytochalasin B binding and immunological reactivity. Proc. Natl. Acad. Sci. U. S. A. 81, 7233–7237.
Dietrich, W.D., Alonso, O., Busto, R., 1993. Moderate hyperglycemia worsens acute blood-brain barrier injury after forebrain ischemia in rats. Stroke 24, 111–116.
Disdier, C., Chalansonnet, M., Gagnaire, F., Gate, L., Cosnier, F., Devoy, J., Saba, W., Lund, A.K., Brun, E., Mabondzo, A., 2017. Brain inflammation, blood brain barrier dysfunction and neuronal synaptophysin decrease after inhalation exposure to titanium dioxide nano-aerosol in aging rats. Sci. Rep. 7, 12196.
Domingues, H.S., Portugal, C.C., Socodato, R., Relvas, J.B., 2016. Oligodendrocyte, astrocyte, and microglia crosstalk in myelin development, damage, and repair. Front. Cell Dev. Biol. 4, 71.
Dutta, R., Mcdonough, J., Yin, X., Peterson, J., Chang, A., Torres, T., Gudz, T., Macklin, W.B., Lewis, D.A., Fox, R.J., Rudick, R., Mirnics, K., Trapp, B.D., 2006. Mitochondrial dysfunction as a cause of axonal degeneration in multiple sclerosis patients. Ann. Neurol. 59, 478–489.
Dutta, R., Mcdonough, J., Chang, A., Swamy, L., Siu, A., Kidd, G.J., Rudick, R., Mirnics, K., Trapp, B.D., 2007. Activation of the ciliary neurotrophic factor (CNTF) signalling pathway in cortical neurons of multiple sclerosis patients. Brain 130, 2566–2576.

Ehrlich, P., 1885. Das Sauerstoff-Bedürfnis des Organismus. Eine farbenanalytische Studie. PhD thesis, Herschwald, Berlin.

Ek, C.J., Habgood, M.D., Dziegielewska, K.M., Saunders, N.R., 2003. Structural characteristics and barrier properties of the choroid plexuses in developing brain of the opossum (*Monodelphis Domestica*). J. Comp. Neurol. 460, 451–464.

Ek, C.J., Dziegielewska, K.M., Stolp, H., Saunders, N.R., 2006. Functional effectiveness of the blood-brain barrier to small water-soluble molecules in developing and adult opossum (*Monodelphis domestica*). J. Comp. Neurol. 496, 13–26.

Engelhardt, B., Ransohoff, R.M., 2012. Capture, crawl, cross: the T cell code to breach the blood-brain barriers. Trends Immunol. 33, 579–589.

Erickson, M.A., Banks, W.A., 2013. Blood-brain barrier dysfunction as a cause and consequence of Alzheimer's disease. J. Cereb. Blood Flow Metab. 33, 1500–1513.

Erickson, M.A., Jude, J., Zhao, H., Rhea, E.M., Salameh, T.S., Jester, W., Pu, S., Harrowitz, J., Nguyen, N., Banks, W.A., Panettieri Jr., R.A., Jordan-Sciutto, K.L., 2017. Serum amyloid A: an ozone-induced circulating factor with potentially important functions in the lung-brain axis. FASEB J. 31, 3950–3965.

Ewing, J.R., Knight, R.A., Nagaraja, T.N., Yee, J.S., Nagesh, V., Whitton, P.A., Li, L., Fenstermacher, J.D., 2003. Patlak plots of Gd-DTPA MRI data yield blood-brain transfer constants concordant with those of 14C-sucrose in areas of blood-brain opening. Magn. Reson. Med. 50, 283–292.

Fainardi, E., Castellazzi, M., Bellini, T., Manfrinato, M.C., Baldi, E., Casetta, I., Paolino, E., Granieri, E., Dallocchio, F., 2006. Cerebrospinal fluid and serum levels and intrathecal production of active matrix metalloproteinase-9 (MMP-9) as markers of disease activity in patients with multiple sclerosis. Mult. Scler. 12, 294–301.

Fantini, S., Sassaroli, A., Tgavalekos, K.T., Kornbluth, J., 2016. Cerebral blood flow and autoregulation: current measurement techniques and prospects for noninvasive optical methods. Neurophotonics 3, 031411.

Faraci, F.M., 2006. Protecting the brain with eNOS: run for your life. Circ. Res. 99, 1029–1030.

Fernandez-Klett, F., Potas, J.R., Hilpert, D., Blazej, K., Radke, J., Huck, J., Engel, O., Stenzel, W., Genove, G., Priller, J., 2013. Early loss of pericytes and perivascular stromal cell-induced scar formation after stroke. J. Cereb. Blood Flow Metab. 33, 428–439.

Fletcher, J.M., Lalor, S.J., Sweeney, C.M., Tubridy, N., Mills, K.H., 2010. T cells in multiple sclerosis and experimental autoimmune encephalomyelitis. Clin. Exp. Immunol. 162, 1–11.

Flugel, A., Berkowicz, T., Ritter, T., Labeur, M., Jenne, D.E., Li, Z., Ellwart, J.W., Willem, M., Lassmann, H., Wekerle, H., 2001. Migratory activity and functional changes of green fluorescent effector cells before and during experimental autoimmune encephalomyelitis. Immunity 14, 547–560.

Fluri, F., Schuhmann, M.K., Kleinschnitz, C., 2015. Animal models of ischemic stroke and their application in clinical research. Drug Des. Devel. Ther. 9, 3445–3454.

Folch, J., Petrov, D., Ettcheto, M., Abad, S., Sanchez-Lopez, E., Garcia, M.L., Olloquequi, J., Beas-Zarate, C., Auladell, C., Camins, A., 2016. Current research therapeutic strategies for Alzheimer's disease treatment. Neural Plast. 2016, 8501693.

Foster, C.A., Mechtcheriakova, D., Storch, M.K., Balatoni, B., Howard, L.M., Bornancin, F., Wlachos, A., Sobanov, J., Kinnunen, A., Baumruker, T., 2009. FTY720 rescue therapy in the dark agouti rat model of experimental autoimmune encephalomyelitis: expression of central nervous system genes and reversal of blood-brain-barrier damage. Brain Pathol. 19, 254–266.

Friedland, R.P., Jagust, W.J., Huesman, R.H., Koss, E., Knittel, B., Mathis, C.A., Ober, B.A., Mazoyer, B.M., Budinger, T.F., 1989. Regional cerebral glucose transport and utilization in Alzheimer's disease. Neurology 39, 1427–1434.

Frolich, L., Kornhuber, J., Ihl, R., Fritze, J., Maurer, K., Riederer, P., 1991. Integrity of the blood-CSF barrier in dementia of Alzheimer type: CSF/serum ratios of albumin and IgG. Eur. Arch. Psychiatry Clin. Neurosci. 240, 363–366.
Gandhi, G.K., Ball, K.K., Cruz, N.F., Dienel, G.A., 2010. Hyperglycaemia and diabetes impair gap junctional communication among astrocytes. ASN Neuro 2, e00030.
Geier, E.G., Schlessinger, A., Fan, H., Gable, J.E., Irwin, J.J., Sali, A., Giacomini, K.M., 2013. Structure-based ligand discovery for the large-neutral amino acid transporter 1, LAT-1. Proc. Natl. Acad. Sci. U. S. A. 110, 5480–5485.
Giannoni, P., Badaut, J., Dargazanli, C., de Maudave, A.F., Klement, W., Costalat, V., Marchi, N., 2018. The pericyte-glia interface at the blood-brain barrier. Clin. Sci. (Lond.) 132, 361–374.
Gold, M., El Khoury, J., 2015. Beta-amyloid, microglia, and the inflammasome in Alzheimer's disease. Semin. Immunopathol. 37, 607–611.
Gold, R., Linington, C., Lassmann, H., 2006. Understanding pathogenesis and therapy of multiple sclerosis via animal models: 70 years of merits and culprits in experimental autoimmune encephalomyelitis research. Brain 129, 1953–1971.
Goldmann, E., 1913. Vitalfarbung am Zentralnervensystem. In: Beitrag zur Physio-Pathologie des Plexus chorioideus und der Hirnhäute. vol. 1. Abh Königl Preuss Akad Wiss, Berlin, pp. 1–61.
Gonzalez-Guevara, E., Martinez-Lazcano, J.C., Custodio, V., Hernandez-Ceron, M., Rubio, C., Paz, C., 2014. Exposure to ozone induces a systemic inflammatory response: possible source of the neurological alterations induced by this gas. Inhal. Toxicol. 26, 485–491.
Goodall, E.F., Wang, C., Simpson, J.E., Baker, D.J., Drew, D.R., Heath, P.R., Saffrey, M.J., Romero, I.A., Wharton, S.B., 2018. Age-associated changes in the blood-brain barrier: comparative studies in human and mouse. Neuropathol. Appl. Neurobiol. 44, 328–340.
Greene, C., Campbell, M., 2016. Tight junction modulation of the blood brain barrier: CNS delivery of small molecules. Tissue Barriers 4, e1138017.
Grytten, N., Torkildsen, O., Aarseth, J.H., Benjaminsen, E., Celius, E.G., Dahl, O.P., Holmoy, T., Loken-Amsrud, K., Midgard, R., Myhr, K.M., Risberg, G., Vatne, A., Kampman, M.T., 2013. Month of birth as a latitude-dependent risk factor for multiple sclerosis in Norway. Mult. Scler. 19, 1028–1034.
Hall, A.M., Roberson, E.D., 2012. Mouse models of Alzheimer's disease. Brain Res. Bull. 88, 3–12.
Hamm, S., Dehouck, B., Kraus, J., Wolburg-Buchholz, K., Wolburg, H., Risau, W., Cecchelli, R., Engelhardt, B., Dehouck, M.P., 2004. Astrocyte mediated modulation of blood-brain barrier permeability does not correlate with a loss of tight junction proteins from the cellular contacts. Cell Tissue Res. 315, 157–166.
Harik, S.I., LaManna, J.C., 1991. Altered glucose metabolism in microvessels from patients with Alzheimer's disease. Ann. Neurol. 29, 573.
Hartz, A.M., Bauer, B., 2010. Regulation of ABC transporters at the blood-brain barrier: new targets for CNS therapy. Mol. Interv. 10, 293–304.
Hartz, A.M., Bauer, B., Block, M.L., Hong, J.S., Miller, D.S., 2008. Diesel exhaust particles induce oxidative stress, proinflammatory signaling, and P-glycoprotein up-regulation at the blood-brain barrier. FASEB J. 22, 2723–2733.
Hassing, L.B., Dahl, A.K., Thorvaldsson, V., Berg, S., Gatz, M., Pedersen, N.L., Johansson, B., 2009. Overweight in midlife and risk of dementia: a 40-year follow-up study. Int. J. Obes. 33, 893–898.
Hawkins, C.P., Munro, P.M., Landon, D.N., Mcdonald, W.I., 1992. Metabolically dependent blood-brain barrier breakdown in chronic relapsing experimental allergic encephalomyelitis. Acta Neuropathol. 83, 630–635.

Hawkins, R.A., O'kane, R.L., Simpson, I.A., Vina, J.R., 2006. Structure of the blood-brain barrier and its role in the transport of amino acids. J. Nutr. 136, 218S–226S.
Hawkins, B.T., Lundeen, T.F., Norwood, K.M., Brooks, H.L., Egleton, R.D., 2007. Increased blood-brain barrier permeability and altered tight junctions in experimental diabetes in the rat: contribution of hyperglycaemia and matrix metalloproteinases. Diabetologia 50, 202–211.
Heidari Nejad, S., Takechi, R., Mullins, B.J., Giles, C., Larcombe, A.N., Bertolatti, D., Rumchev, K., Dhaliwal, S., Mamo, J., 2015. The effect of diesel exhaust exposure on blood-brain barrier integrity and function in a murine model. J. Appl. Toxicol. 35, 41–47.
Hellstrom, M., Kalen, M., Lindahl, P., Abramsson, A., Betsholtz, C., 1999. Role of PDGF-B and PDGFR-beta in recruitment of vascular smooth muscle cells and pericytes during embryonic blood vessel formation in the mouse. Development 126, 3047–3055.
Heye, A.K., Culling, R.D., Valdes Hernandez Mdel, C., Thrippleton, M.J., Wardlaw, J.M., 2014. Assessment of blood-brain barrier disruption using dynamic contrast-enhanced MRI. A systematic review. Neuroimage Clin. 6, 262–274.
Hill, A.E., Shachar-Hill, B., Shachar-Hill, Y., 2004. What are aquaporins for? J. Membr. Biol. 197, 1–32.
Hoglund, R.A., Maghazachi, A.A., 2014. Multiple sclerosis and the role of immune cells. World J. Exp. Med. 4, 27–37.
Hoglund, R.A., Hestvik, A.L., Holmoy, T., Maghazachi, A.A., 2011. Expression and functional activity of chemokine receptors in glatiramer acetate-specific T cells isolated from multiple sclerosis patient receiving the drug glatiramer acetate. Hum. Immunol. 72, 124–134.
Holland, D., Brewer, J.B., Hagler, D.J., Fennema-Notestine, C., Dale, A.M., Alzheimer's Disease Neuroimaging Initiative, 2009. Subregional neuroanatomical change as a biomarker for Alzheimer's disease. Proc. Natl. Acad. Sci. U. S. A. 106, 20954–20959.
Holmoy, T., Hestvik, A.L., 2008. Multiple sclerosis: immunopathogenesis and controversies in defining the cause. Curr. Opin. Infect. Dis. 21, 271–278.
Hotamisligil, G.S., Arner, P., Caro, J.F., Atkinson, R.L., Spiegelman, B.M., 1995. Increased adipose tissue expression of tumor necrosis factor-alpha in human obesity and insulin resistance. J. Clin. Invest. 95, 2409–2415.
Huber, J.D., Vangilder, R.L., Houser, K.A., 2006. Streptozotocin-induced diabetes progressively increases blood-brain barrier permeability in specific brain regions in rats. Am. J. Physiol. Heart Circ. Physiol. 291, H2660–H2668.
Hunt, A., Schonknecht, P., Henze, M., Seidl, U., Haberkorn, U., Schroder, J., 2007. Reduced cerebral glucose metabolism in patients at risk for Alzheimer's disease. Psychiatry Res. 155, 147–154.
Hurtado-Alvarado, G., Becerril-Villanueva, E., Contis-Montes de Oca, A., Dominguez-Salazar, E., Salinas-Jazmin, N., Perez-Tapia, S.M., Pavon, L., Velazquez-Moctezuma, J., Gomez-Gonzalez, B., 2018. The Yin/Yang of inflammatory status: blood-brain barrier regulation during sleep. Brain Behav. Immun. 69, 154–166.
Iannotti, F., Fieschi, C., Alfano, B., Picozzi, P., Mansi, L., Pozzilli, C., Punzo, A., Del Vecchio, G., Lenzi, G.L., Salvatore, M., ET, A.L., 1987. Simplified, noninvasive PET measurement of blood-brain barrier permeability. J. Comput. Assist. Tomogr. 11, 390–397.
Iliff, J.J., Wang, M., Liao, Y., Plogg, B.A., Peng, W., Gundersen, G.A., Benveniste, H., Vates, G.E., Deane, R., Goldman, S.A., Nagelhus, E.A., Nedergaard, M., 2012. A paravascular pathway facilitates CSF flow through the brain parenchyma and the clearance of interstitial solutes, including amyloid beta. Sci. Transl. Med. 4, 147ra111.

Iliff, J.J., Wang, M., Zeppenfeld, D.M., Venkataraman, A., Plog, B.A., Liao, Y., Deane, R., Nedergaard, M., 2013. Cerebral arterial pulsation drives paravascular CSF-interstitial fluid exchange in the murine brain. J. Neurosci. 33, 18190–18199.

International Multiple Sclerosis Genetics Consortium, Wellcome Trust Case Control Consortium, Sawcer, S., Hellenthal, G., Pirinen, M., Spencer, C.C., Patsopoulos, N.A., Moutsianas, L., Dilthey, A., SU, Z., Freeman, C., Hunt, S.E., Edkins, S., Gray, E., Booth, D.R., Potter, S.C., Goris, A., Band, G., Oturai, A.B., Strange, A., Saarela, J., Bellenguez, C., Fontaine, B., Gillman, M., Hemmer, B., Gwilliam, R., Zipp, F., Jayakumar, A., Martin, R., Leslie, S., Hawkins, S., Giannoulatou, E., D'alfonso, S., Blackburn, H., Martinelli Boneschi, F., Liddle, J., Harbo, H.F., Perez, M.L., Spurkland, A., Waller, M.J., Mycko, M.P., Ricketts, M., Comabella, M., Hammond, N., Kockum, I., Mccann, O.T., Ban, M., Whittaker, P., Kemppinen, A., Weston, P., Hawkins, C., Widaa, S., Zajicek, J., Dronov, S., Robertson, N., Bumpstead, S.J., Barcellos, L.F., Ravindrarajah, R., Abraham, R., Alfredsson, L., Ardlie, K., Aubin, C., Baker, A., Baker, K., Baranzini, S.E., Bergamaschi, L., Bergamaschi, R., Bernstein, A., Berthele, A., Boggild, M., Bradfield, J.P., Brassat, D., Broadley, S.A., Buck, D., Butzkueven, H., Capra, R., Carroll, W.M., Cavalla, P., Celius, E.G., Cepok, S., Chiavacci, R., Clerget-Darpoux, F., Clysters, K., Comi, G., Cossburn, M., Cournu-Rebeix, I., Cox, M.B., Cozen, W., Cree, B.A., Cross, A.H., Cusi, D., Daly, M.J., Davis, E., de Bakker, P.I., Debouverie, M., D'hooghe, M.B., Dixon, K., Dobosi, R., Dubois, B., Ellinghaus, D., et al., 2011. Genetic risk and a primary role for cell-mediated immune mechanisms in multiple sclerosis. Nature 476, 214–219.

Jaeger, L.B., Dohgu, S., Sultana, R., Lynch, J.L., Owen, J.B., Erickson, M.A., Shah, G.N., Price, T.O., Fleegal-Demotta, M.A., Butterfield, D.A., Banks, W.A., 2009. Lipopolysaccharide alters the blood-brain barrier transport of amyloid beta protein: a mechanism for inflammation in the progression of Alzheimer's disease. Brain Behav. Immun. 23, 507–517.

Jakkula, E., Leppa, V., Sulonen, A.M., Varilo, T., Kallio, S., Kemppinen, A., Purcell, S., Koivisto, K., Tienari, P., Sumelahti, M.L., Elovaara, I., Pirttila, T., Reunanen, M., Aromaa, A., Oturai, A.B., Sondergaard, H.B., Harbo, H.F., Mero, I.L., Gabriel, S.B., Mirel, D.B., Hauser, S.L., Kappos, L., Polman, C., de Jager, P.L., Hafler, D.A., Daly, M.J., Palotie, A., Saarela, J., Peltonen, L., 2010. Genome-wide association study in a high-risk isolate for multiple sclerosis reveals associated variants in STAT3 gene. Am. J. Hum. Genet. 86, 285–291.

Janelidze, S., Hertze, J., Nagga, K., Nilsson, K., Nilsson, C., Swedish BIO, F.S.G., Wennstrom, M., van Westen, D., Blennow, K., Zetterberg, H., Hansson, O., 2017. Increased blood-brain barrier permeability is associated with dementia and diabetes but not amyloid pathology or APOE genotype. Neurobiol. Aging 51, 104–112.

Janzer, R.C., Raff, M.C., 1987. Astrocytes induce blood-brain barrier properties in endothelial cells. Nature 325, 253–257.

Jessen, N.A., Munk, A.S., Lundgaard, I., Nedergaard, M., 2015. The glymphatic system: a beginner's guide. Neurochem. Res. 40, 2583–2599.

Jiang, Z., Li, C., Arrick, D.M., Yang, S., Baluna, A.E., Sun, H., 2014. Role of nitric oxide synthases in early blood-brain barrier disruption following transient focal cerebral ischemia. PLoS One 9, e93134.

Jing, L., He, Q., Zhang, J.Z., Li, P.A., 2013. Temporal profile of astrocytes and changes of oligodendrocyte-based myelin following middle cerebral artery occlusion in diabetic and non-diabetic rats. Int. J. Biol. Sci. 9, 190–199.

Johansson, P.A., Dziegielewska, K.M., Ek, C.J., Habgood, M.D., Liddelow, S.A., Potter, A.M., Stolp, H.B., Saunders, N.R., 2006. Blood-CSF barrier function in the rat embryo. Eur. J. Neurosci. 24, 65–76.

Joris, P.J., Mensink, R.P., Adam, T.C., Liu, T.T., 2018. Cerebral blood flow measurements in adults: a review on the effects of dietary factors and exercise. Nutrients 10, 530.

Jurcovicova, J., 2014. Glucose transport in brain—effect of inflammation. Endocr. Regul. 48, 35–48.

Kamada, H., Yu, F., Nito, C., Chan, P.H., 2007. Influence of hyperglycemia on oxidative stress and matrix metalloproteinase-9 activation after focal cerebral ischemia/reperfusion in rats: relation to blood-brain barrier dysfunction. Stroke 38, 1044–1049.

Karlawish, J., Jack Jr., C.R., Rocca, W.A., Snyder, H.M., Carrillo, M.C., 2017. Alzheimer's disease: the next frontier—special report 2017. Alzheimers Dement. 13, 374–380.

Kaya, M., Ahishali, B., 2011. Assessment of permeability in barrier type of endothelium in brain using tracers: Evans blue, sodium fluorescein, and horseradish peroxidase. Methods Mol. Biol. 763, 369–382.

Kempuraj, D., Thangavel, R., Natteru, P.A., Selvakumar, G.P., Saeed, D., Zahoor, H., Zaheer, S., Iyer, S.S., Zaheer, A., 2016. Neuroinflammation induces neurodegeneration. J. Neurol. Neurosurg. Spine 1, 1003.

Kerin, T., Volk, H., Li, W., Lurmann, F., Eckel, S., McConnell, R., Hertz-Picciotto, I., 2018. Association between air pollution exposure, cognitive and adaptive function, and ASD severity among children with autism spectrum disorder. J. Autism Dev. Disord. 48, 137–150.

Kettenmann, H., Verkhratsky, A., 2011. Neuroglia—living nerve glue. Fortschr. Neurol. Psychiatr. 79, 588–597.

Kettenmann, H., Hanisch, U.K., Noda, M., Verkhratsky, A., 2011. Physiology of microglia. Physiol. Rev. 91, 461–553.

Kettenmann, H., Kirchhoff, F., Verkhratsky, A., 2013. Microglia: new roles for the synaptic stripper. Neuron 77, 10–18.

Khan, S.S., Bloom, G.S., 2016. Tau: the center of a signaling nexus in Alzheimer's disease. Front. Neurosci. 10, 31.

Kissela, B.M., Khoury, J., Kleindorfer, D., Woo, D., Schneider, A., Alwell, K., Miller, R., Ewing, I., Moomaw, C.J., Szaflarski, J.P., Gebel, J., Shukla, R., Broderick, J.P., 2005. Epidemiology of ischemic stroke in patients with diabetes: the greater Cincinnati/Northern Kentucky Stroke study. Diabetes Care 28, 355–359.

Kleinfeld, D., Mitra, P.P., Helmchen, F., Denk, W., 1998. Fluctuations and stimulus-induced changes in blood flow observed in individual capillaries in layers 2 through 4 of rat neocortex. Proc. Natl. Acad. Sci. U. S. A. 95, 15741–15746.

Kniesel, U., Wolburg, H., 2000. Tight junctions of the blood-brain barrier. Cell. Mol. Neurobiol. 20, 57–76.

Knowland, D., Arac, A., Sekiguchi, K.J., Hsu, M., Lutz, S.E., Perrino, J., Steinberg, G.K., Barres, B.A., Nimmerjahn, A., Agalliu, D., 2014. Stepwise recruitment of transcellular and paracellular pathways underlies blood-brain barrier breakdown in stroke. Neuron 82, 603–617.

Kress, B.T., Iliff, J.J., Xia, M., Wang, M., Wei, H.S., Zeppenfeld, D., Xie, L., Kang, H., Xu, Q., Liew, J.A., Plog, B.A., Ding, F., Deane, R., Nedergaard, M., 2014. Impairment of paravascular clearance pathways in the aging brain. Ann. Neurol. 76, 845–861.

Kroenke, M.A., Carlson, T.J., Andjelkovic, A.V., Segal, B.M., 2008. IL-12- and IL-23-modulated T cells induce distinct types of EAE based on histology, CNS chemokine profile, and response to cytokine inhibition. J. Exp. Med. 205, 1535–1541.

Krstic, D., Madhusudan, A., Doehner, J., Vogel, P., Notter, T., Imhof, C., Manalastas, A., Hilfiker, M., Pfister, S., Schwerdel, C., Riether, C., Meyer, U., Knuesel, I., 2012. Systemic immune challenges trigger and drive Alzheimer-like neuropathology in mice. J. Neuroinflammation 9, 151.

Kuwabara, H., Evans, A.C., Gjedde, A., 1990. Michaelis-Menten constraints improved cerebral glucose metabolism and regional lumped constant measurements with [18F] fluorodeoxyglucose. J. Cereb. Blood Flow Metab. 10, 180–189.
Langrish, C.L., Chen, Y., Blumenschein, W.M., Mattson, J., Basham, B., Sedgwick, J.D., Mcclanahan, T., Kastelein, R.A., Cua, D.J., 2005. IL-23 drives a pathogenic T cell population that induces autoimmune inflammation. J. Exp. Med. 201, 233–240.
Larsson, H.B., Stubgaard, M., Frederiksen, J.L., Jensen, M., Henriksen, O., Paulson, O.B., 1990. Quantitation of blood-brain barrier defect by magnetic resonance imaging and gadolinium-DTPA in patients with multiple sclerosis and brain tumors. Magn. Reson. Med. 16, 117–131.
Leach, M.O., Morgan, B., Tofts, P.S., Buckley, D.L., Huang, W., Horsfield, M.A., Chenevert, T.L., Collins, D.J., Jackson, A., Lomas, D., Whitcher, B., Clarke, L., Plummer, R., Judson, I., Jones, R., Alonzi, R., Brunner, T., Koh, D.M., Murphy, P., Waterton, J.C., Parker, G., Graves, M.J., Scheenen, T.W., Redpath, T.W., Orton, M., Karczmar, G., Huisman, H., Barentsz, J., Padhani, A., 2012. Imaging vascular function for early stage clinical trials using dynamic contrast-enhanced magnetic resonance imaging. Eur. Radiol. 22, 1451–1464.
Lee, W.J., Hawkins, R.A., Peterson, D.R., Vina, J.R., 1996. Role of oxoproline in the regulation of neutral amino acid transport across the blood-brain barrier. J. Biol. Chem. 271, 19129–19133.
Lee, P., Kim, J., Williams, R., Sandhir, R., Gregory, E., Brooks, W.M., Berman, N.E., 2012. Effects of aging on blood brain barrier and matrix metalloproteases following controlled cortical impact in mice. Exp. Neurol. 234, 50–61.
Leibson, C.L., Rocca, W.A., Hanson, V.A., Cha, R., Kokmen, E., O'brien, P.C., Palumbo, P.J., 1997. Risk of dementia among persons with diabetes mellitus: a population-based cohort study. Am. J. Epidemiol. 145, 301–308.
Lengfeld, J.E., Lutz, S.E., Smith, J.R., Diaconu, C., Scott, C., Kofman, S.B., Choi, C., Walsh, C.M., Raine, C.S., Agalliu, I., Agalliu, D., 2017. Endothelial Wnt/beta-catenin signaling reduces immune cell infiltration in multiple sclerosis. Proc. Natl. Acad. Sci. U. S. A. 114, E1168–E1177.
Leray, E., Moreau, T., Fromont, A., Edan, G., 2016. Epidemiology of multiple sclerosis. Rev. Neurol. (Paris) 172, 3–13.
Liguori, C., Romigi, A., Nuccetelli, M., Zannino, S., Sancesario, G., Martorana, A., Albanese, M., Mercuri, N.B., Izzi, F., Bernardini, S., Nitti, A., Sancesario, G.M., Sica, F., Marciani, M.G., Placidi, F., 2014. Orexinergic system dysregulation, sleep impairment, and cognitive decline in Alzheimer disease. JAMA Neurol. 71, 1498–1505.
Lim, J.Y., Reighard, C.P., Crowther, D.C., 2015. The pro-domains of neurotrophins, including BDNF, are linked to Alzheimer's disease through a toxic synergy with Abeta. Hum. Mol. Genet. 24, 3929–3938.
Lu, M., Kuroki, M., Amano, S., Tolentino, M., Keough, K., Kim, I., Bucala, R., Adamis, A.P., 1998. Advanced glycation end products increase retinal vascular endothelial growth factor expression. J. Clin. Invest. 101, 1219–1224.
Lucchinetti, C.F., Popescu, B.F., Bunyan, R.F., Moll, N.M., Roemer, S.F., Lassmann, H., Bruck, W., Parisi, J.E., Scheithauer, B.W., Giannini, C., Weigand, S.D., Mandrekar, J., Ransohoff, R.M., 2011. Inflammatory cortical demyelination in early multiple sclerosis. N. Engl. J. Med. 365, 2188–2197.
Lucero, J., Suwannasual, U., Herbert, L.M., Mcdonald, J.D., Lund, A.K., 2017. The role of the lectin-like oxLDL receptor (LOX-1) in traffic-generated air pollution exposure-mediated alteration of the brain microvasculature in apolipoprotein (Apo) E knockout mice. Inhal. Toxicol. 29, 266–281.
Luissint, A.C., Artus, C., Glacial, F., Ganeshamoorthy, K., Couraud, P.O., 2012. Tight junctions at the blood brain barrier: physiological architecture and disease-associated dysregulation. Fluids Barriers CNS 9, 23.

Lund, A.K., Lucero, J., Lucas, S., Madden, M.C., Mcdonald, J.D., Seagrave, J.C., Knuckles, T.L., Campen, M.J., 2009. Vehicular emissions induce vascular MMP-9 expression and activity associated with endothelin-1-mediated pathways. Arterioscler. Thromb. Vasc. Biol. 29, 511–517.

Lund, A.K., Lucero, J., Harman, M., Madden, M.C., Mcdonald, J.D., Seagrave, J.C., Campen, M.J., 2011. The oxidized low-density lipoprotein receptor mediates vascular effects of inhaled vehicle emissions. Am. J. Respir. Crit. Care Med. 184, 82–91.

Maiti, P., Singh, S.B., Sharma, A.K., Muthuraju, S., Banerjee, P.K., Ilavazhagan, G., 2006. Hypobaric hypoxia induces oxidative stress in rat brain. Neurochem. Int. 49, 709–716.

Male, D., Rahman, J., Linke, A., Zhao, W., Hickey, W., 1995. An interferon-inducible molecule on brain endothelium which controls lymphocyte adhesion mediated by integrins. Immunology 84, 453–460.

Mammana, S., Fagone, P., Cavalli, E., Basile, M.S., Petralia, M.C., NICOLETTI, F., Bramanti, P., Mazzon, E., 2018. The role of macrophages in neuroinflammatory and neurodegenerative pathways of Alzheimer's disease, amyotrophic lateral sclerosis, and multiple sclerosis: pathogenetic cellular effectors and potential therapeutic targets. Int. J. Mol. Sci. 19, 831.

Mannie, M., Swanborg, R.H., Stepaniak, J.A., 2009. Experimental autoimmune encephalomyelitis in the rat. Curr. Protoc. Immunol.Chapter 15, Unit 15.2.

Marcos, B., Garcia-Alloza, M., GIL-Bea, F.J., Chuang, T.T., Francis, P.T., Chen, C.P., Tsang, S.W., Lai, M.K., Ramirez, M.J., 2008. Involvement of an altered 5-HT-{6} receptor function in behavioral symptoms of Alzheimer's disease. J. Alzheimers Dis. 14, 43–50.

McKittrick, C.M., Lawrence, C.E., Carswell, H.V., 2015. Mast cells promote blood brain barrier breakdown and neutrophil infiltration in a mouse model of focal cerebral ischemia. J. Cereb. Blood Flow Metab. 35, 638–647.

Mecocci, P., Parnetti, L., Reboldi, G.P., Santucci, C., Gaiti, A., Ferri, C., Gernini, I., Romagnoli, M., Cadini, D., Senin, U., 1991. Blood-brain-barrier in a geriatric population: barrier function in degenerative and vascular dementias. Acta Neurol. Scand. 84, 210–213.

Miller, S.D., Karpus, W.J., 2007. Experimental autoimmune encephalomyelitis in the mouse. Curr. Protoc. Immunol.Chapter 15, Unit 15 1.

Milo, R., Kahana, E., 2010. Multiple sclerosis: geoepidemiology, genetics and the environment. Autoimmun. Rev. 9, A387–A394.

Minami, T., Okazaki, J., Kawabata, A., Kawaki, H., Okazaki, Y., Tohno, Y., 1998. Roles of nitric oxide and prostaglandins in the increased permeability of the blood-brain barrier caused by lipopolysaccharide. Environ. Toxicol. Pharmacol. 5, 35–41.

Montagne, A., Barnes, S.R., Sweeney, M.D., Halliday, M.R., Sagare, A.P., Zhao, Z., Toga, A.W., Jacobs, R.E., Liu, C.Y., Amezcua, L., Harrington, M.G., Chui, H.C., Law, M., Zlokovic, B.V., 2015. Blood-brain barrier breakdown in the aging human hippocampus. Neuron 85, 296–302.

Morbelli, S., Brugnolo, A., Bossert, I., Buschiazzo, A., Frisoni, G.B., Galluzzi, S., van Berckel, B.N., Ossenkoppele, R., Perneczky, R., Drzezga, A., Didic, M., Guedj, E., Sambuceti, G., Bottoni, G., Arnaldi, D., Picco, A., de Carli, F., Pagani, M., Nobili, F., 2015. Visual versus semi-quantitative analysis of 18F-FDG-PET in amnestic MCI: an European Alzheimer's disease consortium (EADC) project. J. Alzheimers Dis. 44, 815–826.

Mosconi, L., de Santi, S., Li, J., Tsui, W.H., Li, Y., Boppana, M., Laska, E., Rusinek, H., de Leon, M.J., 2008a. Hippocampal hypometabolism predicts cognitive decline from normal aging. Neurobiol. Aging 29, 676–692.

Mosconi, L., Pupi, A., de Leon, M.J., 2008b. Brain glucose hypometabolism and oxidative stress in preclinical Alzheimer's disease. Ann. N. Y. Acad. Sci. 1147, 180–195.

Mueckler, M., Thorens, B., 2013. The SLC2 (GLUT) family of membrane transporters. Mol. Asp. Med. 34, 121–138.
Murphy, A.C., Lalor, S.J., Lynch, M.A., Mills, K.H., 2010. Infiltration of Th1 and Th17 cells and activation of microglia in the CNS during the course of experimental autoimmune encephalomyelitis. Brain Behav. Immun. 24, 641–651.
Nakagawa, S., Deli, M.A., Kawaguchi, H., Shimizudani, T., Shimono, T., Kittel, A., Tanaka, K., Niwa, M., 2009. A new blood-brain barrier model using primary rat brain endothelial cells, pericytes and astrocytes. Neurochem. Int. 54, 253–263.
Nalecz, K.A., 2017. Solute carriers in the blood-brain barier: safety in abundance. Neurochem. Res. 42, 795–809.
Navale, A.M., Paranjape, A.N., 2016. Glucose transporters: physiological and pathological roles. Biophys. Rev. 8, 5–9.
Nishizaki, T., Matsuoka, T., 1998. Low glucose enhances Na+/glucose transport in bovine brain artery endothelial cells. Stroke 29, 844–849.
Niwa, K., Kazama, K., Younkin, S.G., Carlson, G.A., Iadecola, C., 2002. Alterations in cerebral blood flow and glucose utilization in mice overexpressing the amyloid precursor protein. Neurobiol. Dis. 9, 61–68.
Obermeier, B., Verma, A., Ransohoff, R.M., 2016. The blood-brain barrier. Handb. Clin. Neurol. 133, 39–59.
Oldendorf, W.H., 1973. Stereospecificity of blood-brain barrier permeability to amino acids. Am. J. Phys. 224, 967–969.
Oppenheim, H.A., Lucero, J., Guyot, A.C., Herbert, L.M., Mcdonald, J.D., Mabondzo, A., Lund, A.K., 2013. Exposure to vehicle emissions results in altered blood brain barrier permeability and expression of matrix metalloproteinases and tight junction proteins in mice. Part. Fibre Toxicol. 10, 62.
Ortiz, G.G., Pacheco-Moises, F.P., Macias-Islas, M.A., Flores-Alvarado, L.J., Mireles-Ramirez, M.A., Gonzalez-Renovato, E.D., Hernandez-Navarro, V.E., Sanchez-Lopez, A.L., Alatorre-Jimenez, M.A., 2014. Role of the blood-brain barrier in multiple sclerosis. Arch. Med. Res. 45, 687–697.
Ott, A., Stolk, R.P., van Harskamp, F., Pols, H.A., Hofman, A., Breteler, M.M., 1999. Diabetes mellitus and the risk of dementia: the Rotterdam study. Neurology 53, 1937–1942.
Pagani, M., de Carli, F., Morbelli, S., Oberg, J., Chincarini, A., Frisoni, G.B., Galluzzi, S., Perneczky, R., Drzezga, A., van Berckel, B.N., Ossenkoppele, R., Didic, M., Guedj, E., Brugnolo, A., Picco, A., Arnaldi, D., Ferrara, M., Buschiazzo, A., Sambuceti, G., Nobili, F., 2015. Volume of interest-based [18F]fluorodeoxyglucose PET discriminates MCI converting to Alzheimer's disease from healthy controls. A European Alzheimer's disease consortium (EADC) study. Neuroimage Clin. 7, 34–42.
Paget, C., Lecomte, M., Ruggiero, D., Wiernsperger, N., Lagarde, M., 1998. Modification of enzymatic antioxidants in retinal microvascular cells by glucose or advanced glycation end products. Free Radic. Biol. Med. 25, 121–129.
Pardridge, W.M., Boado, R.J., Farrell, C.R., 1990. Brain-type glucose transporter (GLUT-1) is selectively localized to the blood-brain barrier. Studies with quantitative western blotting and in situ hybridization. J. Biol. Chem. 265, 18035–18040.
Patching, S.G., 2017. Glucose transporters at the blood-brain barrier: function, regulation and gateways for drug delivery. Mol. Neurobiol. 54, 1046–1077.
Patlak, C.S., Blasberg, R.G., 1985. Graphical evaluation of blood-to-brain transfer constants from multiple-time uptake data. Generalizations. J. Cereb. Blood Flow Metab. 5, 584–590.
Paz, C., 1997. Some consequences of ozone exposure on health. Arch. Med. Res. 28, 163–170.
Petrelli, F., Pucci, L., Bezzi, P., 2016. Astrocytes and microglia and their potential link with autism spectrum disorders. Front. Cell. Neurosci. 10, 21.

Petty, M.A., Lo, E.H., 2002. Junctional complexes of the blood-brain barrier: permeability changes in neuroinflammation. Prog. Neurobiol. 68, 311–323.
Pfefferkorn, T., Rosenberg, G.A., 2003. Closure of the blood-brain barrier by matrix metalloproteinase inhibition reduces rtPA-mediated mortality in cerebral ischemia with delayed reperfusion. Stroke 34, 2025–2030.
Plog, B.A., Nedergaard, M., 2018. The glymphatic system in central nervous system health and disease: past, present, and future. Annu. Rev. Pathol. 13, 379–394.
Popovich, P.G., Stokes, B.T., Whitacre, C.C., 1996. Concept of autoimmunity following spinal cord injury: possible roles for T lymphocytes in the traumatized central nervous system. J. Neurosci. Res. 45, 349–363.
Poser, C.M., 1986. Pathogenesis of multiple sclerosis. A critical reappraisal. Acta Neuropathol. 71, 1–10.
Pozzilli, C., Bernardi, S., Mansi, L., Picozzi, P., Iannotti, F., Alfano, B., Bozzao, L., Lenzi, G.L., Salvatore, M., Conforti, P., et al., 1988. Quantitative assessment of blood-brain barrier permeability in multiple sclerosis using 68-Ga-EDTA and positron emission tomography. J. Neurol. Neurosurg. Psychiatry 51, 1058–1062.
Proebstl, D., Voisin, M.B., Woodfin, A., Whiteford, J., D'acquisto, F., Jones, G.E., Rowe, D., Nourshargh, S., 2012. Pericytes support neutrophil subendothelial cell crawling and breaching of venular walls in vivo. J. Exp. Med. 209, 1219–1234.
Provias, J., Jeynes, B., 2014. The role of the blood-brain barrier in the pathogenesis of senile plaques in Alzheimer's disease. Int. J. Alzheimers Dis. 2014, 191863.
Quintana, F.J., 2017. Astrocytes to the rescue! Glia limitans astrocytic endfeet control CNS inflammation. J. Clin. Invest. 127, 2897–2899.
Raichle, M.E., Gusnard, D.A., 2002. Appraising the brain's energy budget. Proc. Natl. Acad. Sci. U. S. A. 99, 10237–10239.
Raj, T., Rothamel, K., Mostafavi, S., Ye, C., Lee, M.N., Replogle, J.M., Feng, T., Lee, M., Asinovski, N., Frohlich, I., Imboywa, S., von Korff, A., Okada, Y., Patsopoulos, N.A., Davis, S., Mccabe, C., Paik, H.I., Srivastava, G.P., Raychaudhuri, S., Hafler, D.A., Koller, D., Regev, A., Hacohen, N., Mathis, D., Benoist, C., Stranger, B.E., de Jager, P.L., 2014. Polarization of the effects of autoimmune and neurodegenerative risk alleles in leukocytes. Science 344, 519–523.
Raja, R., Rosenberg, G.A., Caprihan, A., 2018. MRI measurements of blood-brain barrier function in dementia: a review of recent studies. Neuropharmacology 134, 259–271.
Ramanathan, L., Gozal, D., Siegel, J.M., 2005. Antioxidant responses to chronic hypoxia in the rat cerebellum and pons. J. Neurochem. 93, 47–52.
Ransohoff, R.M., 2016. How neuroinflammation contributes to neurodegeneration. Science 353, 777–783.
Ransohoff, R.M., Schafer, D., Vincent, A., Blachère, N.E., Bar-Or, A., 2015. Neuroinflammation: ways in which the immune system affects the brain. Neurotherapeutics 12, 896–909.
Reemst, K., Noctor, S.C., Lucassen, P.J., Hol, E.M., 2016. The indispensable roles of microglia and astrocytes during brain development. Front. Hum. Neurosci. 10, 566.
Reyes, T.M., Fabry, Z., Coe, C.L., 1999. Brain endothelial cell production of a neuroprotective cytokine, interleukin-6, in response to noxious stimuli. Brain Res. 851, 215–220.
Rissman, R.A., Staup, M.A., Lee, A.R., Justice, N.J., Rice, K.C., Vale, W., Sawchenko, P.E., 2012. Corticotropin-releasing factor receptor-dependent effects of repeated stress on tau phosphorylation, solubility, and aggregation. Proc. Natl. Acad. Sci. U. S. A. 109, 6277–6282.
Romanic, A.M., White, R.F., Arleth, A.J., Ohlstein, E.H., Barone, F.C., 1998. Matrix metalloproteinase expression increases after cerebral focal ischemia in rats: inhibition of matrix metalloproteinase-9 reduces infarct size. Stroke 29, 1020–1030.

Romeo, G., Liu, W.H., Asnaghi, V., Kern, T.S., Lorenzi, M., 2002. Activation of nuclear factor-kappaB induced by diabetes and high glucose regulates a proapoptotic program in retinal pericytes. Diabetes 51, 2241–2248.
Rossner, W., Tempel, K., 1966. Quantitative determination of the permeability of the so-called blood-brain barrier of Evans blue (T 1824). Med. Pharmacol. Exp. Int. J. Exp. Med. 14, 169–182.
Rothhammer, V., Quintana, F.J., 2015. Role of astrocytes and microglia in central nervous system inflammation. Introduction. Semin. Immunopathol. 37, 575–576.
Ryu, J.K., McLarnon, J.G., 2009. A leaky blood-brain barrier, fibrinogen infiltration and microglial reactivity in inflamed Alzheimer's disease brain. J. Cell. Mol. Med. 13, 2911–2925.
Sagare, A.P., Sweeney, M.D., Makshanoff, J., Zlokovic, B.V., 2015. Shedding of soluble platelet-derived growth factor receptor-beta from human brain pericytes. Neurosci. Lett. 607, 97–101.
Samuraki, M., Matsunari, I., Chen, W.P., Yajima, K., Yanase, D., Fujikawa, A., Takeda, N., Nishimura, S., Matsuda, H., Yamada, M., 2007. Partial volume effect-corrected FDG PET and grey matter volume loss in patients with mild Alzheimer's disease. Eur. J. Nucl. Med. Mol. Imaging 34, 1658–1669.
Saunders, N.R., Dziegielewska, K.M., Mollgard, K., Habgood, M.D., 2015. Markers for blood-brain barrier integrity: how appropriate is Evans blue in the twenty-first century and what are the alternatives? Front. Neurosci. 9, 385.
Scott, H.L., Pow, D.V., Tannenberg, A.E., Dodd, P.R., 2002. Aberrant expression of the glutamate transporter excitatory amino acid transporter 1 (EAAT1) in Alzheimer's disease. J. Neurosci. 22, RC206.
Segura, B., Jurado, M.A., Freixenet, N., Albuin, C., Muniesa, J., Junque, C., 2009. Mental slowness and executive dysfunctions in patients with metabolic syndrome. Neurosci. Lett. 462, 49–53.
Selkoe, D.J., Hardy, J., 2016. The amyloid hypothesis of Alzheimer's disease at 25 years. EMBO Mol. Med. 8, 595–608.
Sengillo, J.D., Winkler, E.A., Walker, C.T., Sullivan, J.S., Johnson, M., Zlokovic, B.V., 2013. Deficiency in mural vascular cells coincides with blood-brain barrier disruption in Alzheimer's disease. Brain Pathol. 23, 303–310.
Serban, A.I., Stanca, L., Geicu, O.I., Munteanu, M.C., Dinischiotu, A., 2016. RAGE and TGF-beta1 cross-talk regulate extracellular matrix turnover and cytokine synthesis in AGEs exposed fibroblast cells. PLoS One 11, e0152376.
Serlin, Y., Shelef, I., Knyazer, B., Friedman, A., 2015. Anatomy and physiology of the blood-brain barrier. Semin. Cell Dev. Biol. 38, 2–6.
Shabab, T., Khanabdali, R., Moghadamtousi, S.Z., Kadir, H.A., Mohan, G., 2017. Neuroinflammation pathways: a general review. Int. J. Neurosci. 127, 624–633.
Shames, D.M., Kuwatsuru, R., Vexler, V., Muhler, A., Brasch, R.C., 1993. Measurement of capillary permeability to macromolecules by dynamic magnetic resonance imaging: a quantitative noninvasive technique. Magn. Reson. Med. 29, 616–622.
Simpson Jr., S., Blizzard, L., Otahal, P., van der Mei, I., Taylor, B., 2011. Latitude is significantly associated with the prevalence of multiple sclerosis: a meta-analysis. J. Neurol. Neurosurg. Psychiatry 82, 1132–1141.
Skoog, I., Wallin, A., Fredman, P., Hesse, C., Aevarsson, O., Karlsson, I., Gottfries, C.G., Blennow, K., 1998. A population study on blood-brain barrier function in 85-year-olds: relation to Alzheimer's disease and vascular dementia. Neurology 50, 966–971.
Sohet, F., Daneman, R., 2013. Genetic mouse models to study blood-brain barrier development and function. Fluids Barriers CNS 10, 3.

Starr, J.M., Wardlaw, J., Ferguson, K., Maclullich, A., Deary, I.J., Marshall, I., 2003. Increased blood-brain barrier permeability in type II diabetes demonstrated by gadolinium magnetic resonance imaging. J. Neurol. Neurosurg. Psychiatry 74, 70–76.

Starr, J.M., Farrall, A.J., Armitage, P., Mcgurn, B., Wardlaw, J., 2009. Blood-brain barrier permeability in Alzheimer's disease: a case-control MRI study. Psychiatry Res. 171, 232–241.

Stewart, P.A., Magliocco, M., Hayakawa, K., Farrell, C.L., Del Maestro, R.F., Girvin, J., Kaufmann, J.C., Vinters, H.V., Gilbert, J., 1987. A quantitative analysis of blood-brain barrier ultrastructure in the aging human. Microvasc. Res. 33, 270–282.

Strecker, J.K., Minnerup, J., Schutte-Nutgen, K., Gess, B., Schabitz, W.R., Schilling, M., 2013. Monocyte chemoattractant protein-1-deficiency results in altered blood-brain barrier breakdown after experimental stroke. Stroke 44, 2536–2544.

Sugita, H., Kaneki, M., Tokunaga, E., Sugita, M., Koike, C., Yasuhara, S., Tompkins, R.G., Martyn, J.A., 2002. Inducible nitric oxide synthase plays a role in LPS-induced hyperglycemia and insulin resistance. Am. J. Physiol. Endocrinol. Metab. 282, E386–E394.

Sullan, M.J., Asken, B.M., Jaffee, M.S., Dekosky, S.T., Bauer, R.M., 2018. Glymphatic system disruption as a mediator of brain trauma and chronic traumatic encephalopathy. Neurosci. Biobehav. Rev. 84, 316–324.

Surwit, R.S., Kuhn, C.M., Cochrane, C., Mccubbin, J.A., Feinglos, M.N., 1988. Diet-induced type II diabetes in C57BL/6J mice. Diabetes 37, 1163–1167.

Suwannasual, U., Lucero, J., Mcdonald, J.D., Lund, A.K., 2018. Exposure to traffic-generated air pollutants mediates alterations in brain microvascular integrity in wildtype mice on a high-fat diet. Environ. Res. 160, 449–461.

Sweeney, M.D., Sagare, A.P., Zlokovic, B.V., 2018. Blood-brain barrier breakdown in Alzheimer disease and other neurodegenerative disorders. Nat. Rev. Neurol. 14, 133–150.

Tamura, A., Graham, D.I., Mcculloch, J., Teasdale, G.M., 1981. Focal cerebral ischaemia in the rat: 1. Description of technique and early neuropathological consequences following middle cerebral artery occlusion. J. Cereb. Blood Flow Metab. 1, 53–60.

Tei, N., Tanaka, J., Sugimoto, K., Nishihara, T., Nishioka, R., Takahashi, H., Yano, H., Matsumoto, S., Ohue, S., Watanabe, H., Kumon, Y., Ohnishi, T., 2013. Expression of MCP-1 and fractalkine on endothelial cells and astrocytes may contribute to the invasion and migration of brain macrophages in ischemic rat brain lesions. J. Neurosci. Res. 91, 681–693.

Thai, D.R., 2002. Excitatory amino acid transporter EAAT-2 in tangle-bearing neurons in Alzheimer's disease. Brain Pathol. 12, 405–411.

Tofts, P.S., Kermode, A.G., 1991. Measurement of the blood-brain barrier permeability and leakage space using dynamic MR imaging. 1. Fundamental concepts. Magn. Reson. Med. 17, 357–367.

Tofts, P.S., Brix, G., Buckley, D.L., Evelhoch, J.L., Henderson, E., Knopp, M.V., Larsson, H.B., Lee, T.Y., Mayr, N.A., Parker, G.J., Port, R.E., Taylor, J., Weisskoff, R.M., 1999. Estimating kinetic parameters from dynamic contrast-enhanced T(1)-weighted MRI of a diffusable tracer: standardized quantities and symbols. J. Magn. Reson. Imaging 10, 223–232.

Tohidpour, A., Morgun, A.V., Boitsova, E.B., Malinovskaya, N.A., Martynova, G.P., Khilazheva, E.D., Kopylevich, N.V., Gertsog, G.E., Salmina, A.B., 2017. Neuroinflammation and infection: molecular mechanisms associated with dysfunction of neurovascular unit. Front. Cell. Infect. Microbiol. 7, 276.

Tonne, C., Elbaz, A., Beevers, S., Singh-Manoux, A., 2014. Traffic-related air pollution in relation to cognitive function in older adults. Epidemiology 25, 674–681.

Turner, R.J., Sharp, F.R., 2016. Implications of MMP9 for blood brain barrier disruption and hemorrhagic transformation following ischemic stroke. Front. Cell. Neurosci. 10, 56.

Tyler, C.R., Zychowski, K.E., Sanchez, B.N., Rivero, V., Lucas, S., Herbert, G., Liu, J., Irshad, H., Mcdonald, J.D., Bleske, B.E., Campen, M.J., 2016. Surface area-dependence of gas-particle interactions influences pulmonary and neuroinflammatory outcomes. Part. Fibre Toxicol. 13, 64.

Tyler, C.R., Noor, S., Young, T.L., Rivero, V., Sanchez, B., Lucas, S., Caldwell, K.K., Milligan, E.D., Campen, M.J., 2018. Aging exacerbates neuroinflammatory outcomes induced by acute ozone exposure. Toxicol. Sci. 163, 123–139.

Tzivian, L., Dlugaj, M., Winkler, A., Weinmayr, G., Hennig, F., Fuks, K.B., Vossoughi, M., Schikowski, T., Weimar, C., Erbel, R., Jockel, K.H., Moebus, S., Hoffmann, B., Heinz Nixdorf Recall Study Investigative Group, 2016. Long-term air pollution and traffic noise exposures and mild cognitive impairment in older adults: a cross-sectional analysis of the Heinz Nixdorf Recall study. Environ. Health Perspect. 124, 1361–1368.

Uchida, Y., Sumiya, T., Tachikawa, M., Yamakawa, T., Murata, S., Yagi, Y., Sato, K., Stephan, A., Ito, K., Ohtsuki, S., Couraud, P.O., Suzuki, T., Terasaki, T., 2018. Involvement of claudin-11 in disruption of blood-brain, -spinal cord, and -arachnoid barriers in multiple sclerosis. Mol. Neurobiol.

Underly, R.G., Levy, M., Hartmann, D.A., Grant, R.I., Watson, A.N., Shih, A.Y., 2017. Pericytes as inducers of rapid, matrix metalloproteinase-9-dependent capillary damage during ischemia. J. Neurosci. 37, 129–140.

Verbeek, M.M., de Waal, R.M., Schipper, J.J., van Nostrand, W.E., 1997. Rapid degeneration of cultured human brain pericytes by amyloid beta protein. J. Neurochem. 68, 1135–1141.

Verma, S., Nakaoke, R., Dohgu, S., Banks, W.A., 2006. Release of cytokines by brain endothelial cells: a polarized response to lipopolysaccharide. Brain Behav. Immun. 20, 449–455.

Volk, H.E., Lurmann, F., Penfold, B., Hertz-Picciotto, I., Mcconnell, R., 2013. Traffic-related air pollution, particulate matter, and autism. JAMA Psychiat. 70, 71–77.

Wake, H., Moorhouse, A.J., Jinno, S., Kohsaka, S., Nabekura, J., 2009. Resting microglia directly monitor the functional state of synapses in vivo and determine the fate of ischemic terminals. J. Neurosci. 29, 3974–3980.

Wang, Y., Welty, D.F., 1996. The simultaneous estimation of the influx and efflux blood-brain barrier permeabilities of gabapentin using a microdialysis-pharmacokinetic approach. Pharm. Res. 13, 398–403.

Wang, J., Li, G., Wang, Z., Zhang, X., Yao, L., Wang, F., Liu, S., Yin, J., Ling, E.A., Wang, L., Hao, A., 2012. High glucose-induced expression of inflammatory cytokines and reactive oxygen species in cultured astrocytes. Neuroscience 202, 58–68.

Weller, R.O., Subash, M., Preston, S.D., Mazanti, I., Carare, R.O., 2008. Perivascular drainage of amyloid-beta peptides from the brain and its failure in cerebral amyloid angiopathy and Alzheimer's disease. Brain Pathol. 18, 253–266.

Wilcock, D.M., Vitek, M.P., Colton, C.A., 2009. Vascular amyloid alters astrocytic water and potassium channels in mouse models and humans with Alzheimer's disease. Neuroscience 159, 1055–1069.

Wolburg, H., Lippoldt, A., 2002. Tight junctions of the blood-brain barrier: development, composition and regulation. Vasc. Pharmacol. 38, 323–337.

Wolburg, H., Wolburg-Buchholz, K., Engelhardt, B., 2005. Diapedesis of mononuclear cells across cerebral venules during experimental autoimmune encephalomyelitis leaves tight junctions intact. Acta Neuropathol. 109, 181–190.

Wolburg, H., Noell, S., Mack, A., Wolburg-Buchholz, K., Fallier-Becker, P., 2009. Brain endothelial cells and the glio-vascular complex. Cell Tissue Res. 335, 75–96.
Wolman, M., Klatzo, I., Chui, E., Wilmes, F., Nishimoto, K., Fujiwara, K., Spatz, M., 1981. Evaluation of the dye-protein tracers in pathophysiology of the blood-brain barrier. Acta Neuropathol. 54, 55–61.
Wunder, A., Schoknecht, K., Stanimirovic, D.B., Prager, O., Chassidim, Y., 2012. Imaging blood-brain barrier dysfunction in animal disease models. Epilepsia 53 (Suppl. 6), 14–21.
Xia, X., Jiang, Q., Mcdermott, J., Han, J.J., 2018. Aging and Alzheimer's disease: comparison and associations from molecular to system level. Aging Cell 17, e12802.
Xu, Z., Zeng, W., Sun, J., Chen, W., Zhang, R., Yang, Z., Yao, Z., Wang, L., Song, L., Chen, Y., Zhang, Y., Wang, C., Gong, L., Wu, B., WANG, T., Zheng, J., Gao, F., 2017. The quantification of blood-brain barrier disruption using dynamic contrast-enhanced magnetic resonance imaging in aging rhesus monkeys with spontaneous type 2 diabetes mellitus. NeuroImage 158, 480–487.
Zehendner, C.M., Sebastiani, A., Hugonnet, A., Bischoff, F., Luhmann, H.J., Thal, S.C., 2015. Traumatic brain injury results in rapid pericyte loss followed by reactive pericytosis in the cerebral cortex. Sci. Rep. 5, 13497.
Zlokovic, B.V., 2008. The blood-brain barrier in health and chronic neurodegenerative disorders. Neuron 57, 178–201.
Zlokovic, B.V., 2011. Neurovascular pathways to neurodegeneration in Alzheimer's disease and other disorders. Nat. Rev. Neurosci. 12, 723–738.
Zuroff, L., Daley, D., Black, K.L., Koronyo-Hamaoui, M., 2017. Clearance of cerebral abeta in Alzheimer's disease: reassessing the role of microglia and monocytes. Cell. Mol. Life Sci. 74, 2167–2201.

FSC
www.fsc.org
MIX
Papier | Fördert
gute Waldnutzung
FSC® C083411

Kolibri 360 GmbH
Ferdinand-Jühlke-Straße 7
99095 Erfurt, Deutschland
produktsicherheit@kolibri360.de